W9-AFN-720

SHELTON STATE COMMUNITY
COLLEGE
JUNIOR COLLEGE DIVISION
LIBRARY
DISCARDED

True Love and Perfect Union

TRUE LOVE AND PERFECT UNION

The Feminist Reform of Sex and Society

WILLIAM LEACH

Basic Books, Inc., Publishers *New York*

Library of Congress Cataloging in Publication Data

Leach, William, 1944–
True love and perfect union.

Includes index.
1. Feminism—United States—History—19th century.
2. Women—United States—History—19th century.
3. Women—United States—Social conditions. I. Title.
HQ1419.L4 305.4'2'0973 80-50557
ISBN: 0-465-08752-3

Printed in the United States of America
DESIGNED BY VINCENT TORRE
10 9 8 7 6 5 4 3 2 1

To the memory of my father

Thomas Edward Leach

and my mother

Esther Lee Leach

Contents

Part 2

FEMINISM IN THE PUBLIC REALM

Part 3

THE RELATION OF FEMINISM TO SOCIAL PRACTICE

Acknowledgments

I am in debt to many friends who have helped me in different ways. I have learned much from Florence Bartoshesky, Thomas Cole, Faye Dudden, Bonnie Smith, and Mary Young. Sarah Elbert shared her knowledge of American feminism with me and always expressed an enthusiasm for my work that lifted me through some difficult times. She and Elizabeth Fox-Genovese commented carefully on the manuscript, for which I am grateful. Fox-Genovese read the earlier drafts. Her excellent theoretical and historical insights have considerably improved the text.

To Warren Susman and Christopher Lasch I owe a deep debt. Since my undergraduate days at Rutgers, Susman has given me generous and inspiring guidance. I had the good sense to send him the second revision of the manuscript. Acute in his intellectual judgments, he persuaded me to remove several structural weaknesses. Christopher Lasch has read this book in all its forms. I have tried, and more often than not failed, to emulate his unrelenting precision, his clarity, his analytical depth. Unwilling to dictate the direction of my thought, free from the need to create a coterie of imitative students, he has nonetheless significantly influenced my intellectual development.

Lasch also introduced me to my editor, Jeannette Hopkins, who in turn introduced me to Jane Isay of Basic Books. In both cases I had good fortune. Isay's lively commitment to the book gave me the energy to complete it sooner than I had expected. Under Hopkins's superb and exciting editorial supervision, I have suffered and sweated a great deal and learned a great deal about how to write a book. She has improved every page and in every way.

I am grateful to Mary Heathcote for her excellent copy editing and to the women who typed the various manuscript versions of this book:

Jean DeGroat and Lorraine Clark of the University of Rochester; Patricia Camden and Shirley Lawrence of Wesleyan University; and Judy Rickman of The University of the South. They deserve my thanks and my respect. So does the Center for the Humanities at Wesleyan University, Middletown, Connecticut, where I completed the first draft of this book.

True Love and Perfect Union

PROLOGUE

TRUE LOVE AND PERFECT UNION

THIS BOOK deals with the mid-nineteenth-century feminist attempt to create a more perfect union. Often at odds with themselves, feminists struggled to forge a new, rational union between the sexes, a union based on what they often referred to as true love. They demanded that both sexes have equal symmetrical development. Many of them expanded the concept of symmetry to include the social system. Feminists rallied round the principle of no secrets, contending that both sexes needed equal access to scientific and sexual truths. They hoped that this principle would instill harmony into private life, although it often resulted in slander and gossip, a consequence feminists did not intend. Feminists championed equal educational and economic opportunities for both sexes in the public realm. Though many of them still clung to fashionable sentimental perspectives and were caught up in the excitement of a fashionable new world created by mid-century dry-goods palaces and department stores, most feminists demanded equally functional dress in public for both sexes. Many feminists were attached to the romanticism of the pre–Civil War period, but, rebelling against their own impulses, they sought to rationalize passion and to suppress private fantasy. They wanted equal rational sexual pleasure, true, rational love, for both sexes. Such equality would bind, not divide, the sexes.

So, too, feminists labored for greater unity between the private and public spheres, between the individual and the community. Joining other reformers, male and female, they increasingly and often unwillingly turned against individualism, which they thought destroyed social unity, and toward a symmetrical, communitarian system of relations based on the principles of social science. One of the founders of the American Social Science Association, Caroline Dall, exemplified this evolution. A transcendentalist in her youth, she became a social scientist in her maturity. Always conflicted, she sought resolution in a new social harmony. Organized feminists—above all, Elizabeth Cady Stanton—turned to positivist social science; so did other reformers with feminist sympathies, including labor reformers, moral educationists, and progressive spiritualists.

Feminists wanted to build totally rational social and economic arrangements that would harmonize the life of their own class, integrating the private and public spheres, and the two sexes, more closely. They joined other humanitarian reformers to construct a cooperative world view founded on a social science that sought to reconstruct society according to principles of equipoise, cooperation, and organization.

Feminists saw the injurious effects of individualism on the sexual relation all around them after the Civil War. The proof loomed before them in the bewildering growth of the cities. By the 1860s, nine cities exceeded the 100,000 mark, and four hundred cities reported populations close to 2500. Proportionate to the total population, Americans had also witnessed the greatest influx of immigrants into the country in its history.[1] For feminists, this growth had called forth two radically opposed worlds, one of poverty and the other of fashion, both likely to subvert the sexual relation and to destroy forever the natural "harmonies" of a republican society.

From the days of the early republic into the Jacksonian period, the independent producing classes—artisans, affluent farmers, and those men and women linked in some way with mercantile wealth—feared and attacked the "parasitic," fashionable elite, the "riotous livers," on the one hand, and the dependent, thriftless poor, on the other. Such classes constituted threats to the natural "balance of forces" that feminists as well as other Americans believed supported the existence of republican liberty.[2] By the 1860s and 1870s these attacks had assumed a new urgency and a persistence hitherto absent in the earlier period. Not only did a wealthy, nonproducing class and an urban proletariat exist for the first time in almost every major city in the East, but a new political regime, riddled with factionalism and corruption, catered to the needs of these same classes. Intelligent, productive, independent-minded Americans feared the demise of their republican heritage; they feared, in particular, the breakdown of all social bonds.[3]

"The palaces of the rich," the feminist Richard Hume wrote in 1869 for the *Woman's Advocate*, a precursor of the *Woman's Journal*, "as well as the crowded hovels of the poor are the generators of the moral and physical malaise which destroys nations. . . . It is no wonder that marriage is shunned" and "that divorces are common."[4] Everywhere, the New York radical feminist Victoria Woodhull wrote in 1870, "the distance between man and woman, as husband and wife, is gradually becoming less and less the crucial point of attraction of all concerned. Does woman realize that as a sex she is becoming estranged from man?"[5] One hundred twenty-five thousand young single men in New York City alone, the feminist *Hearth and Home* complained in 1871, refused to marry, preferring to live indecent lives and to patronize the "lower and more infamous" forms of "amusement."[6]

Perhaps the most sensational signs to feminists that individualism endangered marriages and the sexual relation was the outbreak of domestic murders—husbands by wives, wives by husbands, fathers by sons—and the eruption of divorce scandals—a veritable "divorce mania," according to the feminist *New Northwest*—in the late 1860s and early 1870s.[7] In New York City two dramas, *Divorce* and *Man and Wife*, played on Broadway simultaneously, to packed houses; and in 1873 the magazine writer Kate Field even adapted Hawthorne's *Scarlet Letter* for the stage.[8]

The feminist press bore continual witness to the numbers of men and women who had lost faith in their marital vows. Feminists often hailed those women who escaped intolerable marriages, but they deplored the circumstances that created the basis for such marriages in the first place. Several feminists complained of the ease with which people married, an ease nourished by still resonant common-law ideas and practices. Anyone at almost any age could marry without license, banns, witnesses, or parental consent.[9]

In their own right, however, feminists seemed to be contributing to, even while they deplored, the breakdown of the sexual relation by claiming for women the legacy of liberalism. Mid-century feminists were liberals. Like most other Americans they inherited the conviction from their parents that the political life was designed to liberate human potential and not to restrain it, that it functioned best only in a negative fashion. The natural social processes of life, if left untrammeled, would spontaneously produce their own laws, and thereby social harmony and unity. One freethinking feminist wrote in an 1869 article called "The Modern Problem Social, Not Political," "We have not to contend for freedom as citizens, or to overthrow sects or establishments. . . . Whatever changes take place with us must be only development and application to the various details of what makes now the inner and animating principle of our national experience."[10] The political "problem," in other words, had already been resolved with the removal of older forms of authority that reformers associated with the very meaning of the political: monarchy, aristocracy, "patriarchalism," and so forth. The one remaining political task was to apply and realize a deeply rooted "animating principle" to the condition of women. Nothing more remained to be done but to grant both sexes the same legal equality, the same suffrage, to protect them in their "interests" and "private property," to secure them equally the right of self-ownership and the right to "sell their labor for whatever price they considered it worth in the open market."[11]

Feminists felt perhaps more contempt for traditional sources of authority than any other group of reformers did. They believed conservative elements had willfully kept women from securing their rights to the liberal heritage. In 1878 the National Woman's Suffrage Association, the most openly radical faction in the mainstream feminist movement, declared, "In her most vital interests woman should no longer trust authority, but be guided by reason."[12] Like abolitionists, they identified the state with masculine dominance, aggression, and appropriation. "The state," wrote Abby May, daughter of the Unitarian abolitionist Samuel May and member of the American Social Science Association, in 1869, "has proven itself void of all positive worth. . . . It has made one man rich at the expense of a thousand."[13]

Feminists increasingly withdrew their allegiance from all established religion, and some made successful efforts to enter the ministries themselves, especially the liberal Unitarian and Universalist ministries, hoping to preach the nonsectarian, humanitarian religion of the New Testament. Others became freethinking spiritualists, and still others—a large proportion of the feminist leadership—joined Free Religion, a rationalist sect that split off from Unitarianism in the 1860s and that attempted to homogenize all religious thought into a Religion of Humanity.

Most feminists berated the church as the major "support of the present subjection of women" and criticized, in one way or another, the reactionary, antifeminist clergy.[14] In 1850 Lucy Stone, later to become president of the American Woman's Suffrage Association, warned her friend Antoinette Brown not to become a minister and pleaded with her to think as a "free spirit." Religion, she wrote, had not only "outgrown" its usefulness, but also had "hitherto bemoaned *women* into *nothingness*."[15] "Our strongest enemies entrench themselves in the church," said Elizabeth Cady Stanton.[16] And Celia Burleigh, the freethinking Universalist minister, claimed in the *Woman's Journal* that "the cause of humanity suffers more from its priests and pharisees than from its publicans and harlots."[17]

Many feminists felt pride in ignoring "all denominational ties" in their meetings and associations.[18] It was only after 1875, when the temperance cause gave a more conservative cast to the women's movement and a new life to organized religion, that such attacks on traditional religious authority were throttled. Many feminists inevitably closed ranks with temperance reformers, while others expediently accepted or simply mourned the fresh intrusion of patriarchal ministerial leadership into the life of their own cause.

At the same time that feminists embraced liberalism and turned their backs on traditional authority, however, they also attacked the possessive individualism that had historically been a part of the liberal tradition. They looked for a new ideology, new forms of authority to organize society.

Some recent historians of feminism have claimed that the mid-century feminist leadership either had no compelling ideological content or had no ideological content at all except as it strove to apply liberalism or natural rights theory to woman's condition or as it reflected the "pragmatic" bias of American reform thought generally. Carl Degler was one of the first to lay out this position when he wrote in 1965, apropos of the feminist movement, that "in America the soil is thin and the climate uncongenial for the growth of any seedlings of ideology," and both Aileen Kraditor and William O'Neill shared this opinion.[19]

Kraditor has done more to shape historical inquiry into the feminist movement than any other historian of feminism. In her important book, *The Ideas of the Woman Suffrage Movement, 1890–1920*, she argued that feminist ideology evolved in the 1890s from an emphasis on natural rights to one based on expediency. By 1900, according to Kraditor, a new generation of feminists no longer appealed to abstract principles of justice to defend woman's right to the franchise; rather, they would use any argument, but above all the argument based on native American woman's virtuous influences, to get the vote for women. Ann Douglas has most recently reiterated this view in *The Feminization of American Culture*.[20]

Kraditor focused on the late nineteenth and early twentieth centuries, when suffrage apparently consumed feminist attention. This book focuses on the period 1850–1880 and specifically on the 1870s, when feminists considered a range of social and economic as well as political questions. My examination of this period (and it does not pretend to be complete) has led me to make certain claims about mid-century feminism that depart from the older views presented by Kraditor and reinforced by Degler, O'Neill, and others. Feminists did not wait until the end of the century to argue from the standpoint of woman's virtue: from at least the 1850s they often used natural rights and woman's moral superiority at the same time in their struggle for women's rights. More important, however, they tried to create a new social ideology, shared by other reformers and prefiguring later Progressivism, that turned against liberalism even while feminists held to it and still retained a loyalty to the concept of woman's virtue. The existence of this new ideology suggests a greater continuity between the old and the new generations of feminists; it should make necessary a reassessment of the later movement during the period Kraditor examined in her book.

I have tried in this study to determine the inner meaning of mid-nineteenth-century American feminism and to place it within a wider strategy of the reformist bourgeoisie to reconstruct American institutional life along rationalist, cooperative, egalitarian (at least for the bourgeoisie), and generally positivist lines. Nineteenth-century feminism should not be viewed as simply "suffragist" nor as a fixed ideology of equal rights, but as a multifarious network of associations and apparatuses that produced an egalitarian ideology determined and limited by its historical matrix and that attempted, in its own right, to shape contemporary historical conditions. Instead of focusing on the feminist application of natural rights theory, therefore, or on the quest for the political franchise, both of which merged with and at times masked new political perspectives, I have chosen to examine the "social problems," or what Elizabeth Cady Stanton called the "more vital questions of this reform." It was in the realm of civil society, the area of political struggle that we ordinarily think of as private, that feminists sought, and in some degree captured, power. It was in this sphere that they established the strongest alliance with other reformist intellectuals of their own class. In this sphere they attacked the possessive individualism embedded within the democratic liberal tradition.

By the liberal tradition I mean the right of the individual, unfettered by oppressive custom, prejudice, law, or public opinion, to determine his or her own destiny. The liberal tradition underlined the moral worth of every individual. By individualism I mean what the political philosopher C. M. MacPherson has called possessive individualism or the right of the individual to proprietorship of his or her "own person or capacities, owing nothing to society for them." According to this conception the "individual was seen neither as a moral whole, nor as part of a larger social whole, but as an owner of himself. The human essence is freedom from dependence on the wills of others, and freedom is a function of possession. Society consists of relations of exchange between proprietors. Political society becomes a calculated device for the protection of this property and for the maintenance of an orderly relation of exchange."[21]

Individualism has always been linked with the liberal tradition, but it has not always had this possessive quality, as MacPherson points out. It has not always been directly tied to the market. In its early Christian form and later in Puritanism and romanticism, individualism recognized the moral and creative worth of the individual. Feminism and post–Civil War reform generally tried to save the liberal tradition by severing it from its possessive individualist element and by reestablishing its relationship with this earlier Christian emphasis, albeit in a secularized and communitarian form.

Most feminists never completely repudiated possessive individualism, but they did attack it, and they did propose a new social vision to counter it. It is within this wider commitment to social rather than conventional political issues, and to the reconstruction of society into a communitarian whole, that the overwhelming feminist emphasis on "social problems" and, more particularly, on egalitarian marriage should be understood. In spite of the many differences among feminists about the nature and tactical importance of the social problems, most feminists sought to make women productive individuals within a scientifically organized, egalitarian society. Most feminists, then, shared a "common theory of marriage."[22] The point of this "new theory"—the heart of nineteenth-century feminism—was "to end the battle between the sexes" and "bring them back together" into an enduring, conflict-free harmony based on shared authority.[23] "Half the family broils," the *Springfield Republican*, one of the most influential newspapers of the day, editorialized in 1871, "are through the very inequality of our present system, its very gift of false supremacy to the husband over the wife. . . . The power of leadership knows no sex . . . There is more harmony, more unity, in equality than in inequality; else all the laws of nature are false."[24] And the *Golden Age* insisted, "We should strive to tear away this thin veil of segregation, and remove every cause for alienation and bring them together. . . . The new era promises to bind men and women more closely than they have ever been." This era promised, moreover, to bind them "on an impregnable basis."[25] "Through this suffrage movement," Isabella Beecher Hooker announced in the same year, "it is hoped to bring the sexes closer together."[26]

Feminists wanted the kind of marriage that would survive and not dissolve through conflict. Jane Croly, Elizabeth Cady Stanton, Harriet Beecher Stowe, and Stephen Pearl Andrews may have all disagreed on the divorce question, but they joined hands over the criteria for lasting egalitarian relationships. As the feminist editor of the *Springfield Republican* put it, the "frequency of divorce in the United States" had greatly illuminated for everyone the "thoughtless manner in which marriage is contracted."[27] "A large social freedom," the New York feminist Lydia Fuller wrote to Victoria Woodhull, "means among other things, right conditions for marriage. These we must have before we can have marriage in any real sense."[28] And Jane Croly in her book on marriage emphasized that "what we need, to correct some of the evils of marriage, is not the liberty to commit the same errors a second time, but the qualifications that will prevent us from making the mistake in the first place."[29]

Feminists, as much as or perhaps even more than their contemporaries, were deeply concerned with the changes taking place in the sexual relation: they consistently brought that relation to public attention and consistently labored for its reform. At the heart of their concern was a belief that the pressures and character of urban and business life and the impact of possessive individualism had created conditions unfavorable to the best interests of both men and women.

In their campaign against possessive individualism, however, feminists also attacked the individual (as they castigated romanticism). They put reason at the service of uniformity, intending to rationalize individual passions and desires. Thus they brought the liberal tradition itself into question. The confusion between the individual and possessive individualism vexed the feminist movement as it did other reform movements. It was a confusion feminists did not completely understand nor clear up. As the most exciting reform of the nineteenth century, feminism often approached a democratic-socialist critique, but more often it suggested the contours of late-nineteenth-century Progressivism.

A historian recently wrote that the "crisis of Reconstruction was a part of the world-wide crisis of the nineteenth-century liberal tradition."[30] Like other postbellum reform movements, feminism experienced and reflected this crisis; it expressed the movement away from the older possessive individualist perspectives of the past to a different world view. At the same time the feminist leadership witnessed another change: it was no longer located in the towns and small cities of the country, but had become based in the large urban centers, drawing on a much bigger pool of feminist-minded women and men. A network of feminists appeared in the major cities—Chicago, Cincinnati, Philadelphia, St. Louis, Boston, New York, San Francisco, Washington, Detroit, and Pittsburgh. These new locations contributed to changes in feminism. Increasingly feminists came to accept the fact that classical liberalism, with its mixture of possessive individualism and the democratic-liberal tradition, no longer met the needs of a new society.

This book focuses on the mid-century feminist attempt to construct a new social ideology departing from the older classical liberal model. The book's scope, however, goes beyond this to deal with two other important matters, neither of which has been satisfactorily explored by historians: first, the central role feminists played in making the reform tradition; and second, the feminist reliance on science and organization as the basis for a humanitarian ideology. I do not believe that feminism was a movement peripheral to or independent of other reform movements. After the Civil War, innumerable, apparently contradictory reform groups, overtly feminist or with feminist sympathies, labored to construct a new world view that offered an alternative to classical liberalism. Even some free-love groups, historically identified with the most extreme form of individualism, contributed to the formulation of a more communitarian system of relations. Abjuring the traditional party system as corrupt and turning their backs on conventional political practice, such reformers did their real business in a social framework. They directed their political attention to schools, asylums, libraries, hospitals, and prisons, to the professions and to the home, and—perhaps most elusive of all to political interpretation—to human biology, to the bodies of men and women, to sexuality.

These groups, male and female, hoped to create new places and new roles for both sexes. More broadly, they participated in an effort to reconstruct the private and public life of the educated bourgeoisie while at the same time securing for it a solid position within institutional and professional structures. They combined the interests of sex with the interests of class.

The historian George Fredrickson has observed that "only in the thought of Lester Ward and the policy-making of John Wesley Powell did the [Civil War] generation suggest the possibility of middle ground (between scientific pessimism and utopian idealism)—a social philosophy grounded in the belief that science, organization, and planning could be enlisted on the side of humanitarian reform."[31] This assessment needs reconsideration, which brings us to a second important aspect of mid-century feminism. Feminists, as well as much of the reform tradition out of which they emerged, did not radically depart from the humanitarian heritage of the antebellum years. The tradition was altered but not repudiated. We see this humanitarianism in the cosmopolitan character of the movement. It based the emancipation of women on cosmopolitan knowledge and experience, mobilizing truths and factual information to prove that the subject condition of women varied from culture to culture and thus had no historically universal validity. Feminist writers of the nineteenth century rarely cited the American experience alone to prove their case, but compared the condition of women in America with that in other countries and cultures—Sweden, Italy, Russia, England, Turkey, Scotland, India, and China.[32] Such cultural cosmopolitanism, such an interest in the general character of women's experience, contributed to the demystification of social life and to the gradual erasure of social differences within the reformist class. It can also be seen as a spur to the emergence of the social and cosmopolitan sciences—the science of comparative religions and mythologies, anthropology, sociology—that seek to investigate the inner life of other cultures and to understand, in particular, their marriage customs, their family patterns, and their sexual behavior. Perhaps this helps to explain why the social sciences have often been viewed as "soft" and "feminine" sciences. It also may explain why feminism and social science have been historically linked in a working, if occasionally confrontational, alliance. Feminism encouraged, even demanded, a secular, humanitarian, internationalist spirit in order to develop as a viable ideology. Religious and political authoritarianism as well as cultural parochialism blocked such a development. "My country is the World," the abolitionist William Lloyd Garrison cried in the 1830s. "My Countrymen are all mankind!"[33] Feminists inherited this cry as a chief legacy. Feminism—indeed, the reform tradition itself—sought to open up all society to the free, leveling dynamic of an intellectual and economic market. It should come as no surprise, therefore, that bourgeois feminists communicated with one another, attended one another's conventions and congresses, read the same literature, and shared a similar ideological ground. From the early conventions of the 1850s to the Industrial Congresses, Purity Congresses, International Associations and Councils, and World Fairs and Exhibitions, interlocking interest and discourse was palpable and indisputable. Flowering within the secularization and democratization of thought and behavior, feminism deepened, extending cosmopolitan processes into the middle and upper-middle classes.

Perhaps the clearest index to the humanitarian tendency of feminism comes in the character of feminist political discourse. Feminists simply did not talk or write politically; they did not employ the language of power, faction, or conflict, nor did they make significant use of political parties or attempt to create a new, genuine political alternative to the established party system. "Let the friends of suffrage no longer look for aid to any existing political organization. Let us cease to be Republicans and Democrats," wrote Henry Blackwell in the *Woman's Advocate*.[34] "Down with Politicians," Susan B. Anthony wrote for the *Revolution* in 1868, "up with the People, has been our cry from the beginning. Our clientele are the oppressed of all classes, all peoples."[35]

This disavowal of conventional political activity resulted from a convergence of several factors. After the war, and especially in the 1870s, the Republican and Democratic parties were organized according to racial, sectional, and religious commitments, ignoring important and divisive ideological questions. Moreover, both parties fell into the hands of powerful business elites.[36] The parties' repudiation of important social and economic issues alienated the feminist leadership, as it did labor and agrarian activists. Although some feminists tried, without success, to create or to join third party movements, their alienation from the established parties tended, more often, to reinforce an already established feminist pattern. This pattern was characteristic of the reform experience as a whole. Feminists spoke always of social vision and of the social organism. When they campaigned to alter the laws in the various states—and especially when they campaigned to get the suffrage for women—they did so as apolitical lobbyists, advocates not for any particular group but for "universal humanity." Politics by its very nature implied struggle between groups and classes for domination and priorities, a struggle for which feminists had only disdain. Like the abolitionists before them and the Progressives of a later generation, feminists hoped to create order, community, cooperation, and harmony from "outside" the conventional political system in which factional struggles and conflicts were inevitable.[37]

"Our politics is corrupt," wrote feminist Moncure Conway. "Our nations snarl and fight; our commerce is systematized selfishness: the ideal life begins when man shuts the door to all these and enters where service is not done for money."[38] "We speak not for ourselves alone," Stanton told the National Woman's Suffrage Association in 1869, "but for all mankind."[39] The health and hospital reformer Caroline Severance modestly described the Boston Woman's Club as "a small group of New England women who seek to create true fellowship of women through an organization founded on no test of creed, or partisan purpose, but on simple womanhood."[40] Women's clubs, wrote the Chicago freethinker Kate Doggett, could "be claimed by no sect, no social clique; eligibility consists principally in the ability to contribute to the good of the whole."[41] And the radical Free Religionist Frederick Hinckley asserted that the woman's movement "is not a rich man's movement, it is not a poor woman's movement, it is a *woman's* movement. It knows no class of woman as against another class."[42] The political meant for feminists that ideal realm above politics where all opposites and differences coalesce into one integrated whole and earthy squabbles do not intrude. "Woman Suffrage," the Boston Unitarian Jesse Jones wrote in the *Woman's Journal* in 1874, "is not primarily a political but a social question; and means a profounder revolution in the whole structure of society, than many advocates seem ever to have dreamed of."[43] In these terms suffrage simply had no political content: it was above all a "social question."

Like the later Progressives, furthermore, feminist leaders tried, though often unsuccessfully, to combine their older humanitarianism with a new ideology based on scientific knowledge and on principles of organization, symmetry, cooperation, and elitist centralization. Science and organization were the new authorities. Feminists of all kinds, from spiritualists to social scientists, turned to the science of preventive health and to physiology, to social Darwinism (but a social Darwinism that emphasized evolutionary progress toward cooperation and harmony), and to Comtian and Spencerian positivism. Some turned to Hegelian rationalism. They drew on the Enlightenment utopian tradition in their quest to remodel the legacy of liberalism.

Part I

FEMINISM AND PRIVATE LIVES

CHAPTER 1

HYGIENE AND THE SYMMETRICAL BODY

"PHYSIOLOGY," a male feminist hygienist wrote in 1874, "is a better guide than religion" to understanding the meaning of our individual lives. "The solution to the problem of women's sphere must be obtained in physiology," said Antoinette Brown Blackwell, the first American female preacher and an influential feminist.[1]

By the 1850s the study and application of hygienic or natural principles vied successfully with and even surpassed religion as a means of rationalizing individual behavior within an increasingly secular society. Even earlier, in the 1820s and 1830s, health reform, retaining a strong perfectionism, had begun to split off from formal religion, and for many reformers, especially those from liberal Protestant backgrounds, health reform had by the 1850s taken on a secular character close in ideological affinity to the Jeffersonian rationalism of the late eighteenth century.[2]

In the course of this evolution from religion to health, feminists also developed a new view of the human body. They called it the symmetrical body. Organized and rational, harmonious and whole, such a body was supposedly immune from the outbreak of internal extremes; it insured women and men from the subversive effects of impassioned, individualistic behavior. By the post–Civil War period, feminists had extended this concept of individual symmetry to include the social and economic system. Individual symmetry equipped the body against the full effects of individualism; social symmetry, however, brought individualism itself into question, suggesting the outlines of a more cooperative social order.

Symmetries of all kinds, from psychological to political, private to public, weave in and out of feminist discourse in the middle of the nineteenth century. They are the fundamental starting points for an understanding of the changing feminist conception of the world.

From Religion to Health

THE PHYSICALITY of the world was seen to mirror the moral law: one had only to study the physical universe or, more important, one's own physiology, to discover the basis for true health, and therefore for true virtue. "The anatomical and physiological construction of man," according to the feminist and principal leader of American spiritualism Andrew Jackson Davis, in the first volume of his *Great Harmonia*, "are designed to receive and elaborate the animating element in nature into an immortal and endlessly progressive soul."[3] "Sickness is a crime," observed Elizabeth Cady Stanton, "since it is an evidence of a violation of some physical law." To her cousin, Gerrit Smith, a wealthy upstate New York abolitionist who anguished over a republican solution of the "woman question," she wrote, "Follow Nature ('be natural'), and the facts of life and the deductions of logic must drive you irresistibly" to the solution of this "tough and tangled problem."[4] Physical exuberance was the key to eternal growth and development. Not grace, but health paved the path to salvation. Better the "skill of an infidel," one reformer wrote in 1866, referring to medical practitioners, "than the piety of a fool." "Let those who have trusted to grace instead of nature," wrote another, "take care of the results of their mistaken course."[5] Not sin but sickness became the required conversion experience, the enforced meditative interlude through which one passed to sanity and well-being in a democratic and capitalist society.

This shift from religion to health, this movement from abstract and otherworldly concepts of truth to a truth hidden within the reality of physical things, in large measure determined the character of mid–nineteenth century reform consciousness.[6] On the personal level the emphasis on health gave great significance to one's own body: what one thought, what one did with one's body set the course, for better or worse, of one's life. The uses of the body, of sexual hygiene, still retain this function in our own time, although sexual fulfillment has replaced sexual purity—or in the parlance of the nineteenth century, hygienic symmetry—as the major proof to ourselves that we have lived and that we have found happiness. On the public level, this nineteenth-century shift in consciousness contributed to directing the reformist mind to the social or "physical conditions" of life, to health and educational reform, sanitation, asylums, prisons. The health of the body and the uses made of it decided whether nations or societies would rise or fall.

Body metaphors have always been a part of American political language. We find them in Puritan, Federalist, and Republican discourse. By the nineteenth century, however, such metaphorical usage had taken on a much wider currency. Reformers described the "divorce mania" as a symptom of an "organic deformity" within the "structure of society," the crowded cities as "organic excrescences," and "all wrongs" as signs of unnatural dislocations and nagging "sickness."[7] "The sickness of our times," wrote John Chadwick, physician, positivist thinker, and champion of women's right to become doctors, "afflicts the social organism."[8] "Great wrongs accumulate," said the feminist leader Matilda Joslyn Gage, "until society becomes sick and in the effort to throw off the disease, terrible social corruptions occur."[9] In this context the battle between rival political systems—between American republicanism and European monarchicalism and, in time, European socialism—metamorphosed into a Manichean conflict between the forces of sickness—crowdedness, civilization, bondage, and disease—and the forces of health—space, nature, freedom, and light.[10]

Feminists relied heavily, and at times almost entirely, on science and hygiene as ideological tools rather than on abstract principles of justice to ply their case for public and private equality for women. We cannot understand feminist ideas on marriage and the sexual relation without underscoring this reliance. Thus many feminists viewed natural rights wholly in hygienic terms. When asked in 1872 what "natural rights" meant, the feminist journalist Theodore Tilton responded: "whatever we need for the *body's welfare* and whatever we need for the *mind's growth*." The feminist doctor Mary Putnam Jacobi said that "happiness occurs where the powers of the individual are in equilibrium" and "it is the only test of right, because that is the natural test of right."[11]

The emphasis on the body's welfare owed its intensity to the outrages that traditional medicine in America, as well as in England and France, had inflicted on women, and also to the preoccupation with prevention of physical disorders in the reformist tradition itself.

Reformers denounced physicians' prescriptions of noxious drugs and sundry mechanical devices that exacerbated disease and even caused death. Feminists detested the doctors who unscrupulously exaggerated the vulnerability of women to sexual disease and who made fortunes from the ignorance of fashionable women. "I blame the medical profession," wrote Russell Trall, a popular water-cure doctor who set up his own coeducational and irregular medical school in New York City partly to correct these abuses. "[Woman] has been misled and miseducated by it; she has been taught that she is naturally more frail and feeble and prone to disease than man; and that she must be dosed and drugged with the most potent poisons for the most trivial indispositions."[12] To save women from such treatment, feminists themselves had to lay claim to the science of health.

Feminist interest in health and physiology also reflected the special stake that feminists had in prevention. The new wave of "civilized" disorders such as nervous tension and exhaustion, headache, insomnia, and neuralgia affected both sexes, but hysteria, doctors claimed, appeared to afflict women unduly. The science of health and feminism so overlapped that many feminists embraced preventive medicine or the "science of life" as the proper professional province for women, a claim they continued to make into the twentieth century. "We should give to man cheerfully," the pioneer woman doctor Harriet Hunt announced in the 1850s, "the curative department, and woman the preventive. . . . She needs to go as physician and investigate habits that are poisoning the fountain of life, inducing precocity of the animal nature and an irritability of the nervous. The female physician must be preventive. She must look upon life through sanitary channels."[13] It was the feminist commitment to preventive hygiene that placed women squarely in the heart of the bourgeois reformist tradition and explained, in large measure, the endurance, the power, and the pervasiveness of the nineteenth-century feminist movement.

The role that women played in the Sanitary Commission, which allocated medical supplies to the Union armies during the Civil War, and in the sanitary, temperance, and social purity movements of the postwar period bears out the importance of this relationship between feminism and the science of health. It sheds light on educated feminists' avid consumption of the scientific and social science literature of the day—especially the works of Darwin, Comte, Buckle, Haeckel, and Lubbock—and even on the popularity of minor "sciences" like mesmerism, clairvoyance, and electromagnetism among the less educated feminists. "As science is more woman's business," wrote Sarah Thayer to Amy Post in 1857, "she will be more sedate and be lifted from the degradation of the slave."[14] Susan B. Anthony told a woman's meeting in 1871 that every step of science was linked with the march of women toward equal rights, and that every step "has been obstructed by the prejudices of the world which had to be overcome."[15] Dr. Clemence Lozier, dean of the New York Medical College for Women and president of the New York City Suffrage Society, advised her former medical students at their annual reunion to study "Darwin, Tindall, Herbert Spencer, Buckle, and Agassiz" as well as "ordinary medicine."[16] Abba Gould Woolson, at a meeting of the Moral Education Society of Boston in 1880, urged that girls "study physiology, natural philosophy, the laws of mechanics, chemistry, the characteristics of a republican form of government, rhetoric and natural sciences, all of which are far more important to the girl than modern languages."[17] Feminists wrote about scientific matters in popular periodicals; they joined scientific societies and associations; they debated scientific questions at women's conventions, congresses, and social science meetings; and they implored all women to rely not only on their "intuitions" and impulses for truth, but also on "rigid reasoning," "impartial scrutiny of facts," and the "scientific study of social conditions."[18]

Feminists readily fused hygienic republicanism with evolutionary science. They knew that only evolving or developing organisms freed from constricting, artificial conditions can find health and fulfillment. No settled notion of absolute justice was persuasive; rather right, virtue, and law were all perceived as vulnerable to functional criteria. In 1872, for example, the *New Northwest*, an Oregon feminist periodical, declared that "every development of science is a step. It is now discovered that constitutions grow and are not made—that is to say, founded on the laws of nature."[19] Another feminist wrote for the *Ballot Box* that "it is now an aphorism that development cannot advance without inherent personal liberty."[20] "Evolutionary theory," Mrs. E. H. West declared in 1879 before the Illinois Social Science Association, "offers the secret to successful living and social evolution affords the hope of future and the rule of progress." Such popularizers of evolutionary theory as Henry Buckle, Thomas Huxley, and Herbert Spencer applied evolution to the condition of women, lending support and substance to feminist arguments. The German popularizer of Darwin, Ludwig Buechner, who journeyed to the United States in the early 1870s, and contributed articles to the *Woman's Journal* and *Woodhull and Claflin's Weekly*, blamed sexual inequality on environmental conditions, and lectured frequently on the positive evolutionary potential in the construction of egalitarian marriages and in the entrance of women as equals with men into the professions and political life.[21]

Social science itself may have emerged in its reformist character because many Americans had come to believe that harmony between the sexes was the most lasting basis for a healthy, biologically advanced social order. Feminist social scientists—among them Lester Ward, Albert Brisbane, John Bascom, Amassa and Francis Walker, Henry Carey, Robert Hamilton, Elizabeth Osgood Willard, and Louis Masquirier—maintained that the laws of the social sciences replicated the laws of all organic phenomena and that, once discovered, they would create objective grounds for the most balanced and orderly social system. Sociophysiologists, they sought to monitor rationally human passions and drives, to channel and constrain the disruptive content inherent in all efforts to attain individual freedom by political means. Pragmatic social scientists (and to a large extent feminists) conceived of themselves as in tune with nature and thus above politics and ideology.[22]

The Ideal of Individual Symmetry

WOMAN'S RIGHTS SUPPORTERS zealously embraced the study of preventive hygiene and physiological science as one of the secrets of marital success as well as of social order generally. Preventive hygiene comprised the educated use of historically commonplace prescriptions: pure air; loose, uninhibiting dress; exercise; unconstrained living conditions with clean, well-ventilated open spaces; and a careful "regulation of the passions."[23] Prevention mapped the way to well-balanced physical harmony, a new physiological conception—new, at least, for the American middle classes—of the human body.[24]

No longer persuaded or mystified by mind-body dualisms, by the conflict between physiology and psychology, or by theories of the innate depravity of man, reformers viewed the faculties and organs of the body as naturally flawless if allowed to implement inherent laws. They also believed that the emotional, intellectual, and physical economy of the human system—what Matilda Joslyn Gage called the "triplicate unity"—would remain healthy so long as all its parts performed in a sustained "natural reciprocity" or in a perfect "balance of power," so long as "every faculty" received "proper cultivation."[25] For feminists no evil existed in the physical world save disease or other conditions that were capable of being changed; equilibrium was to be regained or acquired.[26]

The former transcendentalist Georgiana Bruce Kirby, for example, held fast to this position. The daughter of a shipowner who died at sea without having seen his child and of a mother who gave her very little mothering, Kirby immigrated to Canada from England with family friends, moved to Boston alone in 1837, "stepped out of her class" to work as a domestic and then as a schoolteacher, fell in with the progressive Unitarian intelligentsia, and took part in the Brook Farm experiment of the mid–1840s. There she learned reformist hygienic principles that stayed with her for the rest of her life.[27] In the 1860s, she went to California, married, bore several children, and managed a progressive home that acted as a magnet for young women suffering under the weight of the traditional "domestic system."[28] She lectured widely on rational hygiene. Her pamphlet "Transmission; or Variation of Character Through the Mother," highly praised by feminists, developed the commonly held reformist theory that environmental conditions governed genetic competence in men and women; sexual union could be severed from the stigma of "coarseness" by reconception "as an act of union as pure in its character as the blossoming of the lily or the rose." "Science proves," she wrote, in triplicate unity, "that sin, evil, wickedness mean merely a want of balance among the faculties in themselves good."[29]

Most feminists, from refined and highly educated Unitarians to less cultivated petty-bourgeois shopkeepers and artisans, sought to live hygienic, symmetrical lives. Most looked at the body in positive terms. The feminist novelist Zadel Barnes Buddington Gustafsen wrote in an article that appeared in both the *Woman's Journal* and the *Golden Age*:

> When men and women learn that no physical organs, no physical functions, no passions, are in the least shameful or degrading . . . when they realize that it is the irrational abuse, and never the rational use of Nature, which culminates in evil; when they begin sincerely to attempt insight into the complemented relationship between body and mind, into the laws of adaptation which makes men and women perfectly suited to and absolutely in need of each other—then it will be hoped that men and women will work together not with the heart only but with the understanding.[30]

One reformer, writing for *Alpha*, even defended nudity on the grounds that "naturally no one organ of the body is more obscene than another . . . the exposure of no portion of the human body, *if one were properly educated*, would be considered obscene."[31]

Symmetry shielded the body from sickness: if the educated, positive use of the will ruled the body's natural forces, disease would never appear. Feminists conceived of the body as an organic system of positive-negative, masculine-feminine "valences," or as a system of "centripetal and centrifugal forces" kept in balance by the "galvanic battery of the brain." The best body, in other words, "balanced" or "equilibrated" the maternal and paternal energies.[32] The sexual organs played no greater role within this system or in shaping human behavior than the liver or the heart or the stomach; sexual needs matched in importance and in their harmlessness the need to sleep, think, eat, and defecate. "Woman's sexual processes,"

Nancy Swisshelm wrote for the *Woman's Journal* in 1875, "belong to the involuntary processes of life and are no more startling to a healthy woman than digestion or circulation."[33] "It is established beyond all doubt," Sarah Hackett Stevenson explained in *Physiology of Women*, "that all the functions of the body should be exercised, and that the sexual functions form no exception."[34]

Feminists did not condemn passion. "Repression" of any faculty kills, Lydia Maria Child and Abigail Scott Duniway warned in *Woman's Journal* and *New Northwest* respectively.[35] Like Lester Ward, an important expounder of this tradition, they were unwilling to "damn the stream of human desire." Yet as advocates of balance they did object, and objected passionately, to intemperate use of passion. They sought to "direct" desire "into channels not only innocent, but useful."[36] In this quest for self-control and balance feminists resembled the early Puritans, but, unlike the Puritans, they were born into a tumultuous world already changed and thus felt less fear of spiritual defilement. The early Puritans denounced the temptations fostered by a new capitalist-liberal society taking shape around them that they had helped create. Proponents of social discipline and community, many even rejected liberalism itself. Feminists accepted the liberal order even while they feared its effects on behavior. Both the acceptance and the fear found expression in the concept of individual symmetry.[37]

In 1869 Mrs. T. H. Keckeler, the first woman from the American West to graduate from a regular medical institution and one of the doctors feminists relied upon for hygienic information, in her book *Thaleia* and in a series of articles for the *Herald of Health*, elaborated at length the theory of harmonious development.[38] Every organ and faculty of the body, she said, should be exercised in order to secure the happiness of the individual. No single "power" of the body should dominate other "powers." "All excessive departures from the average," she observed, "partake of the nature of monstrosity. . . . Extremes instead of harmonious balance . . . destroy the progress of humanity, for some line of action would soon gain ascendency to the exclusion of all the other." For example, "The brain must be exercised in various directions or it will lead to monomania or insanity."[39] The Stanton and Anthony paper, *Revolution*, recommended that Keckeler's works "should be on sale in every city in the country."[40]

Antoinette Brown Blackwell agreed with Keckeler. "How can there exist," she asked, "a more fundamental antagonism between individual well-being and the balanced exercise of one function, than between it and any or every other function? They have all grown together in mutual adaptation. A disturbance of one is the disturbance of all."[41]

Free-love feminists as well as the more conservative ones held this view. The free-love radical Alfred Cridge wrote to Caroline Dall in 1871 that nothing should "prevent gratification of *natural* desires, although means may be found to correct unnatural intensifications."[42] Every organ or faculty of the body that broke the "balance of power" by exaggerating its function endangered the body with disease. A tyrannical or overstimulated uterus or penis made the body sick. "There is no *evil* in the contact of sexes," David Croly wrote reassuringly and with the scientific blandness that marked reformist prose on the subject. "Action is itself natural and healthful provided there are no excesses on the part of either of the connectionists."[43]

The ideal of symmetry also became an article of faith for socialists both in the United States and abroad. In 1883 the German feminist August Bebel wrote in *Woman Under Socialism* that the

> mental activity of the human being is the expression of the physiologic composition of its organs. The complete health of the former is intimately connected with the health of the latter. A disturbance of one inevitably has a disturbing effect upon the other. Nor do the so-called animal desires take lower rank than the so-called mental ones. One set and the other are effects of the identical combined organism: the influence of the two upon each other is mutual and continuous. This holds good for man as well as woman.[44]

Bebel believed that "bourgeois society is far from a general acceptance of this maxim," a view that would have met with little contention from American bourgeois feminists of the 1870s.

Rooted in the eighteenth-century organic tradition of common sense that emphasized the importance of social interdependence and harmony, and refined further by utopian thought, the ideal of the perfectly symmetrical human being drew the allegiance of many nineteenth-century Americans.[45] According to a recent historian of Unitarianism, upper-class ministers "frequently led their congregations in the contemplation of the balanced character of the virtuous man . . . If there was a single conception that dominated Harvard Unitarian thought . . . it was the conception of harmony . . . To harmonize, regulate, and balance was the task of the ruling power."[46] No wonder, then, that we find the concept so clearly present in the thinking of such Unitarian feminists as Georgiana Bruce, Caroline Dall, Thomas Higginson, and Caroline Severance, among others. The concept had even greater diffusion at the university level, where almost every college student, male and female, studied the precepts of moral philosophy. More important, the ideal pervaded the works of popular health reformers and physiologists, who, like the Unitarian intellectuals and college students of the day, offered the symmetrical, rational human being as the model of mental and physical sanity. Indeed, this intellectual concept, which required for its working out a reasonable degree of education, leisure, and wealth, appeared to unify upper-class men and women from mercantile backgrounds, genteelly educated professionals, and petty-bourgeois shopkeepers and artisans into one reformist bourgeoisie.

Symmetry as a Social Concept

THE MODEL OF EQUILIBRIUM had obvious feminist implications: women must be fully developed emotionally, intellectually, and physically to secure and keep good health; they must relegate their sexual responsibilities to a rational place within a range of equally important priorities, and thus take on the identity of natural, republican citizenship. "Women will go up," a feminist wrote in 1871 for the *Woman's Journal*, "when rationality is substituted for sensuality in the aim of womanhood." Rationally developed, women could then truly marry men.[47] "We lay down the broad principle," wrote Sarah Hackett Stevenson, "that men and women were made for each other, physically, mentally, and morally." "Only" the "sexual" has kept them apart.[48] In a suffrage speech delivered in Philadelphia in 1871, the liberal minister Celia Burleigh said: "What is demanded now for woman is symmetrical development; hitherto only her affectional nature has been developed—her intellectual kept in the background."[49] Novelist Marie Howland warned of the dangers of violating the law of equilibrium. "It rules everywhere," she said to a National Woman's Suffrage Association audience in 1872, "and the compensation for disenfranchising one half the race is terrible, mask the facts as we may."[50]

From the 1850s onward, feminists attacked men who drank alcohol, ate, labored, fornicated, or did anything else to excess: such men passed on their seriously diseased and unbalanced sensibilities to their children, creating a race "of infirm, deformed, imbecile or unusually diseased persons."[51] In 1852 Susan B. Anthony said flatly, "No woman should consent to be the wife of a drunkard because she must be the medium of the stamping of new forms of immortality with his gross carnal nature."[52] Twenty-five years later Annie S. McDowall wrote for *Alpha* that "a woman who loathes or fears her husband will find that the children born to her while living in a chronic state of terror and loathing, will inevitably be fretful and troubled in their infancy" and will ultimately become "treacherous animals."[53] Feminists warned men to flee those women who would willingly sacrifice their lives to any single function; they warned men further not to subject the unwilling woman to the dominion of a single function. The feminist man married neither "drudge nor ornament." He would not try to transform his wife into a woman chained to "sexual servitude or bodily toil" or to conditions in which "her mind rises no further than the roof that shelters it," nor would he wish his wife to be a fashionable woman who "went out into the street to spend money, or to exhibit its worth when spent on her own highly decorated form."[54] Both drudge and ornament were dependent on others and enslaved, one to excessive labor and child-bearing, the other to love and leisure. Bondage to a single function inevitably engendered disharmony and disease in both wives and children. "Those who live in luxury and idleness," observed Mary Gove Nichols, another water-cure specialist, "and those who live in want and hard labor are the most diseased and wretched." Shaped physiologically by a long history of degradation and oppression, these people created diseased progeny that contributed tragically, in the words of Dr. Ann Preston, to the "degeneracy of the race."[55] The images of drudge and ornament reappear again and again in bourgeois feminist polemic. Mary Wollstonecraft, Margaret Fuller, and the Grimke sisters employed them and feminists continued to use them in the twentieth century.[56] For feminists, the ideal marital choice stood between the two extremes—strong-minded types with republican sympathies who did not consent to be slaves or toys. "Their love of harmony is such," wrote Paulina Wright Davis, "that it drives them from either extreme."[57]

As the historian Linda Gordon has recently shown in *Woman's Body, Woman's Right*, mid-century feminists relied heavily on a "folkloric" hereditarian tradition to bolster their position on marriage, although it "lacked a sound genetic base."[58] Before 1850 an environmentalist Lockean philosophy had dominated the reformist mind, but after 1850 health reformers like Trall and Jackson, the phrenologists Orson Fowler and Samuel Wells, and most other feminists tried to fuse environmentalism with the doctrine of the inheritability of acquired characteristics.[59] All male and female excesses, they believed, had deep and lasting effects on the physical-mental organism and could be transmitted to children. Only men and women free from such conditions could create a race and society of perfected human beings. In the late 1860s and early 1870s, stirpiculture, a new science of animal breeding propounded by such reformers as John Humphrey Noyes of the Oneida Community, along with the antienvironmentalist and antireligious thought of the English eugenicist John Galton, even challenged the dominance of the folkloric tradition. Rejecting the doctrine of acquired characteristics, Galton substituted the "principle of the continuity of germ plasm which asserted that nothing passes between parent and child except that which is contained in the nucleus of the fertilized egg."[60] Charlotte Lozier, Jane Croly, a host of feminist free-lovers, and Charles Eliot, president of Harvard, favored the use of scientific breeding for human beings; in the 1870s and 1880s Elizabeth Cady Stanton spread the message of John Galton wherever she lectured on the science of life.[61]

As Linda Gordon has shown, a strong pessimism pervaded the eugenic thought of this period. Feminists appeared to abandon the optimistic theory of *tabula rasa*. They painted a darker picture of existing social obstacles to progressive change.[62] "They who can give the world children," Stanton wrote in 1869 for the *Revolution*, "with splendid physique, strong intellect, and high moral sentiment, may conscientiously take on themselves the responsibility of marriage and maternity." The rest should be legally prevented from doing so.[63] Ten years later Stanton still held the same position, but by this time she had absorbed Galton: "The law of heredity should exclude many from entering the marriage relation . . . The object of crossing is an important one in marriage. There is a law of vital as well as of chemical affinity which should be observed."[64] By the late 1870s many feminists supported state legislation to block "unhygienic" marriages.[65] Using the latest social science methods, Sara Spencer announced before the Fifth Woman's Congress in 1877 that "our statistics have shown that more than two millions of children have been born and are now living in this country whom the whole world declared unfit to live, born to a heritage of suffering, vice and crime."[66]

On the other hand, feminists never tired of the essentially environmentalist concept that only symmetrically developed men and women, only whole human beings, should marry. This was a deeply rooted principle in the bourgeois liberal tradition and the one with the most feminist promise. Even Stanton, the most Galtonian of feminists, presented her eugenic views from an environmentalist perspective, emphasizing the importance of character and acquired strength as much as or more than genetic competence in both sexes.[67] All feminists believed that only strong, independent, but also tender men and women, who combined in their natures the best virtues of both sexes, could make good marital partners and parents.[68] "Marry your conjugal mate," declared John Cowan, quoting the free-thinker George Stearns, "your personal duplicate, your approximate equal in development and your like in age, temper, intelligence, sentiment, devotion, taste, habitudes."[69] Cowan outlined his views on symmetrical health in his *The Science of a New Life*, probably the best single source for insight into the sexual beliefs of most feminists of the time. Thomas Wentworth Higginson, William Lloyd Garrison, Robert Dale Owen, Lucy Stone, Elizabeth Cady Stanton, and the feminist doctor Alice Stockham swore by it. Stanton used it as a "textbook" in her lectures "to women alone."[70]

A passive, servile, dependent woman devoted only to childbearing and child rearing or an aggressive man obsessed with sexual dominance "destroyed the best qualities of both and the best interests of Humanity."[71] One characteristic statement said:

> *Passivity* of women coordinated with the appetites of men create drunkards. . . . No man may sufficiently emphasize the impossibility of rearing noble children while woman is a serf and a slave to man. . . . Noble motherhood can come only by the emancipation of woman and the strength of character . . . freedom alone will give her. The laws of heredity will not permit the elevation of man from the curse of drunkenness until woman is made free and thereby strong.[72]

To strengthen their polemic on healthy marriages, feminists resorted to their own version of Darwinian theory. For example, one of the most vivacious and gifted of mid-nineteenth-century feminists, Antoinette Brown Blackwell, in her book *The Sexes Throughout Nature*, criticized Darwin for resting his evolutionary case on the "time-honored assumption that the male is the normal type of his species" and the female only a "modification to a special end" and discussed the concept of equilibrium or symmetry in both individual and racial terms from a completely scientific perspective. Although in 1850 she had become the first American woman ordained to the ministry, she threw over the "intolerable" Calvinist tenets of her youth and resigned six months after her ordination. Lucy Stone's warning to her had been prophetic. "The bottom fell out of my theology," she said simply, in a reminiscence recorded by Alice Stone Blackwell.[73] After her marriage in 1856, Blackwell left public life to rear five children; she occupied her spare time reading and thinking about scientific and hygienic literature, devouring Spencer and Darwin and studying social science.[74] By the early 1870s, when Blackwell once again surfaced as a public person, she had become thoroughly immersed in the scientific-rational world view, but without losing her faith in God.

Individual men and women of "this day," Blackwell wrote in *The Sexes Throughout Nature*, display in extreme form opposite sexual traits: higher "sensibility" and self-abnegation in women, "sexual fervor," aggression, and rationality in men.[75] To illustrate conventional male behavior, she used the hubristic male peacock, a bird of a lower order of development and one entirely lacking in the capacity for "parental love." Such traits in men and women reflect the "weight of social conditions" that have "arrogantly repressed" true sexual potential or the "union of equivalents." In the future, and as an outcome of uninhibited evolutionary change, the best men and women will "begin to develop the character of the other" and the best children will "combine in modified form all the qualities of both parents at their best." Thus the "equilibrium of sex" will be restored: men will learn "householding" and nurturant ways and women will acquire "more symmetrical development" and exhibit powers of "abstract thinking and feeling." From an inferior state of uneven and explosive development, the race will progress to a perfect equilibrium of forces, the "cooperative elements" of each sex "balancing the other in their entirety." Unlike some Darwinians, Blackwell preferred to mark the movement toward balance and cooperation rather than struggle and savage rivalry as the true index of the evolutionary progress. "That Nature could have originated otherwise than through the natural creation or adaptation of a *cooperative constitution* of things, coordinating all substances, sentient and unsentient, is, to my apprehension, utterly incredible."

Highly individualistic in its original bias, health reform had taken on a cooperative social dimension quite absent in the earlier Jacksonian period—a period of early capitalist development when most Americans clung tenaciously to laissez faire principles. As the historian William Coleman has observed in a different context and of the earlier era, "the correct use of things non-natural" and the quest for perfect internal balance also identified the health ideology of the late-eighteenth-century French bourgeoisie in the first flush of its power: radically individualist, it repudiated state intervention in matters of health as in all other areas of social and economic life.[76] For the bourgeois, health, like business, represented the domain of individual freedom and stood apart from the general requirements of the political state. Mid-nineteenth-century feminists still abided by this ideology, with its expansionist base and keen allegiance to productive labor, but with some crucial differences. For one thing, because they had come to conceive of society itself in organic ways or as a mirror image of the symmetrical body, they believed that society progressed only as it moved forward as a whole, harmonious and fully developed in its parts. When any part, either sex, any individual lurched out greedily to claim the prizes of life, shock waves threw the social organism out of balance and into disruption. Thus feminist principles of "symmetrical action" or "nature's regimen of work" called into question excessive capitalist accumulation, extreme specialization, and "irrational" intensive demands on men by business "civilization." "The positive and negative forces," as one feminist put it in 1873, "must come to a perfect state of equilibrium in all social and governmental arrangements as well as in the mental and physical organizations of individuals."[77]

Feminists tended to view deviations from the healthy norm—excessive sex, smoking, drinking, eating—in social, not individual, terms. Dr. Mary Putnam Jacobi quoted Auguste Comte: "Even the habit of personal cleanliness becomes elevated from the rank of a simple hygienic precept, by the recollection that it renders each adapted to serve the others."[78] Feminists had come to believe that changing social and economic conditions no longer permitted men and women to follow independent hygienic regimes or to create natural well-balanced wholeness without some community assistance. American reformers, and feminists above all, increasingly turned to the state to administer the needs of public health (the least apparently political of all areas of life) on the grounds that such administration answered universal human needs. By 1875 most feminists had thrown their support behind state and city sanitary reforms, the creation of state boards of health, the introduction of physiological and hygienic education into the school curriculum; and some feminists—most notably Elizabeth Cady Stanton—even proposed that the state should prevent the physically and mentally handicapped from marrying.

The critique based on symmetry, however, went beyond an indictment of nineteenth-century social and economic conditions to become perhaps the first American reformist effort to demystify the body and to rationalize sexuality. The body and sexuality would thus be freed from associations with sin, depravity, guilt, and evil. Sexuality was lifted out of the realm of religion and released from instinctive life into the realm of functional science and rational, educated self-scrutiny. Feminist hygienists initiated what twentieth-century sex reformers have almost completed. They rationalized and demystified sexuality by conceiving of it as a conflict-free natural (though very educable and self-conscious) function equal in importance to and differing only marginally from eating, sleeping, and defecating. By focusing on the mastery of physical processes as the basis for happiness and health, feminist hygienists prefigured the work of Havelock Ellis, Alfred Kinsey, William Masters and Virginia Johnson, and John Money.[79] Like many twentieth-century feminist sex reformers, they attempted to free the body in a democratic setting, while at the same time relying on it as a principal instrument of social control. They too concentrated on the need to create an expressive wholeness that would prevent psychological imbalance or passionate excess, and they too, by elaborating the illusion of personal freedom, contributed to the mystification of the basic and unchanged relationship between the individual and social and economic status—that is, the need to compete alone in an "open" market for material well-being. The nineteenth-century reformist tradition of "wholeness" and sanitary perfection, of revulsion for dirt and disorder, and of contempt for smells, the smell of the body and of mortality—the tradition that has often been called "Victorian"—began to take shape in the late nineteenth century. It did not die in the twentieth century, but came truly into its own to become fully integrated into everyday social life.

The idea of the significance of the body, or, more broadly, the concept of mental, physical, and social health, was the most important contribution of bourgeois women to society. It enhanced their social status and gave them more control over their own lives. The deeply felt words of the Connecticut feminist and member of the famous Beecher clan, Isabella Beecher Hooker, in a letter to John Stuart Mill in 1869, reflected this sense of control and hence of power:

> What father can say "Thou art my child," as a mother can? and through what channels does he count the life beat of his child as his own? To my mind there is more sense of power in this sense of motherhood than in all things else. . . . To create, is to live—to express our own being through another is everlasting youth—and to mould, guide and control this is the glory of existence, its supreme delight.[80]

We find this potential made even more explicit in Elizabeth Osgood Willard's *Sexology*: "Theory and Philosophy are internal and feminine because the feminine or material law is the law of organization on the internal fundamental plane."[81] The intimate connection between women and the private sphere, between women and internal forms of control, between mothers and the bodies of both sexes made it inevitable, in fact, that women would emerge in the social realm as a considerable power, especially in a historical context where conventional and external forms of male authority had become somewhat eroded or threatened. Feminists, it appears, understood this and constructed their ideology, in part, from the elements of such knowledge.

Of course great differences distinguish nineteenth-century feminists from their twentieth-century counterparts. The earlier group wanted to "spiritualize" sexuality, to expunge the "dark drop which circulates in the social organism"; they sought to rationalize or "equilibrate" all bodily functions to free both men and women for active careers geared to production.[82] Because they focused on freeing women as equals with men in the public as well as the private sphere, their critique did not go beyond heterosexual egalitarianism to incorporate other forms of sexual expression. Indeed, they considered self-directed sexual activities—especially solitary masturbation—dangerous pathologies. Sharing the sexual views of their middle-class contemporaries and, above all, seeking to liberate women from the confines of the private sphere, they feared inwardly directed sexual behavior as a potential for "disease" in themselves.

By shoring up individuals with a fabric of undeniably oppressive taboos, by giving contour, as it were, to sexual expression, nineteenth-century feminists retained the ideas of guilt and conflict, although they helped begin the struggle to create the social conditions they hoped would remove them. At the same time they pointed individuals toward productive activity in the public realm. In our own time the existence of sexual deviation, and of the guilt conflict formerly caused by it, has been seriously questioned and often entirely rejected. In a society that now emphasizes the virtues of private life and has increasingly turned away from the withering demands of public life, all forms of sexual behavior have their vocal and persuasive defenders. Deviance is losing its stigma, the individual's capacity to challenge the social order seems to have weakened, and the continuing presence of genuine inner conflicts within women and men has become obscured.

CHAPTER 2

PREPARATION FOR MARRIAGE: SCIENTIFIC KNOWLEDGE AND THE CULT OF NO SECRETS

ON A WARM spring evening in May of 1878, several Boston reform women met at an exclusive private club on Chestnut Street, a narrow and very fashionable street on Beacon Hill, known for its elegant homes and for the wealthy reformers who lived in them. Formerly the location of the Boston Radical Club, the building was now simply a private gathering place for reformers. The socially ambitious and charming Mrs. John T. Sargent presided as a paper was delivered by a Mrs. Wells on the growth of literature dealing with abortion and on the trial of Charles Bradlaugh in England for distributing birth-control information to the public.

Influenced by the famous Boston gynecologist Dr. Edward H. Clarke, who had advised her on these subjects shortly before he died, Wells made two general points: that the "literature circulating everywhere" on abortion was bad and "taught immorality," and that "unhappiness in married life is caused by women having children when they don't want them; and that they ought not to have them." Her argument characterized feminist debate on these questions at the time: bemoaning the horrors of abortion, feminists nevertheless insisted upon limitations on fertility. Wells herself had written for the *Woman's Journal* that there had been "much false modesty—rising from religious creeds and from ancestral ignorance of physiology and anatomy. . . . When we know more of the body and mind we shall use the poetical language of science."[1] The other women assembled on Chestnut Street—including feminists Lucy Stone and Martha Clarke—thought Wells handled her subject with skill, delicacy, and, above all, scientific objectivity. They agreed that some measures should be taken before marriage to prepare young adults for sexual relations and parentage and for decisions on the number of children they would bear. Most believed that mothers should instruct their sons in the ways of sexual control. That failing, Stone suggested, mothers should interview prospective husbands and set them straight on these matters.[2]

The Chestnut Street assemblage symbolized the feminist commitment to the rational Enlightenment principle of "no secrets" that almost every major feminist espoused. This principle was in part a scientific one. All women, feminists believed, should study the scientific facts of life to discover reliable truths upon which to build a better, more lasting social order. No secrets had an equally, if not more important, social dimension. By releasing all knowledge to women, and, above all, by giving them unlimited access to scientific information on sex and physiology, no secrets protected women against the deceit, exploitation, and conflict that inevitably ensued from sexual ignorance. No secrets made private mysteries public property for both sexes. It supposedly formed the foundation for rational decision making and for a more equitable binding consensus within the private sphere. At the same time, however, it exposed the private sphere to public intrusions, catching feminists in a swirl of public revelation that culminated in a reaction to the principle's application.

The Doctrine of No Secrets

ZENA FAY PEIRCE, a Cambridge reformer well known in the 1870s for her cooperative housekeeping scheme and the wife of the founder of pragmatism, Charles S. Peirce, told the New York City Woman's Parliament in 1869, "Bad morals will exist only as long as virtuous women do not choose to look at all they see, to understand all they hear, and to tell all they know." The popular Universalist minister Phoebe Hanaford wrote a letter to the *Woman's Advocate* the same year praising Harriet Beecher Stowe for her bold revelations in *Lady Byron Vindicated:* "Your book shows that 'there is nothing hid that shall not be revealed.' "[3] Isabella Beecher Hooker explained privately to Mary Livermore, the powerful Chicago feminist and coeditor of the *Woman's Journal*, that "it is secret friendships and association with the impure that is dangerous—not a public advocacy of a righteous cause on the same platform," and Julia Ward Howe complained to the freethinking radical Moncure Conway about the "mummery and deceit" of modern life. "It has gone on long enough," said Howe. "A more honest dispensation is coming in."[4]

Some feminists took the principle of no secrets seriously enough to analyze it in their writings. The great feminist doctor Mary Putnam Jacobi delivered a lecture on "The New Moral Safeguard" before the New York City Positivist Society. "Live in the open air!" she exclaimed. "A thing that one is not willing the whole world should know is wrong."[5] A year later she argued this thesis at greater length before the same society. She attacked Christianity and, above all, the Catholic Church for introducing into "consciousness a multitude of new crimes and a new intensity of feeling about these inner crimes." "The antique world," she said,

> trembled before the powers of Nature. After the introduction of Christianity, the world trembled before the more awful terrors of conscience. But as the consolations of Christianity left untouched the evils which had caused pain, and only encouraged the sufferer to bear these with patience, so the fulminations of Christianity were directed so predominantly to the purification of the soul, that both the antecedents and the consequences of crime were relatively overlooked; wrongdoing was not the offense against public order which it had been

> to the Roman law, the crime,—it was the sin which did not need to be public in order to be deadly,—indeed was liable to be more deadly the more it was secret and concealed. Hence sprang up the conception of unspeakable crime.[6]

This remarkable statement displayed the undeniable stamp of the rationalist Enlightenment, and in its sweeping condemnation of Christianity it went much further than most feminists were willing to go, although it captured the substance of feminist thought on this question. Putnam Jacobi hated the Catholic Church. The Church gloated, she believed, over the private conscience, creating painful private sins that could be removed only through private means—the Catholic confessional. In her book, *The Value of Life*, she called Catholicism the "religion of failure, of ignorance" and the institution of private confession an easy and infantilizing escape and a ticket to chronic immorality.[7] Catholicism artificially manufactured "deadly" and "unspeakable" private sins that threatened the public order and then allowed these sins to flourish. She denied the existence of sins, coming close to repudiating the private conscience. Apparently withdrawing personal life in all its variability from the clutches of the puritanical priesthood, at the same time she insisted that the essence of the personal life—its private character—be open to public inspection. "Live in the open air!" she said, and public order will be preserved.

Moncure Conway, an intellectual similar to Putnam Jacobi in many ways, even advised women to remember the moral strengths of Lady Godiva, a lady who indeed lived in the open air. According to Conway, traditional women had worked with their spirits suppressed and ashamed, their passions muzzled, and their influence compelled to "seek hidden and often illegitimate channels." Modern women, however, could no longer afford to repeat these mistakes. They must "face with clear-eyed purity all the facts of nature . . . cast away the mental and moral swaddling clothes transmitted from Eden, and put forth all their powers for the welfare of mankind."[8] In 1883 Lester Ward summed up this evolving reformist position in the first volume of his *Dynamic Sociology*: "Whatever must be done secretly or clandestinely will be done improperly and become an evil though it possesses no intrinsically evil elements."[9]

Allegiance to this principle tended to unify many mid-nineteenth-century feminists who in other areas might have seriously quarreled. Such anarchists as Ezra Heywood, Angela Heywood, Moses Hull, Lois Waisbroker, and Moses Harman pleaded for absolute openness in matters of sex and hygiene; they printed obscenities and accounts of perversions in their publications, and regarded the disclosing of the worst social evils as the first step to removing them.[10] Ezra Heywood, for example, built his reform career on a muckraking passion. Born in Massachusetts in 1829, Heywood prepared himself for the Congregational ministry in the 1840s, but, like many feminist men of his time, he left the church to follow the radical and secular Garrison banner. After the Civil War he held fast to his radicalism. He edited the anarchist journal *The Word*, and disseminated the principles of free love. A rationalist, Heywood sought almost obsessively to disable the irrational power of passion by spreading physiological and sexual truths. "My object in writing 'Cupid's Yoke,' " he wrote about his most notorious statement on the subject of sex, "was to promote discretion and purity in love by bringing sexuality within the domain of reason and moral obligation."[11] In 1872, in the wake of the McFarland-Richardson divorce scandal of 1870 and on the brink of the most sensational scandal of the century, the sexual affair of the famous Congregational minister Henry Ward Beecher with the wife of journalist Theodore Tilton, Heywood looked forward with a rejoicing probably unmatched by any of his contemporaries to what he called the "Great Universal Washing Day, when everybody's dirty linen will be paraded, and when the leaders of sham morality in high places will find it will cost more than $.75 a dozen."[12]

Nineteenth-century feminists were convinced that secrecy poisoned the harmonious relationship between humankind and nature, between the public and private spheres. Real, undistorted facts could emerge only within an open context of equivalence; and only in such a context could men and women formulate lasting and basic laws conducive to health. Mary Davis, the wife of Andrew Jackson Davis and herself a spiritualist and an active feminist, wrote to the New York Free Religious Convention in 1873 that "moral laws can never be derived from authority, but only from a study of the relations of human nature."[13] Secrecy implied an unjust, authoritarian system of relations in which certain groups and classes monopolized truth and power; in which men wielded undisputed authority over women and nature itself held a mystifying dominion over human life. Openness promoted a naturally egalitarian relationship between the sexes and, as in the economic world of the free and open market, led to the discovery of universal moral laws that might form a permanent basis for certainty and order. It also transformed the nature of human relations and of human perception.

On a blustery Boston evening in January 1869, and in the midst of much feminist ferment, Caroline Dall, one of the most important nineteenth-century feminists, held a revealing conversation with another reformer, Martha Goddard, in Goddard's home. Dall considered it so important that she recorded it in her private journal:

> If people could only know the secrets of other lives, how differently they would judge, Martha said at last.
>
> Yes, I said, that would be Heaven—the exact knowledge of facts.
>
> That is a grand thing to say, said Martha. I never heard that before, I always think—well *there* . . . I shall have justice.
>
> And *eyesight*, I replied, that is what I want above all. . . . Indeed, doubt and uncertainty make hell.[14]

The passion for "eyesight" in a growing secular and industrial world—what John Stuart Mill called the "habit of analysis" and Michael Foucault later called the "gaze"—the need to identify and appropriate the very nature and heart of things informed, infused reformist consciousness.[15] It pointed to a new kind of personality that stood dispassionately apart from what it gazed upon and even what it touched, utterly removed from the politics of life, in an effort to obtain the truth. "Intellect without feeling," J. Stahl Paterson, a feminist and an important member of the Free Religion Association, wrote in 1870 for the *Radical*, "has no prejudice: it seeks only truth. Blind, contagious intellect finds no lodgement in such a mind. . . . It is mainly through the just discernment of the relations between things, and the intellectual control of social conditions, that there can be progress."[16]

Scientific Knowledge: The Career of Burt Green Wilder

THE NINETEENTH-CENTURY scientific imagination, the "intellect without feeling," clearly rested on the principle of no secrets, shaped as that imagination was by liberal-capitalist imperatives. By the late 1860s and early 1870s, it had become institutionalized in the curricula of the modern university. Early in the spring of 1867 the great and innovative president of Cornell University, Andrew Dickson White, brought to his progressive faculty a man who would later become one of the legendary and most respected figures in the study of comparative anatomy in America—Burt Green Wilder. A strong-willed and disciplined man who always finished what he started and who loved nothing more than to putter, unflustered, in his laboratory, Wilder devoted his life to the scientific discovery of hidden truths, a passion that incorporated the physical and social life of men and women.

Wilder was born in Boston on August 11, 1841. His parents practiced a Swedenborgian version of Unitarianism and, like so many members of the progressive Boston bourgeoisie who had broken with established religion, had adopted the rejuvenating tenets of homeopathic medicine, Grahamism, and Garrisonian abolitionism. As Wilder put it in his "Reminiscences," his parents "hated oppression."[17] The family interest in all things healthy and healthy-minded determined the course of Wilder's career. He attended the well-known Lawrence Scientific School in the late 1850s and later graduated from Harvard Medical School. He studied with such noted biologists as Jeffries Wyman and Louis Agassiz, and he got the best laboratory training possible in America at the time by enlisting as a surgeon in the Fifty-fifth Massachusetts Volunteer Infantry during the Civil War. As a member of one of the first universities in the country to introduce formally the study of comparative anatomy, physiology, and neurology into the curriculum, Wilder also joined the swelling ranks of temperance and hygienic reformers and, in the 1880s, became the director of the Non-Smoker's League.[18]

Wilder used the laboratory (again, a recent innovation) as a pedagogical device. He employed dissecting procedures that by the end of the century became commonplace in most universities. According to one of his students, Simon Henry Gage, Wilder relied primarily on the laboratory, rather than on "books and scientific papers," to teach his classes. To that extent he would have agreed with Mary Putnam Jacobi when she wrote in 1879 that "it is in the life of the laboratory, where the secrets of nature are not only divined but reproduced, that the true joy of knowledge can best be learned. Certain it is that from the laboratory this joy has been diffused with immeasurably greater rapidity and intensity than ever was the case where the library was the only centre of diffusion."[19] To that extent Wilder's technique pinpointed the shift in biology away from the study of causation and toward the positivistic examination of the phenomena of life.[20] "The last word," wrote Gage, "was always to be found in the actual things and only there. But to interpret the hieroglyphics, that was the question. He opened his laboratory where we could work with the real things and taught us to read in nature's own book."[21]

Wilder's students spent more than a third of their time in the laboratory experimenting with and dissecting a wide range of ordinary and extraordinary specimens: the embryos of dead or killed animals; the brains of the criminal and insane; the anatomical parts of circus animals that Wilder had scavenged in his spare time; and, above all, the easily obtainable and cheaply preserved common house cat.[22] Although Wilder criticized inhumane practices and recommended with other reformers the use of chloroform to execute criminals, he had no reservations about drowning great numbers of cats for experimental purposes. In the name of scientific objectivity and the rational pursuit of truth, Wilder's laboratory stripped the cat of its domestic and sentimental symbolic significance. Antivivisectionists and those faithful to bourgeois sentimentalism railed against him in the 1870s and 1880s, while he reviled them for trying to "abolish the objective method of teaching physiology" and failing to distinguish between painful vivisections and those attended by the use of anesthetics. Death under an anesthetic or even by drowning seemed to Wilder far more humane and defensible (especially from the experimenter's point of view) than death accompanied by protracted pain. By his own account, he never performed sentisection or painful vivisection.[23]

Wilder supplied his students with thousands of cats over the years and never wearied of extolling their virtues as ideal laboratory specimens—a penchant that once prompted Oliver Wendell Holmes to gibe: "I have no doubt that cats feel honoured by the good things you have said of them as subjects for dissection, but I pity you if you find yourself in a room full of ladies anyday."[24] Wilder preferred cats as anatomical specimens because they served as the best introduction to the morphological study of human beings and other mammals. Unlike the "inconveniently large," "expensive," and difficult to get human subject or the endlessly "variable" and noisy dog, the marvelously silent cat did not vary in shape or size. It also "succumbed quickly to anaesthetics"; it reproduced abundantly and could be "easily reared . . . and kept in confinement even in considerable numbers without difficulty."[25] Most important, the amenable anatomy of the cat provided the "best standard of comparison for the study of other mammals."

Among the departments within the increasingly complex and diversified fields of natural history and biology, Wilder studied the branch of homology. With the cat as his point of departure, he sought to compare and determine the corresponding organs and structures in other animals as well as the corresponding parts of the same individual.[26] As a sideline he hoarded over two thousand brains of "educated and orderly" persons, criminals, elephants, and camels; and, like his mentor, Louis Agassiz, collected and advertised for animal embryos from "pregnant females of every breed, in different periods of gestation . . . that the peculiarities of the young at each period of growth shall be ascertained for each breed."[27]

"Science," as the Philadelphia naturalist Grace Anne Lewis told the 1876 Woman's Congress, "has changed all its methods and we now look for likeness rather than difference."[28] This generalization applied aptly to Wilder's work, for although he wanted, in part, to discover the dissimilarities among the inner structures of mammals, more often he sought to uncover structural convergence, resemblance, and equivalence, and to elucidate the harmonious and coordinated perfection of the human system. Like his great rationalist forebears, Cuvier and Milne-Edwards, he wished to understand the laws that made the human body an efficient "workshop" in which each organ and structure skillfully cooperated with every other organ and structure to produce a functioning and organic whole.[29] In the spirit of this quest, and in the manner of contemporary social and natural scientists who studied diseases, criminals, and populations, Wilder projected a uniform nomenclature that would transcend national differences by giving each zoological fact or object "but one . . . exact . . . Latin . . . name."[30] In his most famous book, *Anatomical Technology* (1882), he deplored existing vernacular systems of classification as arbitrary; he called them unintelligible and chauvinistic. Cosmopolitan by nature, scientific facts must no longer be trapped within the linguistic idiosyncracies of national cultures. Trachea, therefore, must supersede "windpipe, weasand and *conduit aerien*," and for the sake of brevity and exactness "*tuberculum biseunium antericis*" must yield to "*lobi optici*." "It is not becoming," he wrote, echoing the thought of Huxley, Haeckel, Darwin, and Agassiz, "that any nation however wise and great should ask all the rest to take their intellectual food with the chopsticks of its peculiar pattern."[31] Furthermore, science must no longer cloister itself in the medical profession. After all, he concluded, quoting Agassiz, "scientific truth must cease to be the property of the few; it must be woven into the common life of the world."[32]

This passion for uncovering "hidden" facts and structural homologies and for giving them exact and appropriate names touched every aspect of physical existence, including human sexuality and hygiene. Like so many progressive physiologists of the post–Civil War period, Wilder tried to write and teach unreservedly and dispassionately about sexual as well as hygienic processes. In 1875 he published *What Young People Should Know, The Reproductive Function in Man and the Lower Animals*, in which he tried to name exactly and to demystify the sexual organs.[33] Like many feminists, moreover, he reduced sexuality to its physiological core and used the concept of equivalence to play down the importance of the sexual organs in the human system. "True love," he wrote, "is related to the physiology of the reproductive organs as is a friendly embrace to the muscular contractions which perform it; or a social meal to the process of digestion."[34] When he wrote of male and female sexual desire he again applied the principle of equivalence. "Sexual desire," he declared, depending heavily on the work of George Napheys, George Beard, and above all William B. Carpenter, "is usually ascribed to the male sex alone. This may have been due to an unwillingness on the part of men to attribute desire to women; at any rate . . . it is a generally admitted fact that there is a correspondence between menstruation in the human female and the condition of heat in the lower animals"; true, many women "have no actual knowledge of a . . . venereal orgasm; but there are others who experience orgasm fully as intense as that which accompanies ejaculation in the male."[35]

In conjunction with these writings and with the full support of Andrew White, who squelched opposition from the Cornell trustees, Wilder initiated a lecture program in 1878—at the request of the male senior class—on "Physiology and Hygiene," a program otherwise known as a "lecture on certain duties of men before and after marriage."[36] Both White and Wilder believed that the "complex and artificial conditions of civilized society" had created "an absolute need of including with a liberal education some knowledge of the body which is to be our companion through life."[37] White himself had asserted before the American Public Health Association in 1873 that college students should be taught physiology from the very first year of instruction: "Prevention is better than cure."[38] Wilder's course flourished for twenty-two years and attracted hundreds of students. One such student, Francis S. Benedict, later wrote Wilder asking him for some pamphlets on physiology: "I wish my brothers to be better acquainted with their bodies than I was before I had the privilege of listening to your lectures."[39]

For Wilder, however, the application of the scientific principle of no secrets did not stop in the laboratory or with myth-infected sexuality. As a progressive Unitarian, Wilder doubtless read and approved of the literature of comparative and Free Religion that underlined the "striking similarities" between different religions and "extracted the grain of truth from the mass of rubbish in each creed."[40] As a faculty member of one of the first universities in the country built to expand industrial potential and foster the economic interdependence of nations, Wilder presumably watched with pleasure the growth at Cornell of the engineering and scientific disciplines, the consolidation of the relationship between scientific technology and industry, and the standardization and rationalization of capitalist procedures.[41] In both economics and religion, no secrets or unimpeded openness supposedly brought to the fore the truly human and rational content of man's social existence.

It should come as no surprise, therefore, that Wilder wanted to weave this principle into the common social life of men and women. In 1873 he launched an assault on secret societies at Cornell that made headlines in major eastern papers.[42] In October of that year a student, Mortimer Leggett, had died as a result of a hazing by his fraternity, Kappa Alpha. Blindfolded and taken into the woods at night by two companions, then instructed to find his way back alone, he mistook the ridge of a gorge for a path and plunged to his death. Because he was the son of the United States Commissioner of Internal Revenue, his death forced the university to carry out a detailed investigation. Wilder played a principal role in the inquest. He studied the evidence assiduously and carefully surveyed the scene of the accident. He pummeled intractable witnesses with an endless series of probing questions. "Some of the witnesses," he reported, "would not have been more reluctant to give information if they had been under the charge of murder."[43] After the inquest the university administration quietly forgot the incident, although it did place a ban on hazing that lasted for several years. Wilder, however, continued to preach and write against such societies and vigorously admonished every incoming freshman class to shun them.

Wilder objected to secret societies—as to secrecy of any kind—largely because they acted against the best interests of men and women and contributed to general social disharmony. Like many scientists and popularizers of science in the mid-nineteenth century—among them Lewis Henry Morgan, Ludwig Buechner, Lubbock, Maine, Haeckel, Buckle, Agassiz, and George Beard—he sympathized with the feminist movement. A feminist influenced by Swedenborg and thus in the tradition of Harriet Hunt, James Freeman Clarke, and Julia Ward Howe, he condemned the "barbaric cruelty" and "repression" that had "crushed" woman's spirit and transmuted sexual "equivalence" into sexual disequilibrium. "The real creed of the future," he wrote in 1870, "is equal but not identical; diverse yet complementary; the man for the woman, and the woman for the man."[44] Healthy sexual relations illustrated the same principle of organic interdependence that Wilder tried to find in the laboratory and elucidate in the lecture hall. He singled out secrecy and its institutional forms, however, as one of the most important explanations for social injustice; he thought it responsible for much marital conflict.

"Confidence," he declared, "should be absolute between husband and wife, and nearly so between mothers and sons, brothers and sisters. . . . Intimate relations cannot be kept up" when wives and husbands "form attachments . . . whose clandestine nature . . . tends almost inevitably to render them pernicious." "If for no other reason," Wilder insisted, "we hail the admission of women into the public institutions of the day as a means of lifting the secret society curse."[45] The integration of known and equal commodities brought public health and stability. Open dialogue between husband and wife secured the bonds of marriage otherwise threatened by secrecy and segregation.

Feminist Application of No Secrets to the Sexual Relation

BURT GREEN WILDER's compulsion to interpose the scientific-rationalist principle of no secrets into the social life of men and women as well as into natural phenomena was a compulsion shared by feminists after the Civil War. Feminists everywhere succumbed to the power of the scientific world view, upholding with Wilder Agassiz's dictum that "scientific truth . . . must be woven into the common life of the world." Edward Bliss Foote, the influential feminist and progressive health reformer, used this dictum as the motto for his *Health Monthly*, while other feminists so approached its spirit that one suspects all of them devoured Agassiz, or if not Agassiz then Spencer, Darwin, and Comte.[46] "Every conceivable subject," the *Woman's Journal*, the most representative of all nineteenth-century feminist papers, pledged in 1872, "is, or will be, brought to the test of the scientific method." The "falsities, shams, the illogical dogmas" and "senseless prejudices" of the past would at last lose their power over the human mind.[47] Everywhere women would be freed "to unlock the secrets of nature" and to extract from these secrets underlying physical and social laws.

It was the business of women, in fact, to gaze deeply into the heart of phenomena and to strip them, like Burt Green Wilder's cats, of all sentimental mystifications. In their organizations and societies, feminists encouraged women to study every science that impinged in any way upon the physical life of men and women: biology—"the surest way of opening sociological studies to the larger class of women destined to occupy themselves with social reform"; anatomy and physiology—"the study of which is our last ground for hope for producing the enlightenment needed"; and sanitary or social science —"a new science . . . whose rays penetrate the darkest recesses and reveal the hidden forces which with such irresistible energy have hitherto devastated the race."[48] Feminist spiritualists even proposed that women pursue careers in medical clairvoyance, because medical clairvoyants had the peculiar gift to "*see* the interior of objects and thus to analyze the nature of disease."[49] In the words of the magisterial Andrew Jackson Davis, the clairvoyant "searched into and looked the facts in the face," "declared impartial convictions," and prescribed cures on the basis of such observations.[50]

Although conventional reflections on Victorianism have often isolated prudery as the hallmark of American social behavior, it has become clear to some historians that "Victorian" feminists and, for that matter, most American reformers, felt little squeamishness over matters of sexuality and sexual hygiene.[51] "It is time," the Reverend Dr. Boynton announced in the *Woman's Journal*, "for women . . . who care for their own sex, to call things by their right names."[52] In 1876 the Reverend Frederick Hinckley, a rationalist intellectual, told the Boston Moral Education Society that sexual questions "have been kept in the closet" for too long. "We put our feet upon all false modesty and silence," he said, "and announce and maintain, as our first principle, brave, free, pure discussion of the whole problem of sex."[53] The appearance of seemingly endless numbers of feminist texts during the middle and late nineteenth century, offering plain common sense analyses of sexual processes in considerable detail and often with the most graphic illustrations, provides undeniable proof that feminists did not fear facing the "facts of life." Many feminist books—notably Mary Studley, *What Our Girls Ought to Know*, Sarah Hackett Stevenson, *The Physiology of Women*, Rachel Gleason, *Talks with My Patients*, Eliza Bisbee Duffey, *The Relations of the Sexes*, and Mrs. T. H. Keckeler, *Thaleia*—can still be read with considerable profit.[54]

Still, many feminists demurred at this outpouring. Caught between an older approach that hid the truth from women or veiled it in sentimental innuendo, and the new one that promised to uncover every secret, they advised caution and good taste. The freethinking radical and Christian socialist John Weiss, in a thoughtful review of Russell Trall's *Sexual Physiology* for the *Radical* in 1866, wrote:

> How to do the thing well is certainly a great problem. . . . Which shall we prefer, an eruption of all the secrets of the physician into . . . woodcuts, every counter strewn with them, and boys and girls invited to premature fancies—or the old ignorance of the sexual relation, the old subjection of woman to the slavery of superfluous child-bearing, with all the disgust, alienation, hidden chagrin, foundered health and spirits, which that brings? We think the alternative lies in telling the truth with greater economy of details. We could say, with greater modesty.[55]

"I love the new wine," Caroline Dall noted in her journal in 1858, "but I do not wish to break all the old bottles."[56] In an otherwise favorable review, Thomas Higginson faulted Burt Green Wilder's explicit *What Young People Should Know* for its "over-information." Wilder "does not come so near," he wrote, "as some others have come to keeping the golden men."[57] The *Springfield Republican* objected in even softer tones, although it did say that "the title is badly chosen, for the information is rather for mature persons than for the young."[58] Wilder did present startling illustrations of the sexual organs as well as a "clear" and "realistic" discussion of the physiology of sex, free, in his words, from the "sentimental glamour which has hitherto surrounded the whole subject." Of his illustrations he wrote, "They are no more objectionable . . . than the organs which they represent."[59] In 1876 the Chicago *Inter-Ocean* praised the sex reformer and social scientist Eliza Bisbee Duffey for calling a "spade a spade" in her *Relations of the Sexes* and concurred that "society" had become "so constituted socially that a better acquaintance with the laws ot nature has become a necessity of modern life." Nevertheless, it warned, this "material should not be read by children."[60] Sexual radicals from Woodhull to Duffey labored to smoke out every hidden sexual truth. Both these feminist groups, however, the timid and the bold, tirelessly ridiculed the social "grundyism," the "mock modesty," and the "cant" associated with the world of sentiment and fashion, agreeing with the quintessential hygienist Andrew Jackson Davis that "American modesty is vulgarity gone to seed" and with Elizabeth Oakes Smith that the "natural processes" should be within the "pale of familiar consideration."[61]

Feminists insisted, in fact, that successful marriages depended on accurate knowledge of physiology and not on simple "revelations of Nature," and that sexual ignorance or, worse, false notions learned in secret, only wrecked chances for marital happiness. One reformer advised parents to give children such physiological and sexual knowledge as would "enable them to decide how and when to marry."[62] "The more we attempt to conceal," observed one of the physiologists popular with feminists, "the stronger is the desire to gratify curiosity. No man or woman should enter the conjugal relation, until they become acquainted with the physical and mental condition of the one with whom he or she asks to become united."[63] The progressive educator Anna Brackett warned that children must not be permitted to "pick up" facts "piecemeal in surreptitious and clandestine ways, as if sex were some horrible mystery which must from its very nature, be covered up from the light of day." She also deplored the practice of expurgating the truth for children or obscuring it with "double entendres."[64] Married life had to be "based on knowledge" gained from educated parents or, better, from the schools. As Julia Ward Howe said in 1876 before the Woman's Congress, "the revelations of Nature" no longer sufficed for those people who wished to marry, but needed to be supplemented by "careful study and effort to know."[65] It was because feminists mistrusted the "revelations of Nature" that they wanted practical training in "the reciprocal duties of men and women" taught in the schools and colleges.[66]

Without scientific facts in this realm, as in any other, the human mind and human marriage would become mired in the chaos and uncertainty generated by modern civilized society. Physiological knowledge, however, would convert excitable youth into cautious adulthood, "impulsive action" and the quest for "immediate pleasure" into "rational desire."[67] It would "*prevent the exercise of the imagination upon the bodily organs.*"[68] Without question feminists hoped that all men and women would develop the detached capacity to observe and study their own sexual behavior, that they would learn to step out of their bodies. They believed that this intellectual stepping out would defuse the body's power to disrupt human relations, thereby marking the first step toward controlling sexuality.

Feminists welcomed and encouraged the increasing numbers of books on woman's diseases and on the hygiene of the sexual relation. They helped to fill the need for expertise. Beginning in the 1840s, the major health and feminist publications frequently carried information on sex and hygiene, while an imposing phalanx of reform-minded women publicly argued for sex education in the home. They sponsored the creation of and spoke before physiological societies; they continually lobbied for the widening of the scientific curricula and implored young women to attend the few institutions open to them that offered scientific curricula.[69] In the 1860s and 1870s the drive to make knowledge of hygiene and the sexual processes readily available to all men and women became dramatically stronger. Major feminist leaders, as well as feminist hygienists, took to the lecture circuits. New feminist-reformist publications added to the burgeoning press that continued to educate the public; genuine feminist physiological societies appeared in Washington, Cincinnati, Brooklyn, Chicago, Philadelphia, and New Haven; and feminists partly realized their goal of bringing scientific subjects into the school curriculum.[70]

Perhaps most important of all, hundreds of women graduated from regular medical institutions in the late nineteenth century, and hundreds of others from irregular, eclectic, hydropathic, and homeopathic institutions. As early as 1865 eight regular female physicians were practicing in Philadelphia.[71] In 1871 the *New York World* reported that more than two hundred women had graduated from various American colleges that year, the "majority" of whom "were physicians." In that year too, thirty women doctors were practicing in Indiana.[72] In Boston, according to the chairman of the Massachusetts Board of Health, Henry Bowditch, fourteen women of the highest caliber were working as physicians in 1880. In 1881 seventy-three were practicing in California, and in 1882 eight were practicing in Topeka, Kansas.[73] After 1870 many women had enrolled in the medical departments of coeducational schools, including the University of Michigan, Boston University, Syracuse, Berkeley, the University of Iowa, Iowa State, the University of Buffalo, and Wooster College, in Cleveland.[74] In some schools—notably Boston

University, Syracuse, and the University of Michigan—more women chose to study medicine than any other subject.[75] Boston University, a homeopathic institution, had three women on its faculty—Drs. Mary Safford Blake, Caroline Hastings, and Mercy B. Jackson—and claimed to give stiffer examinations than any other medical school in the country.[76] By 1880 nearly four hundred women had taken regular medical degrees, and by 1888 more than a thousand women had done so.[77] These developments confirmed the Italian feminist Salvatore Morelli's prophetic statement, in 1858, that "the cause of science and women" was "the new force emerging for human progress and a necessity for our emancipation."[78]

One of the most famous of the woman doctors, Clemence Lozier, followed the evolutionary pattern of many leading feminists: married at sixteen to an invalid whom she ended up supporting, and widowed at twenty-seven with seven sons to care for, she worked for eleven years as a principal of a ladies' seminary, moved rapidly from Sunday-school teaching, tract distribution, the New York Female Moral Reform Society, and abolitionism to an eclectic medical school in Syracuse and, finally, to a very respectable career as both dean of the New York Medical College for Women and an important feminist leader. By 1871 she had an income of twenty thousand dollars a year. In 1875 she was consulting physician to the "Bright Haven" water cure, founded by her feminist son, Dr. Alexander Lozier, in West Orange, New Jersey.[79]

Other women replicated her history, and one of them was Lozier's own daughter-in-law, Charlotte Denman Lozier. Before she died at the tragically young age of twenty-five, Charlotte Denman had achieved a great deal. After the death of her mother in 1857, she took charge of her younger siblings, but still found time to graduate from high school at the age of fifteen. Released from her family duties, she went to the New York Medical College for Women and led a successful fight to get women admitted to Bellevue Hospital for clinical work. Graduating with distinction, she later became professor of medicine at the New York Medical College, bore three children, and was vice-president of the New York Working Women's Association.[80] An example to women like the Loziers was an older pioneer, Mercy Warren. Married in 1823, the mother of two children, and widowed young, Warren opened a school for girls in Portland,

Maine, during the late 1820s. In 1833 she married again, this time to a ship's captain with children of his own. She gave birth to four more children, and was then responsible for a progeny of fourteen. Oppressed by her ignorance of medical science and "family management," and disgusted by the horrors of conventional medicine, she began a study of homeopathy in 1840 and for eighteen years practiced with great success as a homeopathic physician in Plymouth, Massachusetts. When her second husband died and the children were older, this resourceful woman finally took a course in regular medicine at the New England Medical College for Women and graduated with full credentials in 1862.[81]

Women doctors—most of whom had active feminist careers—served the cause of feminism in many ways, not the least of which was to give other women a feminist ideal: autonomous, inexhaustible, self-made women with wonderful reserves of will given over to practical work and the scientific study of life, and dedicated, for the most part, to the hygienic reordering of society. But these women also gave the public hygienic information, dispensing in their writings the ideology of physical-social symmetry to a whole generation of feminist women and men. "Let women who have studied medicine," wrote Mary Studley—herself a graduate of the Women's Medical School of New York and for a time resident physician and Natural Science teacher at the Framingham Normal School in Massachusetts—"teach other women what every woman ought to know."[82] By 1874 Putnam Jacobi could inform the First Woman's Congress that the "existence of women doctors has had an immense influence in dissipating the stupid prejudices which have for so long concealed from women the general physiological knowledge which was most important for them to know."[83]

The work of Dr. Anna Densmore during the late 1860s well illustrated Putnam Jacobi's point. One of the first vice-presidents of the New York City Woman's Club, a champion of hospital reform, and a vocal enemy of abortion—"foeticide"—Densmore introduced a series of "scientific lectures" for women that drew the largest number of women (regularly between 150 and 200) ever gathered in the city for such a purpose. Densmore had two things in mind when she began her lectures on physiology and the sexual relation. First, she wanted to educate women in the "facts of life." "If we are to compel a *radical change* in the wifehood and motherhood," she said, "women should know themselves *thoroughly* in all that pertains to the varying attributes of girlhood, wifehood and maternity. . . . Every woman physician should herself be a teacher, and make it a cordial rule to spread the knowledge she has gained in reference to the prevention of disease and the possibility of imparting better constitutions to our children than is now done."[84] Second, she wanted to institute a plan to train female public school teachers to replace doctors in this pedagogical task. Doctors had other work to do besides public lecturing, while teachers had direct access to the minds of young women everywhere. In 1868 the New York City Board of Education voted by unanimous consent to grant Densmore the right to prepare teachers for such service, and even gave her a large hall on Twelfth Street in Manhattan for her classes.[85]

The Limits to No Secrets: Reaction

IN THE 1870s, the application of the principles of no secrets erupted in an epidemic of sexual muckraking, an epidemic dangerously promoted by the detectivelike probings of the popular press. In 1874, for example, the well-known and respected Boston minister, Cyrus Bartol, publicly dissected the moral character of Henry Ward Beecher, the ministerial head of the wealthiest congregation in America, in a lecture presented before the Free Religious Association. He speculated that Beecher's sentimental theology had vitiated his moral nature, and in making his point left nothing to the imagination. Shocked and angered, Thomas Higginson called Bartol's performance a "terrible personal discussion" and Julia Ward Howe did her "best to rebuke Bartol's slander."[86] A few years earlier, Harriet Beecher Stowe and her personal enemy, Victoria Woodhull, had both, in Woodhull's words, reached into the dark, hidden recesses of "secret sexual practices" and dragged into the "daylight" the "scathed and stigmatized . . . excrementations" of the social order.[87] Woodhull exposed the sexual "crimes" of living public figures, while Stowe dug at the grave of the dead Lord Byron, exposing his incestuous relations with his half-sister. Both women attacked male vice with the aim of destroying the double standard. In the words of Paulina Wright Davis, they "took hold of those men whose souls are black with crimes and who should be torn from their thrones of the judgment of woman's morals and made to shrink from daring to utter one word against any woman so long as they withold justice from her"[88] Stowe's and Woodhull's purpose, however, drove them deeper than any specific assault. They believed that sexual behavior would be purified and made healthy when men and women knew and confronted the "facts of life." "The trouble is," Woodhull wrote in 1873, "that the sexual question is in an entirely abnormal condition: and it requires to be driven out into a normal condition, where healthy action only can exist."[89] "I claim for men and women," Stowe declared, "our right to true history. . . . Let us have truth when we are called in to judge. . . . I have vowed to tell the truth."[90]

Vaunted claims to ethical purposes notwithstanding, Stowe, Woodhull, Bartol, and others came very close to exhausting the limits of the legitimate public exposé. Their revelations, in fact, often bordered on slander, frightening not only their victims but the reformers themselves. And, inevitably, some of the biters were bit. In the early 1870s Elizabeth Cady Stanton sent a series of indignant letters to her friends in which she condemned malicious gossip about Woodhull. "It is a great impertinence," she wrote Lucretia Mott, "in any of us, to pry into her affairs. . . . There is a sacredness to individual experience that seems like profanation to search into or expose."[91] To Woodhull herself Stanton wrote that "it may be a light thing for the press of the country to hold up one frail little woman to public ridicule and denunciation, but this reckless hashing of individual reputations is destructive of all sense of justice and honor among our people and will eventually force on us a censorship of the press."[92] Throughout the late 1860s and early 1870s the members of the feminist New York City Sorosis anguished over the meaning of public and private openness. In February of 1871 they debated the topic "Do we resent Calumy, Deceit, and Treachery because they are untrue or because they harm us?" In April came "When does frankness cease to be a virtue?" and in May, "How far does the moral purpose of the author of *Reginald Archer* justify writing such a book?"[93]

The following year the Sorosis itself fell victim to "obscene innuendo" and "insinuations" in its own midst. Sister slandered sister, fostering a "spirit of detraction" that, in the words of one member, Dr. A. D. French, almost "destroyed the elements necessary to carry out the avowed objects of the Society." French offered a strong resolution against such behavior that passed with unanimous consent and a sigh of relief.[94] Later in the decade such reformers as Charles Dudley Warner, Samuel Bowles, and Julia Ward Howe objected to the unrestrained freedom of the press to investigate the private domain of both the living and the dead. Julia Ward Howe struck out at Woodhull and her paper for "stealing" and publishing "secrets which did not belong to her." Bowles, the editor of the *Springfield Republican*, upbraided Fanny Fern's husband, James Parton, for "rummaging among the soiled linen of the past." In a speech to the American Social Science Association, Warner gloomily pondered the effects on the individual and the community of the contemporary "mania . . . for notoriety in social life," a mania stimulated and fed by the newspapers and magazines.[95]

The tragic case of William F. Channing illustrates with pathetic clarity the power of the public exposé to inflict lifelong injury and unhappiness. An admired Boston physician and nephew of the famous Unitarian preacher William Ellery Channing, Channing was also a utopian reformer and a feminist. He believed in cooperative labor, "perfect organization made up of perfect individuals," and "equal sexual copartnership in the new civilization."[96] A man unfortunately plagued by marital difficulties, Channing was a perfect target for a sensational front-page story. In 1859 the Eastern press hauled him before the public eye when he divorced his first wife, Nancy, after a five-year separation. The papers described all the details of the case; hinted at salacious goings-on; and claimed outright that Channing espoused easy divorce and followed the subversive doctrine of spiritualist affinities. Torn between the desire for silence and the need to defend himself, Channing finally responded with a long letter to the *New York Post*, refuting every charge leveled against him. Above all he felt the need to deny that he had divorced his wife "in order to marry a new affinity." "I sought and obtained this release," he declared, in a confession that probably served to widen the damage to his reputation, "for its own sake, as a matter of personal and social duty." Channing concluded his letter with an attack on the Massachusetts marriage laws, which permitted divorce only on grounds of "physical failure" without recognizing the "mental, moral or spiritual elements as entering into the contract. . . . Marriage is made by law," he wrote, "to be the unalterable contract of bodies, thus depriving it of all the conditions which distinguish human marriage, and degrading it to the level of an enforced animal relation."[97] Channing's parting shot did more to appease his sense of justice than to erase the public stigma. He finally had to leave his home and his Boston practice forever. For the rest of his life he felt a burning hatred for the "Puritan system, which still haunts New England like a nightmare and which taints all New England blood. . . . This arrogance of God's province," he wrote Caroline Dall in 1872, "of taking care of the truth of history and of judging the earth is detestable to me in all its branches, I loathe the moral scavenger more than the physical."[98]

The very reformers, however, who denounced the gossip of petty minds and the tyranny of the popular press also championed woman's right to act like moral Godivas, to live in the open air, and to delve dispassionately into the inner secrets of the natural world. Even while they tried to keep separate the private and public lives of men and women, mid-nineteenth-century feminists, by the very nature of their discourse, inevitably blurred the distinction. At the same time that the members of Sorosis forswore excessive frankness among women, they also discussed among themselves sexual physiology, legalized prostitution, abortion, and infanticide; they supported the "laudable custom" of men and women dining together, working together, and "interchanging ideas" together in public "upon a footing of perfect equality." They resolved in favor of "equal interests, equal affections, and equal rights" within marriage.[99] Thomas Higginson winced as he listened to Cyrus Bartol coldly dismember Henry Ward Beecher's sex life, but he had written earlier in the *Woman's Journal*: "What we need now is open instead of secret influence, the English tradition instead of the French."[100] Warner, Bowles, Stanton, and others inveighed against the system of newspaper espionage, yet they felt equal scorn for what Burt Green Wilder called the secret-society curse. Stanton appealed to women to raise their voices against secret societies, "for they not only undermine the family but the state and church;" and Warren Chase of the *Banner of Light* advised young men to "keep out of secret societies and make the world your home."[101] Theodore Tilton described secret societies as "absolutely unnecessary, objectionable, and repugnant;" and Bowles, Warner, and George William Curtis identified them with what Curtis called the "baby-house" of the "juvenile state of civilization."[102] William Channing suffered profoundly from the moral slurs heaped upon him by public opinion, but no feminist ever surpassed him in his commitment to open dialogue between the sexes and to the power of science to redeem mankind. In the same letter in which he expressed his contempt for the moral spies of society, he described to Dall his devotion to the scientific method and love for scientific labors. "Certainly," he wrote, "it is a supreme pleasure to drag out of the depths some new treasure of human use."[103] Whatever their reservations, therefore, feminists not only occasionally indulged the principle of no secrets by muckraking in public, they also installed it in the heart of the sexual relation.

Feminists could not protect themselves from the often subversive use of the principle of no secrets; nor were they able to prevent the reaction that set in within society at large and within the feminist movement itself in the late 1870s and early 1880s against the reformist attempt to inform the public on sexual-hygienic matters. As John Weiss might have said, the thing had not been done well. Old lines were being too boldly crossed and the difference between scientific truth and pornography badly blurred. Prudery was reasserting its ugly tyranny. In 1878, the year of the Chestnut Street meeting in Boston, a heated controversy exploded in Cincinnati over the use of Burt Green Wilder's *What Young People Should Know* in the classrooms of Woodward High School. In January of that year the principal of this coeducational school, Dr. W. H. Massey, discovered one of his English teachers, Edward O. Vaile, teaching Wilder's book to his male students. Massey hauled Vaile before the Board of Education and furiously reprimanded him for exceeding his authority as an English teacher, called Wilder's book a "quack doctor's piece of trash," and recommended a return to the "separate sex" system of education. Humiliated, Vaile demanded a "public examination" of his case, which the board denied him. He also argued that Wilder's book was "fit" reading for girls and boys and warranted a place on the library shelves (where he had in fact already put it, although students needed the teacher's permission to read it). The case soon became a battle between those board members and teachers who considered themselves committed to good teaching and coeducational instruction, and those—Massey, other board members, and parents—who were not. One member of the board argued that the attack on Vaile really represented a veiled "attack on the high schools." Surprisingly enough, the board, led by a progressive president, A. T. Goshorn, exonerated Vaile—an excellent teacher and a highly educated man—but it forbade him to teach anything beyond his own subject and withdrew *What Young People Should Know* from the library.[104]

Feminists tried to keep the distinctions clear between "science" and pornography, *What Young People Should Know* and the *Police Gazette*. Of E. P. Miller's book on the "abuses" of the sexual function, the *Revolution* wrote that it was not an indecent book like the ones that "stare from the newsstand." All you would find in it was "an invaluable aid in arresting the tide of immorality that threatens to engulf us."[105] Even the radical sexologist John Scudder, of the Eclectic Medical College in Cincinnati, author of two of the most sexually explicit books ever to go to press in the nineteenth century and an obvious source for some of Woodhull's ideas, in 1874 condemned "this vicious traffic in obscene literature which is widely distributed throughout the country."[106] It was not enough, however, that reformers tried to draw a clear line between the scientific and the salacious. As early as 1872 New York authorities imprisoned Victoria Woodhull and her sister, Tennessee Claflin, both for publishing "obscene" literature and for slandering a local stockbroker, Luther Challis, in their *Weekly* (really an attack on Henry Ward Beecher). A united front of Comstockers and inflamed clergy launched an assault that year on such previously unscathed figures as Orson and Lorenzo Fowler, Walt Whitman, and John Humphrey Noyes. Noyes found himself in bitter exile in Canada in 1879, and by the end of the decade such feminist periodicals as the *Physiologist* (edited by Dr. Sara Chase), *Truth Seeker*, *Alpha*, *New Age*, and *Dr. Foote's Health Monthly*, among others, were suppressed or threatened with suppression.[107] Feminists of the National Woman's Suffrage Association sharply opposed attempts to muzzle some of these publications and in several cities organized protest demonstrations against censorship.

Such progressives as Octavius Frothingham, Parker Pillsbury, James Parton, and Robert Ingersoll defended the publication of the *Truth Seeker* and openly excoriated legal authorities for imprisoning its editor, D. M. Bennett.[108] On the other hand, many feminists remained silent and some even supported Anthony Comstock's crusade.

The cult of no secrets identified the interests of many middle-class women and men. In this respect these women and men seemed to fall in line with the traditional American hostility to the secret life. Secrecy, supposedly, had no place in an open, democratic society. Yet many other Americans with different (but not necessarily opposing) interests failed to rally to the banner of no secrets. The farmers of the National Grange and the workers of the Knights of Labor, for example—large organizations that, ironically, were the first of their kind to admit women to membership—functioned under a veil of secrecy in the 1870s and 1880s. A very different sort of secret society, the Ku Klux Klan, also flourished at the same time.[109] For these groups secrecy acted as a unifying force, fostering a shared sense of community and purpose. For feminists, however, no secrets, not secrecy, was a binding as well as a liberating power.

In practice, the principle of no secrets was a double-edged weapon. On the one hand it was intended to free women from entrapment in the private sphere, to harmonize and rationalize the life of both sexes, to bring the private life of women into more open relationship with community life. On the other hand, no secrets removed the barriers between private and public life, exposing the former to direct invasion and control by the latter. In an effort to separate the private life from older forms of control, feminists supported conditions by which new controls might be imposed. In particular, feminists contributed to this new development by insisting upon the installation of sexual hygiene in the school curriculum. Home instruction also played an important role, feminists believed. More often, however, for reasons of general effectiveness, they argued for the displacement of private by public methods of sex education.

CHAPTER 3

A HEALTHY MIND IN A HEALTHY BODY: EQUAL PHYSICAL DEVELOPMENT AND THE COEDUCATIONAL EXPERIENCE

IN UPSTATE NEW YORK, in a beautiful region dotted with slate-bedded glens and glacial lakes, one of the most influential of feminist doctors conducted her well-known Elmira Water-Cure in the 1870s. Rachel Gleason offered her female patients a hydropathic therapy of hot and cold baths, massages, wet compresses, and healthy diet; she lectured them on physiology and reproduction and disparaged their emotional and physical excesses. In her big gymnasium patients in "pretty uniforms" exercised to music every day.[1] Like other female and male reformers, Gleason believed that no woman was "inherently weak or delicate, only restricted by fashion, sedentary domesticity, constricting dress, and bad diet."[2] She and her reformer patients dedicated themselves to restoring women to their rightful physical powers.

Feminists matched their plea for scientific knowledge for both sexes with promotion of equal physical development from childhood on. Because morality itself, they believed, depended on health, vigorous development was imperative for both sexes. "I should suppose," declared Faith Rochester, "that they ought to be brought up alike. There is nothing in their physical functions to lead me to suppose that one needs different diet, clothing, experience."[3] "Boy and girl as they are, with the same physical and mental needs," according to Eliza Bisbee Duffey's popular book, *No Sex in Education*, "nature calls imperatively for an equally active life for both."[4] Mary L. Booth, a gifted historian and feminist journalist who once hoped to publish a "Woman's Journal" under the motto "Equal Rights for All Mankind," wrote that

> there is a degree of distinction made by mothers between the male and female children which is not warranted by the natural differences of sex. This artificial difference strengthens with the advance of age, and thus in the same family it is often observed that the males are vigorous and the females delicate, though both were born with equal strength of constitution.[5]

Booth edited one of the most important fashion magazines of the time, *Harper's Bazaar*, yet she was a persistent defender of the equal physical development of men and women.

Accompanying injunctions and schemes to establish the physical health of individual women on a par with men came the belief, among a large sector of the feminist community, that the coeducational institution was the healthiest form of educational experience for men and women. The theme of healthy bodies took a larger institutional form. Coeducational colleges and universities expressed the new, post–Civil War emphasis on symmetrical institutions as well as on symmetrically developed individuals; these colleges integrated the individual within a larger institutional complex, while promising to develop every faculty—intellectual, physical, social, and moral—according to the symmetrical principle. Individual physical health and coeducational training would further prepare the sexes for more congenial relationships, more cooperative friendships and marriages.

Healthy Bodies

HEALTH IDEOLOGY, with roots in the antebellum period, cut deeply into reform thought from the 1850s into the later decades of the century. A network of water cures, irregular health establishments, and schools, often founded and staffed by feminists, disseminated the gospel of physical health for both sexes. Gleason's water cure, Russell Trall's Hygeio-Therapeutic College in New York City, Dio Lewis's Normal Institute for Physical Education in Boston, and James C. Jackson's palatial Hygienic Institute in Dansville, New York, were preeminent among them.

Gleason's gym was modeled on that of the gymnastic fanatic Dio Lewis, who had opened his Normal Institute in 1862; it was devoted almost exclusively to the strengthening of female bodies and to the science of preventive medicine. At its first commencement exercises Lewis declared his faith: "I came many years ago to think somewhat seriously of that ounce of prevention which is worth tons of cure. Want of muscular exercise was one of the most obvious defects of our physical life."[6] Two other men, Russell Trall and James C. Jackson, played important roles in shaping the direction and content of reformist thinking on health. An invalid in his adolescence and early adulthood, Trall decided to reject allopathic medicine, with its emphasis on cure and drug medication, for the new religion of preventive medicine. He was "fully satisfied that the whole system of drug medication was false in philosophy and absurd in science . . . and that the only truly remedial agents were those materials and influences which possessed *normal* relations to the vital organism—air, light, water, food, temperature, exercise and rest, sleep, passional influences, electricity, etc."[7] He was convinced that men and women were engaged in an "organic war," "defensive struggle" with disease. Organic wholeness was, in contrast, the "sunlight" realm. He equated the rise of the science of health with the emancipation of women and he backed every nineteenth-century feminist position, including the right of a woman to own her body and her consequent right to practice birth control.[8] In the 1850s he built his Hygeio-Therapeutic College, a coeducational institution that attracted hundreds of women and that, a decade later, had graduated far more women than men.[9]

Jackson, Trall's nearly exact contemporary and possibly the most admired water-cure physician of his day, was the son of strict Presbyterian parents of "first class rank in society." He suffered under a tough educational regime supervised by his father and religious drilling imposed by his mother. His parents never bothered with his diet, sleep, or dress. They opposed games and kept him from play. "Instead of treating me according to the laws of my being," he wrote, "they undertook almost immediately after my birth to bring me under rules of their own creation and which had no other origins except in their own intellectual and religious belief. . . . I never knew anything about childhood—its joys or its sorrows."[10]

"My father never called me a child," Jackson wrote. "He longed to have me become a man, and set his genius to work in the institution and establishment of all kinds of expedients to have me leap from my young life into the life of an adult." Jackson's mother displayed more tenderness, "her maternal instinct being strong and her love for me very intense," and her son was therefore propelled "into more frequent and intimate relations" with his mother than with his father. It was a female-centered relationship characteristic of the personal histories of many feminist men in the nineteenth century.[11] Jackson regarded his father as a strict "puritan" who "took him away from his mother" and as a "chronic invalid" overwhelmed by disease. A locally celebrated physician, the father died of "abdominal dropsy" when his son was seventeen. Jackson's mother, in contrast, was magnificently endowed with health and practical intelligence as well as love. Criticizing her only for excessive piety, Jackson described her as a "woman with a very superior organization . . . the handsomest woman of all the region," gifted with perseverance unmatched by "any human except myself." Jackson came to equate disease with men and health with women.[12] His vision of women as indomitable was confirmed by his older cousin, Adelaide Clarke, a fervent abolitionist, a "great walker," "a candid and magnanimous woman" with "great powers of conversation."

By the early 1840s, abandoning both the pieties of his mother and the diseases of his father, Jackson fused their strengths in an imaginative and inverted reconstruction—the "religion of health." His own career combined the celebrity and professional status of his father with the glorious health of his mother. He published and helped to edit, with Dr. Harriet Austin and others, the journal *Laws of Life*, standard reading for all reformers. He founded the Dansville Sanatorium, which he preferred to call "Our Home Hygienic Institute," a successful monument that survived until a fire destroyed it in 1882. One of thirty-five such health institutions, Dansville included doctors, nurses, and patients of both sexes. During the summer season as many as four hundred people flocked to this therapeutic utopia.[13] Reformers including John Hooker, Isabella Beecher Hooker, Celia Burleigh, and Robert Dale Owen came for treatment. "I have rarely enjoyed a summer more," wrote John Hooker in 1880.

> It was one of absolute idleness, which had generally been intolerable to me, and of very great social enjoyment, the other patients being very largely people of education and intelligence, and often clergymen and other professional men, and many among the women being teachers. . . . I made some friendships here, especially among the women I met, that I shall always cherish.

When Dansville burned down, Hooker memorialized its passing in sentimental verse:

> Old home of the best sanitarian science!
> You dear old rookery!
> We mourn you for all your dear old memories
> Of a life ideal;
> And for all your hate for nonsensical flummeries
> And things false and unreal.[14]

Life at Dansville was simultaneously exhilarating and relaxing. "Love," declared the *Laws of Life* in 1873, "dwelt in the Institution and has largely controlled the activities of its helpers and patients."[15] To Jackson, "sickness" was the root cause of family disruptions and "perturbations of communities," and he treated it by "natural" methods—diet, baths, exercise, massage—and educated his patients "to preserve and maintain health" after they left the sanatorium. He did not attack disease with the rigid puritanical frenzy that marked the style of many early health reformers. He disliked gloominess in dress and wore "light-colored" instead of black clothing, velvet caps instead of "stove-pipes."[16] His approach to sex and the sexual relation was profoundly rational and functionalistic, not repressive. The

Dansville Prospectus gave central place to the following "psycho-hygienic" principle:

> There is not a passion nor an appetite nor a propensity in human nature which if properly trained and disciplined might not be made so to express itself at proper times, and in proper ways, as really to add very much to the beauty of personal character. It is only when it has no education, and when the organ expressing it is in a diseased state, that its expression becomes either undesirable or deplorable.[17]

Educated passion, rationalized through a balanced system, would result in perfect health.

Jackson applied this method in utterly feminist ways. Not only did he believe that both sexes should fully understand each other's "sexual structures" in order to "calculate" rationally the best marital choice, but he also advocated the equal physical development of men and women with greater conviction than any other contemporary reformer. "There is no physiological reason why the bodily constitution of girls should not be as good as boys. . . . The organic forces in her are the same as they are in him in every respect but one—the reproductive organism is different."[18] Dansville institutionalized feminist health ideology. Its staff members wrote for feminist periodicals, attended feminist conventions, and delivered papers at the woman's congresses.

University Coeducation as Requisite for Successful Marriage

ALONG WITH a sure grasp of scientific principles and healthful physical development, university coeducation represented, by the late 1860s and early 1870s, a third prerequisite for successful marriage. Coeducation had been implemented to some degree during the antebellum period in secondary and primary schools as well as in many normal schools, private academies, and religiously affiliated colleges.[19] Because an extreme sexual division of labor held sway during this time, however, and because most middle-class women inevitably sought education solely for domestic purposes, coeducation at the college and university levels never took on a truly egalitarian character. In the late antebellum period the tide shifted somewhat and reformers made the first pioneering efforts to create egalitarian coeducational institutions. In the early 1850s, for example, educational reformers joined with "machinists and manufacturers" to back the People's College movement in upstate New York.[20] Precursors of Cornell, the manual labor institutions that emerged from this ferment were designed to "instruct the *whole man*." Their curricula included systematic instruction in both agricultural science and the various branches of manufactures and the mechanic trades, plus instruction in the liberal arts. The most outright coeducational and reformist of these ventures was Central College, in McGrawville, New York. Incorporated in 1851, it accepted both blacks and women on a basis of complete equality with white men. The trustees called Central a school for "humble folk" who respected labor, who detested fashion and "the doctrine of exclusiveness and caste." Feminists and abolitionists like Samuel May, Gerrit Smith, Elizabeth Stanton, Amy Post, Lucy Stone, Amelia Bloomer, and others committed time and money to the life of this college. A campaigner for coeducation throughout the 1850s, Susan B. Anthony called the Central program a "glorious plan."[21] Wherever genuinely coeducational institutions with scientific, utilitarian curricula appeared, feminists could be found in the vanguard of support.

Another pioneering experiment in egalitarian coeducation was the Raritan Bay Union, in Perth Amboy, New Jersey, organized under the guiding genius of Theodore Weld in the 1850s. Like many abolitionists who left the fray of public agitation, Weld had turned his hand to the more pacific but not less important work of social reconstruction and institutional reform. The Raritan Bay Union was a utopian community backed by journalists Albert Brisbane and Horace Greeley and financed by the wealthy Quaker merchant Marcus Spring; in 1853 its trustees invited Weld to become director of the community's school. The new community tried to implement, though without much rigor, the principles and practices of Charles Fourier: the application of capitalist procedures for the benefit of all, the rejection of individualism for association, the creation of community rather than "isolated families," and the introduction of a serial system to develop harmoniously and "naturally" all the human faculties and propensities. "We hold," wrote Albert Brisbane, "that, instead of chains, man requires freedom; instead of artificial order through coercion, man requires Divine harmony that comes through counterpoise. . . . We intend to make organic provision for the entire circle of human wants, for the entire range of human activity."[22]

Weld's school, which he called Eagleswood, reflected these principles and continued to do so in the late 1850s and early 1860s, long after the Bay Union itself had dissolved. Like the progressive manual labor reformers of upstate New York, Weld fashioned a well-balanced curriculum related directly to the "duties of life, practical and professional," with courses in agriculture, the physical sciences (geography, physiology, anatomy, hygiene, natural philosophy, chemistry, zoology, and so on), and the industrial arts as well as courses in music, mathematics, history, drawing, ancient and modern languages, political economy, and the English language.[23] His program of balanced physical growth, gymnastics, and calisthenics, and his healthy hygienic regimen mirrored the balance of his curriculum. Opposing "artificial order through coercion," he exhorted students to evaluate and judge their own characters according to the standards of "perfect equilibrium" and "symmetry" and to practice private forms of social control; excesses, all "impure, indecent, impracticable," and "reckless" deviations from the norm, were antisocial, "tending to counteract the processes of nurture in others."[24]

Weld's ardent devotion to egalitarian coeducation completed a trinity of balance: balanced education, balanced bodies, and a balanced sexual community. "To restrict each sex to schools exclusive of the other," he wrote, "is to ignore a law of reciprocal action vital to the highest welfare of both. In recognition of this law, we receive pupils, youths of both sexes. Their education together, under wise supervision, gives symmetry to mental and moral development."[25] Believing in coeducation as the "true way," he taught his male and female students the same subjects in the same classes; he let them stroll and converse, hike and play together in "little companies," although he discouraged "solitary walks."[26] Such advocacy of an unpopular cause drew the praise and the children of feminists. Elizabeth Smith Miller sent her children to Eagleswood. The Pennsylvania feminist Martha Wright enrolled all her children there in the 1850s. "If only," she wrote her husband,

> I had been born one thousand years or so later. The more I see of Mr. Weld's system of education, the more pleased I am, that it is the right method to develop all the powers, and am anxious that we may be able to continue our children with him for several years, until they are old enough to continue the culture begun there—provided the checks and balances all work well. . . . His theory is beautiful and we can only hope it may prove the right one.[27]

Weld's experiment in coeducation, like other early manual college ventures, did not last long. Sickness in his family forced him to move to Massachusetts, cutting short the brief career of Eagleswood in 1862. The manual labor colleges in New York State failed because of inadequate capital investment.[28] The pattern set by these schools came to full and enduring fruition only in the coeducational upsurgence of the post–Civil War period, a mark of the reconstruction of social life in the North. By 1867, for example, twenty-two coeducational colleges and universities had appeared, and by 1872 there were ninety-seven. By 1873, moreover, 70 percent of commercial colleges had gone coeducational and 60 percent of secondary schools had opened their classrooms to both sexes.[29] Nevertheless, most of the coeducational institutions of higher learning ranked in quality far below male schools (and, later, below such female schools as Vassar and Smith)—a situation that would remain relatively stable, in spite of greater coeducational growth, into the late 1880s, when only eleven of the two hundred and seven coeducational institutions received accreditation.[30] Still, Cornell, the University of Michigan, Wesleyan, Boston University, and the state universities of California, Wisconsin, Iowa, Illinois, Vermont, and Kansas represented landmarks in the history of egalitarian coeducation and real advances in equal educational opportunities for women.

Three important and converging trends produced these changes. First, federal land-grant-college legislation in 1862 fostered agricultural and scientific university education, and led to a revolution in educational thinking and in university curriculum in the late 1860s. Responding to the rapid expansion of the West and the enormous industrial production of the post–Civil War period, with its consequent demands for specialized technical and scientific training, schools like Cornell, the University of Michigan, and Wesleyan abandoned the fixed classical curriculum of Greek, Latin, mathematics, logic, and moral philosophy, and, in its stead, installed a flexible, elective system. Throughout the 1870s the curriculum widened to include biology, physiology and hygiene, chemistry, physics, astronomy, geology, engineering, and agricultural science; the laboratory was introduced for the first time as a principal device for classroom instruction.[31] The traditional small faculty with a president who taught psychology, ethics, political economy, international law, and moral philosophy disappeared; the president's teaching chores were divided among a staff of specialists.[32] The *Springfield Republican* saw in these reforms the toppling of the old theory of the "ill-proportioned and self-made man" and the emergence of a new theory of college education that "lays the foundation for a broad and symmetrical superstructure."[33] Before 1873, schools were paternalistic, focused only on education; they offered few prizes and required strict classroom attendance. After 1873, in an effort to encourage greater student "independence," they gradually became more "communal" and "fraternal;" they expanded dramatically the extracurricular activities that geologist William North Rice of Wesleyan called the "side shows" of university life in contrast with the "entertainments in the man tent"—student publications, the Young Men's Christian Association, debating societies, clubhouses, fraternities, and, above all, athletics.[34]

Second, at this juncture between two educational traditions, egalitarian coeducation took its first important leap. In the minds of such college presidents as Andrew Dickson White of Cornell and John Vleck of Wesleyan, the progressive advancement of women and its embodiment in egalitarian coeducation represented the true index of the scientific advancement of the race.[35] Third, the significance of the relationship between curriculum reform, scientific progress, and the cause of women was clear to feminists; they helped to shape and direct the character of this relationship.

Like Cornell and the University of Michigan, Wesleyan University, in Middletown, Connecticut, illustrated quite clearly during the early years of coeducation the relationship among science, coeducation, and feminism. A Methodist Episcopal school but with a decidedly non-sectarian orientation, Wesleyan dispensed with the Latin and Greek requirement in 1869 and in 1873 introduced the elective system more completely than any other New England college except Harvard.[36] Three men led the way in this transformation: John Vleck, professor of astronomy and president of the university in the early years; William North Rice, a geologist and evolutionist; and Wilbur Atwater, a German-trained chemist who was the son of a Methodist minister.

Deeply influenced by German educational reform, Vleck almost single-handedly changed Wesleyan's curriculum. Rice performed the important service of making Darwin palatable to a generation of men and women who, like himself, refused to renounce completely their religious beliefs.[37] Of this triumvirate, however, Wilbur Atwater was the most famous and influential. "I am enthusiastic in teaching and propagandizing science and scientific truth and want to work where that can well be done,"[38] he wrote Rice in 1872. Atwater built and equipped the first laboratory at Wesleyan in 1873, and three years later, after much pleading with the Connecticut legislature, he brought to Middletown the first agricultural experimental station in the United States, a laboratory patterned after a German model and designed to develop chemical fertilizers and to experiment in new agricultural methods. Atwater dreamed of creating an international scientific community that would transcend all political frontiers. He wanted "to make wonder and dread" the "obedient servants" of humanity and "to remove the inconvenience of *space* and *time*."[39] To hasten the day, he developed a rational and uniform nomenclature of nitrogenous compounds of animal and vegetable substances; conceived of a plan to build international networks of experimental stations; and classified, coordinated, and correlated agricultural knowledge for national as well as international consumption.[40]

Wesleyan became the first eastern college to institute egalitarian coeducation; it admitted four women and fifty-one men to its freshman class in 1873. Although the administration set no formal quotas, it continued consistently to admit on the average two to six women for the next seventeen years,[41] and by the 1890s the number of women students was larger. Wesleyan, like Cornell and the University of

Michigan in the early years, made no distinctions between the sexes except in the matter of housing and had yet to impose a system of "espionage" or "surveillance" over the comings and goings of its students.[42]

Both Vleck and Rice were feminists and both viewed coeducation as a feminist experiment; Atwater apparently never expressed any opinion on coeducation, but, appointed to the faculty the year Wesleyan turned coeducational, he symbolized the link between egalitarian coeducation and science. Feminist and antifeminist students and most faculty alike connected coeducation to "woman's rights" agitation. As one male feminist wrote for the student newspaper: "In these days when the question of woman's rights has become a prolific subject of discussion, when she is demanding a position of legal and social equality with men, it may not be impertinent to inquire on what grounds man refuses to admit her claims."[43] Another male student wrote for the student magazine in the same vein: "To our University belongs the glory of having dared to test the system first among Eastern colleges."[44] These male students were only a small minority of supporting voices, however; the rest ceaselessly ridiculed coeducation from its inception to its termination at Wesleyan in 1909. (Coeducation did not return to Wesleyan until 1972.) These antifeminist men succeeded in excluding women from the junior and senior class offices and, for the most part, from the college newspaper and magazine. In an orgy of college football competitions in the 1890s, they looked upon coeducation as a blight on Wesleyan athletic prestige.[45] But despite this characteristic student campaign, the administration and most of the alumni sustained the coeducational principle in its egalitarian form for nearly forty years, and hundreds of women benefited from that commitment.

The emergence of coeducational colleges and universities in the late 1860s and early 1870s created a climate in which feminists might rally round the coeducational principle. But by this time there were also some colleges for women only. In the 1850s, some states chartered women's colleges—Elmira Female College in New York, Ohio Wesleyan Female College, and Pennsylvania Female College among them. In the 1860s and 1870s such noted institutions as Mount Holyoke, Smith, Vassar, and Wellesley appeared, and from about

1875 all of them developed curricula equal in rigor and quality to that of the best male schools.[46] During this period, however, most feminists took the coeducationist, not the separate education, position.

Theodore Tilton and Moses Coit Tyler championed the cause at Michigan; Samuel May, Susan B. Anthony, Matilda Joslyn Gage, and Elizabeth Cady Stanton gave their warmest support to White at Cornell; Lucretia Mott did much to bring the reform to Quaker Swarthmore; Lillie D. Blake made a crash, unilateral attempt to induce Columbia to go coeducational; and a united front of feminists and neo-feminists—including the Unitarians James Freeman Clarke, Julia Ward Howe, and Caroline Dall—nearly persuaded Unitarian Charles Eliot to convert Harvard into a coeducational school.[47] These reformers demanded that women receive precisely the same education as the men students, that they study the same curriculum (preferably scientific, utilitarian, and nonsectarian), have the same teachers, and "contend together for the same rank and honors."[48]

The argument they put forward to justify these demands rested largely on hygienic-scientific grounds. Though often thrown on the defensive by hostile critics and fatalistic physiologists who claimed that coeducation would cripple woman's feminine appeal and health and disable her maternal functions, the coeducationists wielded their own battery of physiological weapons and cited their own doctors to prove that woman's health and appeal would blossom and mature, rather than wither and decay, within a coeducational system. Feminists knew that menstrual disorder and, to use spiritualist Andrew Jackson Davis's quaint expression, "General De-bility" did not strike naturally active women, that is, country women, peasants, Indians, or women struggling to develop themselves fully and harmoniously.[49] No, sickness struck those women who secluded and segregated themselves in dreamy, claustrophobic domesticity, who "exposed themselves at the opera," who "devoured food at Delmonico's," and who "used drugs" —chloral and opium—"to deaden pain," and those women, finally, who lingered unhealthily on their sexuality and "prostituted body and soul . . . that the passion of men might be satisfied."[50]

Most feminists belabored a widely held reformist theme that segregated private schools removed men and women from the "real," progressive world into a backward-looking world infected with the "varied restraints of modesty," and that the character of the curricula offered by these schools—one fashionable, the other classical—underlined this unreality. These feminists felt certain that sexual segregation in the schools kept the flame of the sexual imagination alive far beyond healthy psychological limits, stimulating "morbid sensitiveness," "hurtful reveries," and a "false consciousness of sex." Trapped in seclusion and left only to the insubstantial fantasies of their minds, male and female students floundered in visions of one another as knights and maidens, gods and goddesses—thus rendering themselves, in real life, disappointed adversaries of the other sex. Persistent dwelling on sex in segregated cloisters turned the sexes in upon themselves, fostering premature "sexual tension" and "unnatural," diseased sexuality—"diseases of the body, diseases of the imagination, vices of the body and imagination"—which, in turn, produced extreme forms of antisocial behavior, stirring up a "vortex of crime."[51] "Every school for young men," declared the *Golden Age* in 1874, "is more or less infected with vices that students do not outgrow or recover from; and every lady's boarding school is more or less perilous to its inmates."[52] Thomas Higginson even suggested that this private life so captivated young men that they would not or could not, in the end, leave it.[53] Stanton bluntly objected to the educational segregation of the sexes, an objection she held for the rest of her life:

> Restricting the social intercourse of the sexes adds undue zest to their meeting. . . . If the sexes were educated together we should have the healthy, moral and intellectual stimulus of sex everquickening and refining all the faculties, without the undue excitement of senses that results from novelty in the present system of isolation.[54]

Acting largely on this conviction, Stanton sent two of her sons to Cornell.[55]

Egalitarian coeducation, in contrast, erased the intellectual and moral differences between the sexes in a simulated natural-family context that "made potent social forces subject to rational influence."[56] In a coeducational setting, wrote one reformer, self-centered sexuality and "sexual polarization" disappear.[57] Let "nature have her rights," wrote Higginson, with the usual feminist assurance that the "facts of life" must be squarely faced and with the deep nineteenth-century conviction that "nature" directed through proper channels never threatened anyone:

> As they are born to interest each other, we think it safer to let this work itself off in a natural way; to let in upon it the fresh air and daylight, instead of attempting to suppress it. In a mixed school as in a family, the fact of sex presents itself as an unconscious healthy stimulus. It is in the separate schools that the healthy relation vanishes and the thought of sex becomes a morbid and diseased thing.[58]

In common with such older educational reformers as Unitarians Horace Mann and Samuel Gridley Howe—in fact, in the manner of a whole generation of reformers who wished to socialize American penal, medical, and educational institutions—Higginson combined the "natural way," "family," and the "healthy relation" in a typical feminist ensemble; he believed that the same education for both sexes could easily be achieved if men and women encountered each other daily, not as sexually excited protagonists, but as equal brothers and sisters within one common family, in an environment, in other words, that replicated the sexless, warm, and cooperative climate of the home.[59] Feminists thought of a coeducational institution as a rather elaborate water-cure establishment in which no conflict, no sexual "polarization," no politics, no sectarian controversies intervened, an institution in which the cooperative, fully integrated human being could develop freely. Thus a new generation of scientifically trained, nonpolitical men and women could "go forth to bless the nation."[60]

At the heart of the coeducational rationale lay the conviction that everything "one-sided" and dangerous happens in a segregated sexual world where everything is hidden. In a coeducational world nothing one-sided or dangerous happens, because everything is seen or revealed. In the early years of coeducation, this belief occasionally resulted in a free, open discourse between male and female students that troubled college administrators. In 1874, at one coeducational school in Wisconsin, several reformist women and men met to discuss "delicate subjects." One boy read a paper on tight lacing, treating at length the damage done to the female reproductive organs, its effect on menstruation, and on childbearing. "Wholly unembarrassed," the girls applauded his paper. "Nothing could be more timely," said one of the most gifted. The administration, however, objected strongly to these exchanges, put a stop to them, and compelled some of the brightest female students to resign from the school.[61]

The belief was that open exchange between the sexes promoted, rather than subverted, more "pure" and less conflicted sexual relations. Take, for instance, Lita Barney Sayles's essay on "Friendship Between the Sexes," which appeared in the *Revolution* in 1870. Inspired by an article the feminist free-thinker Francis Abbott had written for the *Index*, Sayles criticized the limits imposed on the social intercourse of the sexes by a "honey and molasses" marital system that not only violated the principle of full growth but also permitted only "nauseous" and "emetic" friendships between the sexes. Sayles even found "equal and perfect" marriages wanting, because they isolated the sexes from other rewarding social contacts, cutting them off from further "rounded" development, from "comprehensive knowledge" and "experience" of the complexities and "calamities" of the world, and from a "charitable" and compassionate awareness of "what the world needs." "Are we only to live in each other by pairs?" she asked. "This will not do." Good marriages must be supplemented by good "bread and butter," "practical," and "platonic" friendships. "We need" them "just as much; for humanity is a vast brotherhood, and we must continually give and receive to keep in sympathy with it." Besides, such friendships, such deeper growth for both sexes, will "bring" wives and husbands "nearer together. . . . As it is, each fears to trust the other with their interior loves and aspirations."[62]

Sayles was a respectable feminist, an influential member of the National Woman's Suffrage Association and the New York City Sorosis and, later, of the Association for the Advancement of Women, and certainly no free-lover. Nevertheless, she advanced an argument here that could have easily been turned against her by both timid feminists and antifeminists. The fact is, many feminists agreed with her in public and private, although apparently few of them ever projected the position as a major one.[63] In a coeducational polemic, or within an institutional setting, however, feminists could safely pose some variant of Sayles's prophetic argument, albeit without taking it so far. They could in effect transmute the polemic into a marital one. "Much of the mutual ignorance before marriage," the Unitarian feminist John Weiss observed in a lecture on marriage before the New England Woman's Club in 1872, "would be modified by a mixed education of the sexes—the dangers of which are greatly overrated." "In order to make early marriages even tolerably hopeful," he wrote two years later in the *Woman's Journal*, "this coeducation would be necessary;" at least then the sexes would be "thoroughly known to one another."[64] "If men and women are to marry at all," declared another reformer, "I do not know any place presenting superior advantages for that thorough preliminary acquaintance" than a coeducational school.[65] Elizabeth Stanton made the most of this position. No one more than she believed that coeducation improved possibilities for better marriages. "In opening all high schools and colleges to girls," she said in her address on marriage and divorce before a meeting of the National Woman's Suffrage Association in 1870, "we are giving young men and women better opportunities of studying each other's tastes, sentiments, capacities, characters, in the normal condition. In everyday life," she went on, "in the recitation room and in the play-ground, the real character reveals itself, and more congenial marriages will be the result of these early and free acquaintances, far different from those under the artificial stimulus of fashionable society."[66] Thomas Higginson maintained optimistically, before the American Social Science Association in 1873, that "if anything is certain in our school system, it is, that the sexes, once united in school, are united forever."[67] Higginson's optimism was not altogether misplaced. At Wesleyan, for example, 50 percent of women alumnae who married between 1872 and 1898 married Wesleyan men; and at the University of Michigan a high percentage of women who studied medicine married University of Michigan men in the same field.[68]

CHAPTER 4

SEXUAL OWNERSHIP AND THE RATIONALIZATION OF SEXUAL DESIRE

"WOMAN must have the courage to assert the right to her own body as the instrument of reason and conscience, and the fulfillment of the function of motherhood subject to no authority but the voice of God in her soul."[1] Thus Lucinda Chandler, one of the leading lights of moral education in America and an influential feminist, expressed a central feminist goal: women's control of their own bodies. Her words expressed another goal as well, and one fully compatible with the principles of no secrets and complete, balanced development. Chandler sought to rationalize sexual desire, to detach it from an excessive individualism, to domesticate it, to make it harmless and beneficent.

The feminist leader Matilda Joslyn Gage summed up this position with feminist elan on July 4, 1876, at the National Woman's Suffrage Convention in Philadelphia: "Without the control of one's person, the opportunities of the world, which are the only means of development, cannot be used."[2]

If married women were to make cooperation work, they had to

grapple with some basic necessities of woman's existence: the control of fertility, the limitation of conception, the mastery and alteration of sexual behavior in marriage.

The Feminist Consensus on Birth Control

STATEMENTS about limiting conceptions appear with growing frequency after the Civil War. In his 1866 review of Russell Trall's *Sexual Physiology*, which advocated smaller families, John Weiss wrote: "American women have too many children, and have them too often when every physical condition imperatively calls for repose and immunity."[3] Every major feminist periodical of the 1860s and 1870s carried articles arguing for a reduction in the number of births, for "quality" children, and for "instruction that would enable women to regulate the number of their progeny according to their capability of providing for them."[4] Feminists did more than plead for birth control; they placed their arguments within a larger rationalist reassessment of sexuality.

A feminist birth-control polemic had already entered feminist discourse to a minor degree in the late antebellum period, in part in response to the vocal, often embarrassing, preachings of free-love feminists—Mary Gove Nichols, Julia Cranch, Stephen Pearl Andrews, and others—at feminist conventions and in free-love periodicals. Free-lovers utterly deplored "bondage to the passions," although the public often confused their denunciation of passion with an invitation to promiscuity. They differed from the more moderate feminists only in their public opposition to legal control of sexual behavior and in their defense of the most "well developed conscience." Free-lovers embarrassed "respectable" feminists; still, they brought the subject of birth control boldly into the public arena. They marked the outer limits of debate and often pushed the more timid to think in terms of birth control and to speak openly on the matter. Thus even while Lucy Stone distanced herself from the likes of Mary Gove Nichols, she still longed to speak out for "self-ownership" at the conventions. She did so, for the first time, in 1853.[5] Martha Wright may have seriously disliked all the talk by Stephen Pearl Andrews about "unwilling maternity" at the 1858 convention, but she equally disliked the prospect of excluding him from a "free platform." She preferred his speech to the "interminable prosy essays" of other feminists, and believed in self-ownership herself.[6] Elizabeth Cady Stanton had no qualms at all. "Every article you write," she wrote Nichols, "hits the nail on the head. I like you vastly."[7]

The influence of the free-love vanguard, however, does not account for feminist acceptance of the validity of birth control. Much broader change created the conditions for this feminist consensus. For instance, as demographic studies have shown, general fertility rates had dropped from 7.04 children per family in 1800 to 4.24 children by 1880, indicating widespread use of birth control. The drop was caused also by such related changes as the separation of production from the family, the increase in male and female literacy, career demands placed on middle-class men, the growth of commodity production, and a rise in the standard of living—all tendencies that could make comfortable living and larger families incompatible. Also, there were the subtle rebellions of "sentimental" wives on grounds of health against the sexual demands of their husbands, and, finally, a companionate system of marriage that produced a new autonomy for women within the home.[8] Reformers were quite aware of the decline in fertility, of the rising incidence of birth control, and of the many reasons for the existence of both. The social scientist Francis Walker, in a paper on "Some Results of the Census" delivered before the American Social Science Association in 1873, attributed the "causes of retardation" to the "habits of life . . . which generally, if not invariably, connect themselves with, if they do not arise out of, certain conditions of industry and settlement. The habits of life to which I refer are the restrictions of the marrying class: the procrastination of marriage within that class, and the careful avoidance of family increase."[9] Mary Putnam Jacobi reported the same change in the same year to the Woman's Congress in New York. "The diminished fecundity of marriages," she said, "although in individual cases often a misfortune or the consequences of misfortunes or even crime, corresponds on the whole to a diminished necessity for rapid increase of population." This "testifies to the increased difficulty of maintaining families according to our higher standard of living." The reformist editor of the Chicago *Inter-Ocean* wrote of reproductive practices in the 1860s that "there was petulance as to motherhood in every woman's face and babies were welcome only in exceptional cases. There were homes built on new models where wife and husband were to be all and all, and no other individuality was ever to come between them to mar their pleasure."[10]

By the early 1870s, when the major feminist groups were based in the cities, reformers were brought into direct and daily contact with both the "fashionable" urban elites and the working classes. Many feminists sympathized with the plight of women trapped in the fashionable world; after all, many came from fashion themselves and emulated the fashionable styles. They tried to recruit wealthy, fashionable women into the feminist ranks. But, on the whole, feminists joined other reformers in an outcry against the behavior of fashionable people. They protested the wastefulness of the fashionable and their scandalous indifference to the poverty around them. Most important, feminists were disturbed by their sexual and procreative attitudes—their reliance on abortion and the use of unseemly contraceptive devices. Most feminists associated contraceptive controls with the immoral propensities of fashionable people who lived "animalistic" lives, questing after pleasure. Such people scorned those "human" facts that feminists identified with the public realm—work, discipline, production, personal fulfillment. Pleasure itself was not bad, but in feminist eyes fashionable people made it the sole priority, thus throwing public, productive life into a dark light. Only in the twentieth century, when work and fulfillment had begun to split apart and when most Americans had ceased to view the public life as the true focus for "human activity," did this feminist position change. By the 1920s the private life had taken on "human qualities;" pleasure had lost its stigma. Artificial contraception then found acceptance in both feminist and nonfeminist circles.[11]

The feminist view of working-class women, and of all women who lived without the prospect of remunerative marriages, contrasted markedly with their view of the fashionables. They did not scorn poor women, although some feminists—Anna Dickinson, for example —feared the immigrant working classes with their "prolific" birth rate and the potential for racial contamination.[12] Many feminists felt compassion for the women of the working class who worked hard, silently endured the "tyrannies" of their husbands, bore more children than they could care for, and often resorted to infanticide or abortion as the only means of birth control. Eleanor Kirk, who toured the streets of Lower Manhattan, mingling with working-class women and studying working-class neighborhoods at first hand, wrote ironically:

> What did you say? tired taking care of the baby? tired of your offspring? Oh! It is well to qualify a little. Not tired of the child, but so worn out with its screaming and fretting; so exhausted with walking the floor nights; that what? you almost wish it had never been born! You must never, my dear, give utterance to such statements as these. But you are desperate! Tommy hadn't half finished teething when this one was born? What of that? Wives must learn to regard these trifling additions with saint-like equanimity.[13]

In their novels and periodicals feminists often did not reproach working women for relying on abortion, even while they "deplored" it as a "horrible crime."[14] At the First Woman's Congress, Stanton even excused infanticide "on the grounds that women did not want to bring moral monsters into the world."[15]

Moral Education and the Career of Lucinda Chandler

FEMINIST WOMEN began to organize societies within which they might safely, fully, and openly discuss every aspect of marriage and sexuality without intimidation from men. Such groups, appearing in the guise of moral education societies in the early 1870s, engaged their memberships almost entirely in a single-minded dialogue on sexuality, marriage, and birth control. The channeling of feminist activity into such single-issue organizations had proceeded rapidly by the early 1870s, offering undeniable proof not that feminism had split into a myriad of conflicting units and thus lost its power, but that the feminist movement had grown so extensively that it had begun to specialize, not to splinter. It produced an interlocking network of organizations that complemented rather than contradicted each other. The women of these organizations examined other major feminist questions, but their specialized interests consumed their time. Dress reform societies dealt with fashion, radical clubs and liberal leagues with religion, suffrage societies with the franchise, the Association for the Advancement of Women with economic and professional life, the Woman's Christian Temperance Union with temperance, and the moral education societies with birth control and sexuality. Each group fed into and reinforced the other. With the possible exception of the WCTU, which drew for its membership on a vast community of Christian women, all these organizations shared a common commitment to the construction of a new, secular, rational, and scientific egalitarian society. They shared a common membership and even an interchangeable leadership: Stanton and Gage presided over the NWSA conventions and spoke everywhere in the late 1870s and early 1880s on moral education and Free Religion.[16] Abba Gould Woolson, Mary Safford Blake, and Caroline Severance, all of Boston, who served at various times as presidents of the Boston Moral Education Society, worked for dress reform and joined the AAW and the Free Religion Movement.[17] Severance was president of the New England Woman's Club; Woolson and Blake were elected to the executive committee of the First Woman's Congress.[18] Caroline Winslow, Sara Spencer, Anna Garlin Spencer, Lucinda Chandler, Elizabeth Boynton Harbert, and Augusta Cooper Bristol advocated moral education, belonged to the NWSA, attended Woman's Congresses in executive capacities, and supported Free Religion.[19] What emerged then, after 1870, was not a divided movement but a complex of feminist programs working toward a common goal: the construction of a social ideology bearing on every facet of the sexual relation.

The object of the Boston Moral Science Association was the education of young women in life's laws, marriage, maternity, and "social intercourse." In 1873—the year of the founding of the Ladies' Social Science Association, which would become the Association for the Advancement of Women—the Moral Science Association evolved into the first moral education society in America.[20] Other societies emerged in Philadelphia, New York, Chicago, and Washington, all initially focusing on marriage and the reform of the sexual relation. As David Pivar has shown, these societies had small memberships but dynamic leaderships.[21] The Washington Woman's Club was reorganized as a moral education society and published *Alpha*, the main organ of moral education. The prospectus of *Alpha* read as follows:

> We advocate a change in the present marriage laws, which gives woman bound soul and body to her master, for his use and gratification—such a change as may be necessary to secure to her by law, the ownership of her own body, that she may hold it in all purity and freedom from use for what man terms "physical necessity"—that children may be born under better conditions and educated to understand their physical natures and know how to control them; to understand wherein lies the secret of health and happiness.

In a letter to Isabella Beecher Hooker appended to a copy of the prospectus, Emma Wood stated *Alpha*'s purpose. "We still work on," she wrote, "with eyes open and watching on every side, and indeed there is need for work, so long as woman's personal sovereignty is not assured. That must now be the end and aim of our endeavors and we feel that the clouds are breaking—the night is fast passing and the morning will soon come."[22]

Of all the moral educationists, Lucinda Chandler was perhaps the most persevering and the most original. She was faced with frustration and tragedy early. At thirteen she was forced by a spinal malady to abandon her formal education at the St. Lawrence Academy. When she married she had only one child, a boy who drowned at the age of three. Chandler's widely circulated tract, "Motherhood: Its Power Over Human Destiny," powerfully influenced feminist thinking. Vineland, New Jersey, where she lived, was one of the most active reformist communities in the country at the time. Progressive spiritualists, the Friends of Human Progress, Free Religionists, and "Liberalists" of all kinds lived in or near Vineland—among them the hygienist Susan Flower; the dress reformer and social scientist Mary Tillotson; the industrial reformer Dr. Lucinda Wilcox; the sex reformer Eliza Duffey; the naturalist Mary Treat; the social scientist Augusta Cooper Bristol; and such pioneer feminists as Margaret Pryor, Mary Ann McClintock, and Elizabeth McClintock Phillips, all formerly of Seneca Falls, in upstate New York. Like Chandler, these women studied "political science" and social science in the 1870s and sought to replace a competitive with a cooperative capitalist system. In 1874 they established the American Free Dress League, and in 1878 the

Woman's Greenback Club.[23] When Chandler left Vineland temporarily for Boston, she held parlor meetings in the style of Margaret Fuller and, according to her biographers, "achieved the purpose of her heart, the organization of a body of women who were pledged to work for the promotion of enlightened parenthood and to an equal and high standard of purity for both sexes."[24] This organization, of course, was the Boston Moral Education Society. In spite of her health, Chandler often chaired society meetings and lectured on moral education throughout the country. She delivered one of the opening papers at the First Woman's Congress: "Enlightened Motherhood—How Attainable."[25] She also became absorbed in the study of political economy, wrote articles on finance, the land question, and industrial reform. Like Elizabeth Osgood Willard, who developed ideas in *Sexology* on sexual equivalence very similar to hers, Chandler also backed the cause of labor reform. She sought an "equitable economics" and "free access to Nature's gifts."[26]

Chandler became a vice-president of the National Woman's Suffrage Association and, in Chicago, founded the Margaret Fuller Society, to advocate both the education of women in the principles of government and the reform of the economic system—a sign, once again, that feminists always considered private sexual issues in relation to larger public questions.[27] Before the Fuller Society Chandler analyzed the United States Constitution from a populist perspective. She denounced unjust accumulation of wealth, demanded heavier taxation of the rich, and proposed the nationalization of railroads, highways, and communications.[28]

In the early and middle 1870s Chandler wrote for virtually every major feminist periodical. *Woodhull and Claflin's Weekly* serialized her long tract on motherhood, and the *Revolution* published her articles on marital equality and motherhood. Apparently disparate journals opened their pages to her, a fact that reflected the unanimity among feminists on the necessity of moral education.[29]

Chandler's writings, along with those of Eliza Bisbee Duffey, John Cowan, and Georgiana Bruce Kirby, may be taken as the major feminist texts on birth control of this period.[30] Chandler held many ideas that were indistinguishable from Stanton's. She considered woman's right to suffrage "indisputable," demanded woman's right under the law to the control of her own earnings and property, argued for equal access to education and the professions, advocated an equality based on "equivalence of the different and opposite qualities and labors of men and women," and believed that sexual life within marriage needed to be completely transformed.[31] She also agreed with Stanton that, in Chandler's words, "the state must be in its legislation and its political operation a supplement to the integrity and moral purity of the righteous home, or it will inevitably disintegrate and become a destroyer of the home."[32] But unlike Stanton, Chandler singled out birth and sex control as the pivotal reforms. For her, equality under the law represented only "*one* feature of the adjusting process" and, moreover, not the major one. Woman's control of her own fertility and of the sexual relation was the most important reform, the key to equality and advancement, the solution to the "irrepressible conflict between the sexes that resulted from the *dis*order of our 'regulations.' "[33]

Chandler's program rested on two pillars: "self-ownership" and the rationalization of sexual desire. "The only true state of things . . . is when the woman, as all females in the animal kingdom, has control over her own person, independent of the desires of her husband."[34] Self-ownership meant that woman, not man, would decide when, where, and how the sexual act would be performed. She would no longer be subject to the unpredictable and erratic male "lust of dominion and appetite;" she would no longer yield to "unwarrantable intrusions upon her personal sanctity" or be treated "simply as an instrument for multiplying the species."[35] It also meant that woman, not man, would determine when children would be conceived and how many. Planned births would replace "undesigned and unhallowed parentage."[36] "The imposition of the office . . . of motherhood upon women in the 'old fashioned marriage,' regardless of her wish or fitness, has directly laid the foundation of the present instability and inharmony in marriage relations."[37] Chandler believed so strongly in the principle of self-ownership that she wanted it fixed in the law; she joined the moral educationists of Washington in an attempt to repeal the law of coverture in the District of Columbia and to give every woman the "legal . . . custody and control of her person in wifehood to govern according to her wisdom and instincts the maternal office and protect her child as well as she may from the dangers of selfish passion, alcoholism and vice."[38] Viewing all women as mothers in both a spiritual and a physical sense, Chandler wanted the law to recognize this and place motherhood on a par with fatherhood. "The law," she declared, voicing a sentiment shared by most feminists, "perpetuates one of the errors of barbarism which science has exploded and which experience is constantly disproving, viz., that the father alone is the creative power."[39]

Legalization of self-ownership was only the beginning. If women were to control their own fertility, moral education was required to raise both sexes to an equal level of purity, to rationalize or render "natural" and harmless male sexual desire, and to prevent disease. As the instrument of "reason and conscience," mothers, above all, had the power to "guide the surging passion of masculinity into . . . channels of enlightenment."[40] Chandler advised mothers to familiarize themselves with "all the scientific knowledge . . . which involves the relations and uses of sex," all the facts of "physiology" and the menstrual cycle, and "the responsibilities of parentage—and which effect the fundamental relation of life, marriage—the intelligent understanding of which is central to purity."[41] Then the mother was to transmit her knowledge, through the "potent" influence of her "language," first to her son. He was to learn to respect marriage as a "spiritual relation" and not as a means of "gratification;" to "master his passions" and wait upon the "law of the wife;" and, as a husband, to "prepare" carefully and methodically for the "sacred office of parentage." No man, Chandler believed, should engage in sexual intercourse by "accident" or impulse, but only after "premeditating desire" and only after he and his wife had developed all their faculties to an equal level of health and balance.[42] These "equilibrated" conditions not only considerably reduced occasions for intercourse; they also preserved the "ante-natal" state of the wife from disruption and thus prevented the transmission of disease to the children.[43] Finally, the mother was to teach her son about the virtues of continence and restriction of sexual intercourse to the purpose of procreation. He would learn that intercourse is not a "sexual" act, but a cooperative one—planned, rational, harmonious, efficient, and ultimately productive.

A daughter would learn from her mother, especially, how to mute her sexual attractiveness so as not to stimulate the fantasies of men. She would discover, in other words, the moral advantages of dress reform. If girls would only "discard all prominent devices to distinguish sex," Chandler declared, they would "take one of the most efficient steps toward the emancipation of women."[44] The mother was to share with her daughter the facts of the menstrual cycle and of sexual physiology, knowledge so fundamental that Chandler was willing to relinquish the teaching of it to the schools. In a paper she read at a Chicago women's meeting in 1872, Chandler proposed the "introduction into our public schools of a psycho-hygienic department, in which those portions of the human anatomy, which have been heretofore ignored, shall be taught, together with their functions, and also the necessity of keeping not only the body but the mind free from disease."[45] She wanted such instruction made mandatory for graduation. High-school girls, she said, "must have all knowledge which relates to the function of maternity as a part of the study which will admit them to graduate from school."[46] The progressive Free Religionist Frederick Hinckley said the same thing, more forcefully, before the Boston Moral Education Society in 1876. "It is of the first importance," he declared, "that boys and girls, especially boys, should be given a thorough knowledge of their physical structure, and the use and abuse of the sexual function. This is a work for parents and teachers infinitely more important than the teaching of mathematics or language."[47]

Like many mid-century feminists, Lucinda Chandler looked upon male social-sexual behavior with alarm, especially in its prevalent irrational forms—the delirium of masturbatory orgies, uncontrolled marital "carnalism," and the "furious" expression of "selfish propensities."[48] Here was not only an attack on sexual passion but also a repudiation of the patriarchal and romantic-individualist traditions of the past. Here also was a vision of a new social order, a vision untouched by the bleak influence of social Darwinism. Passion and instinct, according to Chandler, were not in themselves abnormal or antisocial; indeed, in their "natural," "normal" state, they promoted "cooperation," "peace and goodwill." "Moral laws," she wrote in the *Revolution*, "are not arbitrary and antagonistic to the welfare and happiness of the individual and community;" rather, "they coincide with the orderly and healthful exercise of the appetites and needs of the physical and with the principles of best government in the State."[49] To put it succinctly, "physical and mental culture" were not at warfare with one another, but inherently harmonious and "unified."[50] Historical conditions of male dominance and female subjection had destroyed this unity and harmony by transmuting normal male "appetites" into aggressive and "selfish" impulses, and by lifting the male "individual" above the "human welfare." Only the "*equal* freedom of women and men in the marital relation" would correct this disequilibrium. Only when "there is no subordination or domination" in marriage would "each sex fulfill the offices and functions ordained to each *according* to the *order of nature*, and not according to the will or selfishness of one party."[51] Only sexual equality would restore harmony and domesticate male passion.

Sexual Passion: Benign and Harmless

BY THE 1870s, self-ownership and the rationalization of sexual desire had become the stock in trade of feminist thinking on birth control. As Linda Gordon has observed in regard to self-ownership, the "whole feminist community" believed that women should have full possession of their own bodies and that they, not men, should have the "right to choose when to be pregnant."[52] Most feminists were also at one on the question of sexual rationalization. The belief that all births should be planned; the contention that sexual intercourse should take place under perfect eugenic conditions of symmetrical health, preferably in daylight when the imagination could not be easily activated, and only for procreative purposes; and the demand for male "continence *in marriage and out of it*"—on all these matters, most feminists were, in varying degrees of rigor, unified.[53] Samuel Byron Britten, the feminist editor of the *Banner of Light* and husband of the famous spiritualist Emma Hardinge, summarized the essence of this rationalist position in an article on "gestation." "The demands of this essential law of our being," he wrote, "will never be duly respected so long as the generation of human life is left to accident, sudden caprice and unconquerable passion. Millions of unwelcome children are forced into the world."[54] In effect, feminists fixed the limits of sexual expression, organized and channeled it, thereby reducing occasions for sexual intercourse.

The idea of passion as an essentially harmless and "natural" human emotion would remain a stable component of bourgeois-reformist thought into the twentieth century. Like Chandler, most feminists began their polemics against passion with an almost romantic appreciation of its power. Thus in a speech filled with hyperbole Augusta Cooper Bristol revealed to the 1873 Woman's Congress in New York the chilling spectre of prolonged conflict between the power of instinct and the power of morality:

> We are living in that dread time when the beast and the angel within the human have closed in deadly conflict, each contesting for Sovereignty. . . . We have so long known the moral history of human life to be one imperfect, yet ever constant endeavor to eliminate the animal from our instincts and passions, that we never dreamed of the dread grandeur and fearful tenacity of its death struggle. We never dreamed of the vital strength of that dark drop which circulates in the social organism.[55]

Bristol proceeded, however, to unfold a vision of a cooperative and "unified" future in which heterosexual passion would harmonize with the best interests of society. Feminists, of course, believed that this transformation waited upon the abandonment of patriarchy, which would permit the equal symmetrical development of both sexes in marriage, each faculty checking the other to create a "healthy" ensemble in harmonious balance—a bourgeois reformist conception that reappears again in the twentieth century as the "integrated personality." The "elimination of the animal" from passion also depended upon the beneficent workings of the evolutionary process that would liberate men and women from the "childhood" of the "race"—"an age of impulse, passion and appetite"—into a "higher grade of conscious existence" characterized by a "regard and care for others, instead of the self."[56] The effects of marital equality, feminists believed, tended to disable the power of passion. "Sexual passion," the *Springfield Republican* declared in 1871, elaborating a view that looked backward in time to the abolitionists and forward to the Progressives, "aggressive in man, is dormant in woman. So long as she must come to him to beg her bread, to flatter her vanities, to feed her weaknesses for dress or indulgences of any kind; and just to the extent of such dependence, man will dictate the sway of sexual passion and the degree of its indulgence."[57]

Some feminists—notably the dour Alphaists—agreed that women had less intense sexual drives than men and that rationalization might, in fact, remove the "passionate feeling" in both sexes. Caroline Winslow was apparently convinced that marriage could proceed very nicely with little or no intercourse at all, while Eliza Duffey allowed the "nerve and brain workers" of the world (she was more generous with the "muscle workers") a maximum of "intercourse once a month," although she thought "six times a life" was sufficient to sexual needs.[58] Reformist rationalization generally did not repudiate sexual pleasure for either sex, although it sought to reduce occasions for it, and to channel and organize it. The writings of the important physiologists George Napheys and William Carpenter, who were favorites with feminists, made this position quite clear. Both advocated birth control and a limitation on offspring.[59]

A moderate feminist, Napheys believed that the mental faculties of men and women were different but "not unequal," and that the advance of "civilization" would make both sexes "absolutely equal before the law." More to the point, he wrote in his book, *The Physical Life of Women*, that both sexes were very similar anatomically and that both had similar sexual needs. "In the vast majority of women," he declared, "the sexual appetite is as moderate as all other appetites. . . . It is a false notion and contrary to nature, that this passion in a woman is a derogation to her sex."[60] The feminist doctor Sarah Hackett Stevenson borrowed heavily from Napheys's book in her *Physiology of Women* (1881). Octavius Frothingham described it in a review for the *Herald of Health* as "pleasant and instructive" and thought it was "high time such a book were written."[61]

William Carpenter, the famous Scottish-trained English physiologist and comparative anatomist, and the brother of Mary Carpenter, the equally famous philanthropist, prison reformer, and active member of the British Association for the Promotion of Social Science, shared Napheys' views on male and female sexual response.[62] While denying in strong words the virtues of "excessive" sexual indulgence, Carpenter maintained that both sexes needed sexual release, and that female desire was particularly strong during the "period of menstruation" and "by no means wanting at other times."[63] Other contemporary physicians, including the Americans George Beard and Horatio Storer and the renowned English physician Henry Maudsley, agreed with Carpenter, observing even more forcefully than he the importance of sexual intercourse within marriage for the physical and emotional health of both sexes. "Sexual repression," wrote Beard, "is a disease of great and increasing frequency and importance."[64] Maudsley even claimed that women "suffer more than men from the entire deprivation of sexual intercourse. . . . In our view the affections constitute the entity of life itself, and their normal exercise is essential to health."[65]

Carpenter had more influence than perhaps any other physician in mid-nineteenth-century medical circles, and several feminist doctors and reformers drew some of their ideas on physiology from his books.[66] Among the feminists influenced by Carpenter, Lester Ward illustrated most clearly the developing reformist perspectives on heterosexual sexual pleasure. His first and most important expression of these concepts appeared in his *Dynamic Sociology*, a book profoundly Comtian in orientation.[67] Central to his thinking was the notion that "desire" represented the motive force in "all action." "One great law permeates the whole sentient world," he asserted, "that before there can be action, there must be desire, that action only can be performed in obedience to desire." Like Lucinda Chandler, Antoinette Brown Blackwell, Sarah Hackett Stevenson, Elizabeth Cady Stanton, and others, Ward looked upon desire, passion, and feeling in both sexes within marriage as essentially beneficent and healthy forces that tended in a "natural" fashion to "harmonize" with the general "social welfare." He even went so far as to say, without fundamental conflict with liberal reformist thought, that "many propensities now regarded as wholly bad will be found capable, by the proper adaptation in the circumstances, of producing great good. . . . All desires are alike before Nature—equally pure, equally respectable. All are performed with the same freedom, the same disregard for appearances. Nature knows no shame." And again: "Everyone has a right to seek the gratification of any and all of his desires, so long as in so doing he does not prevent others from exercising the same right. If society interferes, it should be from a scientific or sanitary, and not from an abstract moral, point of view."

What Ward objected to was the contamination of a natural relation by imaginative fantasies, "mysteries, deep hidden secrets," "illusions, deceptions and paradoxes," and the failure of rational intellect to "organize feeling" into "channels of human advantage." Ward located in the imagination, and in its handmaiden secrecy, the principal causes of chronic conflict between the sexes. He even suggested that sexual intercourse would become ever more healthy, ever more pleasurable, and ever more egalitarian, as it lost its dependence on "modesty" and "prurience" and obsessive privacy. "Whatever must be done secretly and clandestinely," he wrote, "will be done improperly and become an evil, though it possess no intrinsically evil elements."[68]

The fact is, Ward associated secrecy and imagination with religious "monomania," "internal reflection," "fancy," "pure theory," "ideal conceptions," and "prevailing theories of human right" that had historically "impoverished the mind," led to "derangement and delirium," and even "damned the stream of human desire until it finally burst over all its barriers and whelmed everything in the ruin of political revolution."[69] Only relations founded on the "knowledge of things," on "stern objective realities" and "physical fact," and not on "appearance" and "thought," were truly "pleasurable," progressive, and "natural." For Ward, as for other mid-century feminists, the goal of a feminist "social science" was to "strip all physical realities of their masks" and "drag them before the light." "Once caught," he insisted, "they have been tamed, trained and put to man's service" and "directed into channels not only innocent but useful." Once tamed, pleasure and feeling could be embraced without reservation. "The human body is a reservoir of feeling, which, when wholly unobstructed, is all pleasurable. . . . Checked in its natural flow it becomes pain." Ward, however, attacked the imagination—the very thing that produced passionate desire. In this he was at one with the reformist intellectuals of his generation.

The positive feminist approach to heterosexual sexual pleasure can also be understood in terms of the symmetrical concept cited by Bristol. As Stanton said in 1870, "There is a science in pleasure as well as in profit and in its best, its truest sense, it is the harmonious development of all our faculties."[70] Most feminists thought "marriage should be consummated only between *physiologically* perfect men and women with full growth of every organ," and most of them relegated the sexual organs to a status equal in importance to the other bodily processes, insisting that only the "imagination" acting on the organs tipped this balance.[71] Feminists, however, stumbled over contradictions in this theory. They exaggerated the fury of the male sex drive and of the maternal side of woman's nature, marking the former as the enemy of life and the latter as the true source of virtuous social and political order. This hyperbole resulted, in part, from the politicization of biological-sexual reality: feminists often equated the male sex drive with uncontrolled individualism, and maternal feeling with a cooperative society. Feminists eulogized the maternal virtues but thought they could be shared. Most of them hoped to see men acquire such virtues as secondary sex characteristics or introduce them into their own parental behavior, thus leading to symmetry.

"The two most important feminine virtues are tenderness and purity," wrote Elizabeth B. Harbert, journalist and social scientist. "The two most important masculine virtues are courage and knowledge. Humanity will not be perfected either in individual character or social destiny by the greater separate development of these in the sexes, but only in their balanced diffusion in both, making woman wise and courageous, and man tender and pure."[72] Although feminists distinguished between an insatiable male appetite and a more refined, considerably less insatiable female appetite, most envisioned a time when men and women would behave in precisely the same sexual way. Take, for example, the friendly colloquy between the influential, progressive sexologist Dr. Edward Bliss Foote and the adamant moral educationist Dr. Caroline Winslow, published in Winslow's journal, *Alpha*, in 1881. Foote rejected strict sexual continence as a healthy approach to sexuality; he did not limit sex to procreation and advocated preventive birth control. He seemed to grant sexual expression a preeminent place. Winslow took the opposite stance, denigrating sexual expression, although she did not deny men and women the right to pleasure in the act of procreation—indeed, she thought it a sign of health—and demoting the sexual organs to a place below intelligence and morality. Yet Winslow and Foote both disparaged precocious and intemperate "exercise" of the organs; both wished to systematize or industrialize sexual activity, limiting it to "due season and under legitimate circumstances;" both believed that the "vital force" should be "harmoniously distributed;" both insisted upon the relationship between private health and public welfare; and, above all, both yielded to the logic of the symmetrical concept. Winslow wanted both sexes "emancipated from the thralldom of passion." Nevertheless, she said: "I know that the legitimate use of all the functions of the body or endowments of the soul tend to promote happiness, secure sound health and prolong life."[73]

Rationalization did mean the harnessing of passion and the "organization of feeling," but not the renunciation of physical pleasure. Even in our own time, when varieties of sexual experience increasingly find unstigmatized outlets, society still continues to deny sexuality genuine sensuous expression. Nineteenth-century feminists were determined to make men and women cooperative and harmonious human beings and to subordinate reproductivity in the private sphere to productivity in the public. They chose female self-ownership and the rationalization or sublimation of desire as the means of reaching these goals. Opposed to individualism and to its manifestation in competitive capitalism, they nevertheless brought to bear on the sexual relation the full weight of industrial technique.

These feminists longed to create a workable egalitarian marriage that would free both sexes for productive work—even to the point of rejecting the centrality of the family and of maternity for women. Physiological functions that women and men did not share were, correspondingly, inferior. The appeal of liberalism continually undercut arguments based on the differences between sexes. "Motherhood," two women doctors told the Sixth Woman's Congress, "is not central" to the lives of women.[74] Eliza Duffey, in her *Relations of the Sexes*, seemed to capture best of all the underlying direction of the feminist ideology of marriage:

> I utterly deny that marriage, in the present state of the world, is instituted solely for the perpetuation of the human race. . . . Nor is it the duty of married people to have children at all, if according to their own best judgement, they would do an injury to these children by bringing them into the world. . . . Men and women were made for themselves. . . . Though they will develop self, it will not be for selfish purposes, but because they can through their own individuality, best benefit mankind. In marriage they will see greater opportunities for self-development and harmonized action.[75]

CHAPTER 5

THE VINDICATION OF LOVE

"WE SAY LOVE IS BLIND," wrote Emerson in 1860, "and the figure of Cupid is drawn with a bandage round his eyes. Blind: yes, because he does not see what he does not like: but, the sharpest-sighted hunter in the universe is Love . . . finding what he seeks, and only that." Ten years later Moncure Conway, who had devoured Emerson's writings in the 1840s, paraphrased a line from Pascal, "Love, from so long having bandaged eyes, will be all eye."[1] One short but revolutionary decade separated these two opinions. Emerson spoke for the American romantic era then dying away; Conway spoke for the more realistic and more rationalist post–Civil War period.

Sentimental Love and Sexual Stereotypes

CONWAY'S WORDS also articulated the language of feminism. In the latter half of the nineteenth century love itself was transformed by feminist reform. Feminists had turned away from romantic love as blind, passionate, seductive, the bearer of conflict. In a sexually unequal society, romantic love threatened the interests of women. So did sentimental love that portrayed women as idealized objects and the passive recipients of masculine affections. In their place feminists put a rational, symmetrical, and egalitarian love based on a knowledge that made no room for ideality, passion, or fantasy.

Sentimental love, with its exaltation of sexual differences, was the consequence of two overlapping lines of development: the appearance of an economic sexual division of labor, most blatant in the antebellum period, and the transformation of American Protestantism from a relatively grim, static, and authoritarian Calvinism to a more gently remissive, fluent, feminized, and democratic evangelicalism.[2] By the 1830s, especially in the settled urban regions, manhood had come to signify selfish individualism and rational or functional behavior; womanhood had come to mean stable domesticity, nurturant piety, and infantile impulsiveness.[3] After 1850, when womanhood became clearly linked with the consumption of commodities and manhood with production, domesticated religion rooted itself in the world of fashion, selflessness in sentimental narcissism. The periodical press, novels, and poetry popularized these norms, although sexual behavior never neatly reproduced them.

Contemporary American attitudes toward the two most important Christian holidays, Christmas and Easter, exemplified the romance of religion with female fashion and commodity consumption. Many Americans identified Christmas with maternal giving and Santa Claus with the mother. In 1857 the spiritualist Andrew Jackson Davis described such maternal behavior in his autobiography, *The Magic Staff*. In the Christmas holidays of his childhood, Davis's stocking, hung by the chimney, was stuffed with a perplexing combination of objects—toys, fruits, candy, potatoes, and straps. Davis asked his mother to explain what they meant, although the good things, he knew, came from her and the bad from his father. "My dream was realized at once. A benevolent smile pervaded her countenance, as she answered. "Yes, Jackson, I put in everything but the potato and the strap."[4] In the Easter holiday the giving mother was merged with the narcissistic wife who spent her husband's money on herself. Woman converted fashion into a religious ritual. In 1872 the *New York World*, one of the most powerful Democratic papers of the time, lectured its readers: "Easter and new spring bonnets have long been synonymous terms, and as religion yearly becomes more fashionable the relation becomes closer, until it is difficult to decide how much of Easter there is in the bonnets or how much of the bonnets in Easter."[5]

In sentimental love man yielded up his selfish and aggressive self and succumbed to the passive, maternal spirit of woman, finding in the shelter of her redemptive influence his true humanity. Conversely, sentimental love occurred when woman became a reified testament to male power. Such sentimental love rested on a double motive. By personifying women as spiritual mothers while at the same time denying them a reliable, consistent source of power within the home, sentimental love continually threatened marriage with sexual disaster. For women it meant repression and narcissism. For men it meant a basic alienation from women. The sentimentalization of women as mothers made sexual expression between the sexes difficult and at times impossible. The transformation of the home into a haven infused with feminine virtues of frailty and passivity cut to the core of male narcissistic vulnerability: it generated fears of impotence that could be met only by escape from the home or by the assertion of power within it in often sadistic ways.[6] In a sermon of 1865 Henry Ward Beecher expressed the potential conflict inherent in such love. "The life of ordinary men, under Christian influences," he said, "is a conflict and ever-broken compromise. Men are perpetually at conflict with themselves. They are often at a loss to know which they are, what they are. And are they to take the measure of themselves, of their feelings and their passions, by those tremendous surges to which they are subject?" Only under the influence of "fair and gentle wives, star-like and dove-like," could men regain possession of their "souls."[7]

Sentimental love offered compensations that lent validity. In a world without order, it clarified sex roles; in a world without solace, it provided for most men and women the only consistent source of emotional security. In the minds of men, at least, it preserved, even while it denied, the compassionate power of the mother. As historians of the family have recently noted, sentimentalism also helped raise the status of women: they held a new authority over children and received a new respect from husbands.[8] It answered the needs of a Protestant bourgeoisie burdened by the pressures of atomistic and competitive individualism and, like the traditional revival meeting, tended to remove the guilt and anxiety that accompanied the accumulation of wealth and material comfort.

Remarkably remissive, the ethos of sentimental love gave to conspicuous and ostentatious consumption its earliest and perhaps fullest justification as in the name of the spirit man struggled to get his fortune. "Among the appropriate uses of wealth," Henry Ward Beecher advised *Herald of Health* readers in 1867, "may be mentioned that of enobling and dignifying the household. . . . I love to see a man who makes a paradise about his home. . . . I am not a believer in poverty. . . . It is right to indulge these things."[9] Uncritical and reflexive, sentimental love attempted to make palatable and endurable the conflict that the extreme sexual division of labor engendered.

Feminists were drawn to sentimentalism even as they labored to create the conditions that would destroy it. In the late 1860s and early 1870s, for example, almost every meeting of the first woman's club in the United States, New York City Sorosis, began with a recitation from some sentimental verse or fiction, a form of prayer to open the proceedings. On May 16, 1870, Thomas Bailey Aldrich's maudlin poem, "Baby Bell," introduced a session devoted entirely to a discussion of equal rights within marriage.[10] Feminists published sentimental fiction and poetry in their periodicals—in *Una*, *Woman's Journal*, *Alpha*, *Revolution*, *Woodhull and Claflin's Weekly*, and *Golden Age*—and themselves added to the surfeit of sentimental literature. Caroline Dall, Abigail Scott Duniway, Phebe and Alice Cary, Thomas Higginson, Harriet Beecher Stowe, and Theodore Tilton, among others, wrote sentimental novels and short stories and almost everybody, male and female, wrote poetry. Although both sentimental love and what may be described as feminist-rational love characterized bourgeois experience in a new urban, individualistic society of wealth and abundance, sentimental love underlined the bourgeois commitment to personal well-being and comfort and continually intervened in the feminist effort to reform and rationalize social behavior.

Many feminists lived out their lives in a tangle of rational reform and sentimental impulses, as Theodore Tilton's career illustrates. A journalist at twenty for both the *New York Tribune* and the *New*

York Observer, Tilton, through the direct intercession of Henry Ward Beecher, joined the staff of the most influential religious newspaper of the age, the *New York Independent*, as managing editor at twenty-one.[11] As a youthful Garrisonian Tilton was strongly committed to the bourgeois reform principle of the common humanity of all men and women. After the Civil War he focused on the theme of the relationship of union with equal rights—union of the states, churches, and races. The union of the sexes he considered the most crucial question of the day; he insisted that men and women be equals in the family, the school, and the state.[12] The plea for sexual "togetherness" indicated the new frame of mind in educated circles after the Civil War.[13] Like other reform publications of the time, Tilton's *Independent* and, later, his journal *Golden Age* marked the emergence of the rationalist-scientific perspectives that looked forward in time to the Progressive era. "Write and let me print, the freest words you ever penned," Tilton implored the rationalist James Parton, husband of Fanny Fern.[14] More pagan than Christian, the *Golden Age* leaned heavily toward rationalist religion.[15]

Although the rationalist climate of opinion shaped Tilton's reformism, a materialistic and remissive sentimentalism characterized his thought on social questions and, above all, on love. Tilton lived in a city that by the time of the Civil War had been transformed from a remote rural province with pigs running in the streets into the fourth-largest financial and commercial center in the country. By 1860 there were some twenty millionaires living in Brooklyn and many of them attended Beecher's Plymouth Church.[16] Tilton knew fashionable society intimately; many of his friends, both feminists and nonfeminists, were wealthy; and no one enjoyed the emoluments of success and fame more than he. "I believe," he wrote in the *Herald of Health*, "in eating the fat of the land."[17]

Tilton owed much of his good fortune, as well as the moral justification for it, to Henry Ward Beecher. When Beecher went to England in 1863, Tilton sent him letters full of endearments and sentimental declarations of affection. "Your private letters," he wrote, "are like so many kisses. . . . Send some more! My love multiplies for you everyday." "I toss you a bushel of flowers and a mouthful of kisses!" "I never knew how much I loved you till your absence. I am hungry to look into your eyes."[18] The two men kissed when they met and when they parted, and on one occasion Elizabeth Tilton discovered Beecher in her husband's lap discussing the Sermon on the Mount, "both apparently in the pleasantest of moods—Mr. Beecher rose as I entered—I met him with a kiss—he then sat down again on Theodore's knees."[19]

Beecher exposed Tilton to the watered-down remnants of the evangelical Protestant world view and smothered him in the doctrine of Christ's eternal love for mankind. In E. L. Godkin's words, Beecher had freed religion of "anything in the smallest degree disciplinary, either in the shape of systematic theology, with its tests and standards, or of a social code, with its pains and penalties." Beecher and the Brooklynites he shepherded had replaced older forms of male authority and "self-control" with materialistic self-indulgence and sentimental remission. "A large body of persons has arisen," Godkin wrote perceptively, "under the influence of the common schools, magazines, newspapers, and rapid accumulation of wealth, who are not only engaged in enjoying themselves in this fashion, but who believe that they have reached all that is attainable or desirable by anybody, and who therefore tackle all the problems of the day." Such conditions prefigured American social life in the twentieth century. A new middle and upper-middle class of men and women in the major industrial cities sought to understand and order the opportunities that wealth, education, and the breakdown of older forms of political and religious authority had disclosed. "The result," Godkin averred, "is a mental and moral chaos."[20]

Tilton's religious thought replicated Beecher's, and he also learned from Beecher all he would ever know about love. Beecher's novel *Norwood* described his ideas on love between the sexes. Five years after it was published Tilton wrote an island romance called *Tempest Tossed*, which virtually mirrors *Norwood*.[21] For Tilton, modern society gave men and women only the certainty of a love that transcended both time and space. "I think," says a character in *Tempest Tossed*, "there must be a golden chain made of invisible links, hidden in the sunlight, reaching around the world to bind each soul to its mate."[22] Tilton's novel also sets up sentimental sexual stereotypes. The men are "fierce and strong," rational, outwardly directed; the women are inwardly directed, passive, fashion-conscious, vain, and at the same time supremely maternal "nest-builders," furnishing men with an endless source of love.

Tilton tried to live the sentimental fiction he wrote. He considered his wife alternately a "Queen" and "a Birdie perched on a bough." The Civil War to him was a "romantic story." He himself was a heroic "knight errant of the downtrodden." All phenomena, living and dead, were joined in an organic network of affinities.[23] He addressed other men as "father," "beloved father," "bishop," and "mentor"; endlessly vain himself, he appealed to the vanity in other men, and although many men and women perceived the manipulative twist behind the sentiment, most willingly succumbed.[24]

Sentimentalism, however, could not protect Tilton from personal calamity. Its regressive character made him powerless to change the pitiful course of his life. It destroyed his marriage. His dependence on Beecher proved his nemesis; Beecher no less than Tilton was subject to the shifting, morally unreliable vagaries of sentimental narcissism. Beecher seduced Tilton's wife, Elizabeth, and in the uproar that followed, Tilton lost his position at the *Independent* and set up his own short-lived reform publication in New York, the *Golden Age*. When Tilton publicly and correctly charged Beecher with alienating his wife's affections, he, not Beecher, became humiliated. Protected by his rich and powerful congregation, Beecher escaped punishment, while Tilton exiled himself to Paris.[25] Nevertheless, Tilton continued to believe for the rest of his life that love between human beings gave the only meaning to life. Even after years of personal humiliation and exile, he returned like an insatiable child to that central theme. He fled personal chaos into sentimental fiction and, after 1875, turned entirely to a timeless and decadent imaginary world of "golden-haired girls," and "downy-bearded youths."[26] He populated his poems with mythological nymphs, penniless peasants, noble knights, Greek goddesses, and Viking kings, all of whom lived only for love, "Love, stronger than the triple Fates, Love, strongest of the strong."[27] In an unpublished epic poem, his hero enters the tomb of an unnamed Egyptian Pharaoh whose voice he hears calling out for the love of his lost queen. After countless stanzas of uninspired couplets Tilton concludes, "Heaven both willed that love shall be fulfilled/Hence love—and love alone—this is the thing that makes a man king!"[28]

Romantic Love and the Passionate Imagination

ROMANTIC LOVE differed radically from sentimental love. In Europe it emerged in full force as part of a wider romantic and radical individualist reaction to the failure of the Enlightenment and the collapse of the French Revolution. It rejected the stereotypes of the sentimental tradition and the rationalistic, symmetrical perspectives of the reformist bourgeoisie.[29] In America, romantic love appeared (although less dramatically) with the emergence of economic individualism during the late eighteenth and early nineteenth centuries. The American romantic lover resembled his economic counterpart, the risk-taking entrepreneur who, though contemptuous of material wealth for its own sake, gloried in invention, achievement, and power. Such love became ever more possible as the social, economic, and religious patriarchal traditions of the early eighteenth century began to break down.[30]

To a great extent the psychological consequence of unresolved Oedipal tensions, romantic love embodied a passionate longing to regain the sexual pleasure of the pregenital childhood years. It entailed the often painful and uncertain experience of possessing or being possessed, of losing the self or "falling in love."[31] Romantic love rested on fantasies of the Other (any sexual Other) that arose from needs within the self; it projected those needs on the Other. Sentimental love remained fixed in a static regression with the object of desire shrouded in a fetishistic aura; romantic love also cloaked the object of desire in a magical and bejeweled aura, but as a first stage before a new level of "crystalization" that would deepen the intensity of mutual possession.[32] Unlike either rational or sentimental love, romantic love required conflict, tension, resistance, or, broadly speaking, "evil." Without these "obstructions," wrote Denis de Rougement in *Love in the Western World*—though he warned against the dangers of romantic passion—"there is no romance."[33] Nor could the exponent of romantic love hope for the fulfillment of desire or for escape from the lonely frontiers of self. Presumably, romantic love could lead to extreme obsession with self and the passing of youth; bitter regret at failure to achieve satisfaction; self-destruction and renunciation of the world as meaningless; the rational repression of passionate desire; richly sublimated works of the private imagination; or moving out of the private sphere to transcendence and an active public engagement with the political, social, and economic realities of life. Romantic love might also combine all of these things in various ways.[34]

In the second volume of *Democracy in America*, Tocqueville, like de Rougement, admonished men and women to steer clear of romantic entanglements, but he also noted how social and psychological obstacles form romantic passion:

> When a man and woman wish to come together in spite of the inequalities of an aristocratic social system, they have immense obstacles to overcome. After they have broken down and eloped from the ties of filial obedience, they must by a further effort escape the sway of custom and the tyranny of opinion; and then, when they have finally reached the end of this rough passage, they find themselves strangers among their natural friends and relations; the prejudice which they have defied separates them.

Tocqueville pointed to a greater danger to society in the political potential implicit in the consummation of such love:

> One should also not forget that the same energy which makes a man break a common error almost always drives him beyond what is reasonable, that to enable him to dare to declare war, even legitimately, on the ideas of his country and age means that he must have something of violence and adventure in his character, and people of this type, whatever the direction they take, seldom achieve happiness or virtue. That . . . is the reason why, even in the case of the most necessary and hallowed revolutions, one seldom finds revolutionaries who are moderate and honest.[35]

Tocqueville understood the potential relationship between romantic love and romantic revolution. He did not say, however, that Americans suffered, like Europeans, from class divisions and the need to transgress them. According to him, Americans enjoyed an incomparable freedom in courtship, although a stringent sexual division of labor dominated marriage. Tocqueville missed the real class divisions in America because they were obscured by a pervasively held egalitarian ideology and by greater social and economic mobility, but he did observe how public opinion placed constraints on individual expression. Widely diffused in a democratic context, these constraints were more difficult to see and to resist than clear European class patterns. Sentimentalism was woven into these constraints and supported by a deeper rationalism hostile to romantic impulse and passion.

The author of *The Sexes Throughout Nature*, Antoinette Brown Blackwell, in her only work of fiction, an unsentimental, semiautobiographical novel called *The Island Neighbors, A Novel of American Life*, projected her social vision upon an island, as Tilton had done. But unlike Tilton, she used the island symbol to illuminate social reality.[36] Writing thirty years after Tocqueville, Blackwell was aware of American class divisions and exemplified a rationalist perspective that countered the sentimentalism Tocqueville thought unchallenged. As a feminist, Blackwell had basic sympathy with rational love, although in this novel she appeared to embrace romanticism.

Her book describes a long summer vacation taken by an affluent middle-class Boston family on a remote, idyllic, unidentified island, a "primitive niche" somewhere off the coast of Massachusetts. The vacationers include an invalid factory owner and his class-conscious wife, anxious to get her children married well; their four children; and an Irish maid and companion. They have come to the island to get the health and pleasure their Boston lives deny them. The island contrasts in every way with the world the family has left behind. In noisy, industrial Boston they lead a "tread-mill" existence and must contend with "invalidism" and nervous tension, the constant need to observe, for the sake of their children, the differences of "class and breeding," the extremes of wealth and poverty, "fashionable ladies" and dirty street gamins. On their island they discover a rural enclave of self-reliant sailors and farmers untouched by industrialization. Everyone seems to share equally in the island's bounty; time is measured not by the clock but by the seasons. A sensuous "gypsy" spirit takes full possession of their lives. "This island is a jewel in the rough," declares the Irish maid, "and for one I hope they never will spoil it by any grinding-down, polishing process." "I'm with you there, Margaret," says the eldest son, "I never want to come here again, if this island becomes a fashionable watering-place. It would be like changing a nice little country girl into a fashion-plate."

From the moment the Irish maid, the main character, steps upon the island shore, she is torn between her allegiance to her employer and the promise of island life. An ardent suitor appears, a sailor-farmer. She and her suitor are passionate, moody, sensuous, and self-reliant. Blackwell heightens the sense of burning passion by supplying him with an abundance of red hair and by dressing him in an "unusually flaming red shirt."

Their love becomes disorienting and intense. In the words of Emerson, they "see reminders" of each other 'in every beautiful object" and yield to the "maia [deception]" and "magnetic tenaciousness of an image . . . that mocks and instructs the soul."[37] The suitor struggles to protect his independence but cannot disable the power the woman's image has over him. He dwells on the contradictions of her nature and appearance: her moody introversion and her "helpfulness for others"; her indifference to fashion, her courage and her independence, and her dependent status; her "pleasant and honest face" and the lingering signs on that face of childhood smallpox. ("The beloved," wrote Stendhal in the most important book ever written on romantic love, "may have the marks of smallpox, but is no less beautiful to the lover."[38]) He fears yet "yearns utterably" to surrender himself to her compassionate love, and thus to make himself whole; he longs to save her from the unhappiness of her dependence, and thus to make her whole as well. The maid finds herself obsessed by the image of her lover and bewildered by its contradiction. Seeing him on the road by accident, she experiences an "undefined terror"; in her room alone she conjures an image that "flames into defiant red"; and observing him from afar in his red shirt, she feels his image "burn into her very soul and yet she could not turn away." "She was," Blackwell writes, "under the spell of fascination; but she could no more have told whether the sensation was most pleasure or pain, than the little bird could tell you what it feels when it is fluttering under the magnetism of the relentless serpent," although "the young man certainly felt anything but serpent-like." She has strength and resources of her own; in her excitement and despair of her love she contemplates leaving her employment for a more "independent, self-sustaining position" as a typesetter or as a clerk in a telegraph office in Boston, but her lover's power over her casts her into a dependent position; she sees his power both as demonic and as salvation from her "parasitic" existence.

Blackwell throws up obstacles to the ultimate realization of their love, among them her employer's "benevolent despotism." In his attempt to keep her bound to his family the employer tells the suitor privately that his maid has no interest in him. "She was humbly born; but we have raised her in social standing, till we regard her as a friend—almost an equal; and we have far other plans for her than settling her permanently on this very retired, small island." The attempt to separate them by suggesting a class difference simply intensifies their passion. To use Stendhal's term, the separation nurtured by doubt furthers the "crystallization" of love.[39] As she tumbles "headlong" into the "boiling depths" of the sea, the suitor declares his love by rushing to save her. In the end, the family returns to Boston and the couple remains behind to build a new home "high up on the hills," denying the world outside and completing the scenario of romantic passion.[40]

Blackwell's *The Island Neighbors* can be interpreted as an indictment of social life in industrial America. Presumably, to Blackwell the island represents freedom, independence, health, and sensuous fulfillment; Boston is sickness, class stratification, and repression. The inchoateness of Blackwell's distinctions, however, blunts their critical power. She attaches no moral significance to her tale; it is simply "the normal outgrowth of a restful mood," with "no more thought of a moral in it than there is in the play of children or the friskiness of all young animals." If we had to choose which of the two worlds meant more to her, Boston would probably be the choice. For a feminist, Boston, and not the island, would represent a new world of professional opportunity and marital equality. In the novel she rebukes the sailor-farmers for "leaving their wives year after year puttering in the kitchen and nursery" while they enjoy the experience of the world. In modern "civilization," she declares, "men and women will go hand in hand in all progress."

Blackwell's sympathetic rendering of romantic love contradicts the intellectual thrust of her other, more important feminist essays and books. The feminist works outline a rational and sexual egalitarian system of relations. Such a system would end class conflict; it would untangle the romantic and narcissistic entanglements that prevented women from enjoying equality with men. Blackwell attacked extreme forms of industrial specialization that fragmented human "wholeness" and converted men and women into "live machines," divorcing them from their "feelings." She proposed a "natural" (really a highly rational) solution. Her "highest hope for humanity, male and female," she wrote for the *Woman's Journal*, lay in the perfect "balance of *thought* and *feeling* and *bodily action* daily three abreast."[41] Blackwell had a romantic history herself, and her love for her husband is partly depicted in her tale of the two island lovers. Yet her romanticism did not end in self-doubt or social withdrawal; it gave her the imaginative energy to conceive of its political and social antithesis. The early romanticism of Elizabeth Cady Stanton, of Caroline Dall, and of Mary Putnam Jacobi, among others, may have carried them toward the same rationalist goal.[42]

Blackwell's island easily melds with the memory of the old time, of pastoral America, as it does with the more bourgeois concepts of vacation, park, retreat, and water cure. It is a sensuously volatile world that cannot be integrated into an entire system of relations. Both the servant and the passion, two important sources for social conflict, remain on the island, separated from the life of industrial society. In this sense, rationalistic, industrial society cannot tolerate romantic passion. It must abolish it in order to exist; passion represents a threat to its survival.[43] This interpretation, I think, does some justice to Blackwell's book, although it does not do equal justice to the complexity of Blackwell herself, a complexity displayed in part by her overwhelming enthusiasm for her Irish servant. Blackwell threw her own lot, perhaps unwillingly, with rational love: at heart she appreciated, and probably experienced, the joys of romantic passion, a fact that gives her book a special poignance.

Feminist Attacks on Sentimental and Romantic Love

HOW MEN and women choose to love determines, in part, how they act in the public realm. Romantic love potentially promised bold and, as Tocqueville noted, even revolutionary behavior; sentimental love obscured existing reality and supported the status quo. These forms of love are separated here for purposes of analysis, but both could coexist uncomfortably or even tragically in a single consciousness or at different points in the lifetime of a single person. Both could have appeared in the company of the other form of love, rational love, which feminists sought to isolate as the only love capable of advancing the interests of women. Middle-class feminists waged an ambivalent war on sentimental love, most feminists adhered to a high standard of female purity, and few denied or wanted to deny the importance of woman's maternal nature. Indeed, mid-nineteenth-century feminism took much of its power from the tradition of female nurture. Many feminists wrote poetry, short stories, and novels (the literary embodiment of the sentimental legacy) for a living, but at the same time gained considerable satisfaction from the fact that they were competing successfully with men. "Among the most striking facts in the present condition of women," Paulina Wright Davis wrote in 1853, "is her eminence as a novel writer. The novel writer of this day is to the public what the bard was of old; more than preachers, more than legislators, he moulds the thought, he sways the feelings of the common people. To this sphere, wide and free and influential as it is, woman is at the last freely admitted."[44]

Yet feminists struggled to repudiate the sentimentalization of love that justified the extreme segregation of the sexes and shackled women to the household. They considered the sexual spheres dysfunctional, harmonizing nothing in their existing forms. They believed that the outside world constantly intruded in the shape of male dominance, male needs, and male lust, making a mockery of woman's salvational role and threatening her safety. "The 'Conflict of the Ages,' " wrote Caroline Dall in 1867, "has penetrated to the heart of almost every household."[45]

Feminists used the novel to portray the inevitable miseries that resulted from the sentimentalization of the sexual relation. Lois Waisbroker, Ella Giles Ruddy, Marie Howland, and Elizabeth Boynton Harbert wrote such novels, but Lillie Devereux Blake's *Fettered for Life; or, Lord and Master* (1874) exceeded them all in the sensational rendering of male brutality in marriage.

The novel depicts two kinds of marriage, reflecting ornamental and domestic-maternal sentimentalism, and may have been drawn from Blake's own experience. She traces the psychological evolution of a fashionable, feminist-inclined woman from the time of an arranged marriage to the day of her suicide. Prevented from pursuing an independent career by a husband who wants only to adorn his household with her beauty as well as her money, she sinks into despair and leaps to her death in the sea. "In a moment," writes Blake, "she stood swaying and trembling on the edge of an overhanging cliff: then she stretched her arms and with a strange wild cry, sprang into the clamourous sea." The fashionable version of sentimental marriage was bad enough, but in Blake's opinion the marriage that locked women in a submissive, spiritual maternalism was even worse. It made it possible for a man to vent savage rage against the world through his wife. In the most graphic terms Blake describes the marriage of a drunken man named Blodgett (bludgeon), steeped in the corruption of politics, and a gentle woman who consumes "sentimental novels" and pathetically mothers her husband even in the midst of her own degradation:

> He raised one hand and seized her hair, then lifting the other hand, his face glowing red with passion, he dealt her a heavy blow across the face. . . . He struck her again and again. . . . At first, the poor creature replied with wild appeals for mercy, but these died away presently, and there was no sound as he flung her from him to the floor. . . . He kicked the prostrate form more and more, his heavy boots making the strokes almost murderous.

Blake concludes this passage by denying its fictional character. "Day after day our police records are full of the accounts of the wounds, the hurts, the death-blows, that women receive from brutal husbands." Like so many other feminists, Blake took liberties of expression in a novel that she would have never taken in public affairs. The daughter of a wealthy North Carolina slaveowner, John Devereux, and of Elizabeth Edwards, granddaughter of Jonathan Edwards, she was a beautiful woman with large dark eyes and a reputation as the local "ball-room queen"; but when her husband, a Philadelphia lawyer, suddenly committed suicide, she was left to support herself. She took up a career as a novelist and journalist, moved to New York, and married a man who shared her growing interest in feminism and in the ideas of Huxley, Spencer, and Darwin. She joined the suffrage movement and thereafter gave her life to that cause. A member of the National Woman's Suffrage Association, she edited a woman suffrage department of the *New Era* and, in the 1870s, served as president of both the New York City and New York State Suffrage Associations.[46] By her own account, she had opposed the introduction of the "marriage and divorce question" into public feminist debate. Let the "friends of a change in the existing laws of marriage," she said in 1871, organize their own society and cease "embarrassing" and "hampering" the suffrage movement with their fevered complaints. By 1874, however, she had apparently changed her mind.[47] The heroine of her novel repudiates traditional versions of marriage. Moreover, she will marry no man unless he permits her "to follow out my own career in life."

For many feminists the ideal woman was without sentiment—"tender" but "without weakness," "trusting" and "modest" but without "credulity" and "prudery"; "good" and "self-respecting" but without "pietism" and "conceit."[48] "We have women enough sacrificed to this sentimental hypocritical prating about purity," wrote Stanton to Paulina Wright Davis in 1872. "This is one of man's most effective engines for our subjection."[49] Elizabeth Gay, wife of the abolitionist Sidney Howard Gay, attacked the French historian Michelet for peddling the "sentimental pietistic stuff he calls religion" and tried to persuade Caroline Dall to "take up" Michelet's book on love and "damn it as it deserves. It's the greatest insult to human nature, to men as well as women, the greatest insult to God ever perpetrated by a man calling himself a moralist and benefactor of the race." Dall herself thought the book "needs an answer—only a woman *could* answer it and no woman ever will."[50]

Feminists reproached female novelists for offering women false ideas of love as well as for broadcasting unrealistic conceptions of the outside world. Stanton, for example, searched for realistic fictional depictions of woman's condition and thought she found one in Fanny Fern's *Ruth Hall*, which went far "to prove," she said, that the "common notion" dispensed in novels "that God made woman to depend on men" is a "romance and not a fact of everyday life." She exhorted other women to imitate Fern and "divest themselves of all false notions of justice and delicacy and give the world full revelations of their sufferings and miseries."[51]

Caroline Dall wrote that the novel "has lately passed into the hands of women," and as such persists in doing "infinite harm, by drawing false distinctions between the masculine and feminine elements of human nature, and perpetuating, through the influence of genius often *intensifying*, the educational power of a false theory of love."[52] Eliza Duffey, a popular feminist author as well as a member of the NWSA, presented her own assessment of the modern novel at the Fourth Woman's Congress in 1876:

> Women who have had deep heart histories, but are totally ignorant of the life and the people around them, are capable of writing very tender and touching things; but their writings produce no effect, since they fail to comprehend the affairs of the world as they are, and consequently fail to know how to cope with evils, the very existence of which they are sometimes ignorant. There are floods of this class of literature deluging the world; doing its little good, perhaps, to other women, who live in the same isolated, ignorant world; but doing more harm, since it not only narrows the perception of those who are affected by it, but gives a character to the whole of "feminine" literature.[53]

Duffey beseeched women novelists to discard this "feminine" style and practice what she called true "womanly" mastery of their materials. She, and other feminists like her, did not wish to relinquish what control women had over the modern novel—a principal and powerful vehicle, after all, for the articulation of women's domestic role—but they did wish to free it from an inwardly directed, sentimental vision. They wished, in fact, to use the novel against the novel itself.[54]

If feminists showed some restraint in their attacks on sentimental love, or, more exactly, on women who wrote sentimental fiction, they showed less in their critique of romantic love. Romantic love, "love at first sight," "falling in love," "young love" blinded young men and women; it offered no content beyond a fleeting, unpredictable, perhaps pernicious sexual attraction. Constructed out of ignorance and appeals to vanity, and woven of infantile fantasies, romantic love ignited the immature mind with enflamed and unhealthy desires. "Falling in love," wrote one reformer, "sees nothing but itself and its own desires. . . . Where two young persons are thrown together, their passions are liable to burn themselves out and leave but cinders of their possessors."[55] "Falling in love," wrote another, "is often degradation; as persons who meet in convulsive embraces may separate in deadly feuds."[56] In contradistinction to such love, wrote George Stearns, the wealthy feminist free-thinker and former Garrisonian abolitionist, in his book *Love and Mock Love; or How to Marry to the End of Conjugal Satisfaction*, "true love inspires no jealousies, perpetuates no murders, suggests no suicides, induces no miscarriages, creates no family wars, and warrants no selfish lusts."[57] Both the *Revolution* and the *Woman's Journal* exhorted women and men to shun the deceptive dangers of romantic love. "Most of those who love," according to the editor of the *Revolution*, Laura Curtis Bullard, "are, or fancy themselves to be, in love with each other. Each imagines the other to possess those qualities which would make a life spent together delightful to both; and this expectation makes the disappointment, when it comes, the harder to bear." The love resulting from such a need is a "counterfeit passion" fed by "the stir of the senses, the novelty of the experience, by power over one another" and by susceptibility not to the love object so much as to love itself.[58] "Young persons," observed the *Woman's Journal*, "who are so blinded by love that their judgement is rendered torpid . . . are soon and sadly undeceived by the experiences of married life; and such matches are most miserable."[59]

Feminists may have helped produce conditions hostile to the existence of romantic love. Julia Ward Howe could declare, in one of her many lectures before the Concord School of Philosophy, that "falling in love" had become an "obsolete deity." It "has gone so entirely out of fashion," she remarked to her audience, "that a woman my age may be excused from asking whether any one of these present has even a dim idea of what such an experience might be."[60]

In the late 1860s and early 1870s many feminists followed Stanton's earlier imperative to "divest themselves of all false notions of justice and delicacy" and wrote books and novels laying bare the hazards of romantic love. Epes Sargent's *The Woman Who Dared* (1869) was typical, though written in blank verse. The ever faithful male feminist from Maine, John Neal, called it "a conclusive answer to most of the pettifogging objections that are urged against woman being allowed to have dominion over herself."[61] Epes Sargent was the brother of the equally feminist freethinking Unitarian minister John Sargent, and they were descended from a venerable line of Massachusetts merchants. Epes made a respectable living for himself as a journalist and an editor of the *Boston Transcript*, and devoted the last years of his life to the study of scientific spiritualism. His career as a journalist and spiritualist has attracted little attention from historians, and his feminism and his curious little book have attracted none at all.[62] *The Woman Who Dared* was one of the earliest and most elaborate defenses of female marriage proposals in nineteenth-century feminist literature. It deals with the feminist career of a professional artist who single-mindedly seeks "self-culture and self-advancement" and who scoffs at the sentimental convention that men, not women, should woo and propose marriage. "How bleak and void my Future," the heroine declares, apropos of the prejudice against women proposing, "if I stand

> Waiting beside a stream, until some Prince—
> . . . Appears, and jumping from his gilded boat,
> Lay heart and fortune at my idle feet!
> Ye languid daydreams, vanish! Let me act!

And she does, securing for herself a husband, a baby, and her profession at the same time. She also packs a pistol to protect herself from male predators. Attacked in the woods by three men, she shoots and wounds all three. "She kissed the pistol," Sargent writes without a trace of humor, "that had been her mother's/Wiped it, and reverently put it by." *The Woman Who Dared*, however, directs its principal fire against intense feeling and romantic passion. Linda's parents warn her in a double homily on marriage:

> Jealousy and love were never yet true mates; for Jealousy
> Is born of selfish passion, lust or pride.
> How men and women cozen themselves with words; and let
> their passions
> Fool them and blind, until they madly hug
> Illusions.
> Have you a loving heart, and would you feed it
> On what the swine have left—mock it with lies?
> —Passion may lead to Love, but it may lead
> Away from Love, but Passion is not Love;
> It may exist as Hate; too often leads
> Its victims blindfold into hateful bonds,
> Under the wild delusion that Love leads.
> Love, even when abandoned,

Feels pity and not anger for the heart,
(But Passion, selfish, proud or murderous,
Seizes the pistol or the knife, and kills.)

No feminist ever launched a more controversial, vindictive, or passionate broadside against romantic love than Harriet Beecher Stowe in *Lady Byron Vindicated.*[63] At the time she wrote it, in 1869, she was fifty-eight years old. Like Julia Ward Howe, Clemence Lozier, Isabella Beecher Hooker, and to some degree Lydia Maria Child, Stowe was fully into an openly acknowledged feminist phase.[64] She had just read Mill's *Subjection of Women* and written her friend Sara Parton that "Mill's book has wholly converted me—I was right in spots before. Now I am all clear."[65] Under its spell she began to introduce unmistakable feminist themes into her fiction (notably *Old Town Folks* and *My Wife and I*), and seriously mulled over a decision to assume the editorship of the *Revolution* after Stanton's retirement.[66]

Stowe wrote *Lady Byron Vindicated* ostensibly to clear the name of Lady Byron from the obloquies of her English critics. These critics had assailed Lady Byron for ruining her husband's life with her puritanism, her heartlessness, and her hostility to him after the failure of their marriage. According to Stowe, a profound modesty prevented Lady Byron from defending herself against her enemies and it was not until the end of her life that she confessed privately to her friends the awful truth that had kept her silent: her husband, Lord Byron, had committed incest with his half-sister, Augusta. Thus Stowe took it upon herself to do for Lady Byron what Lady Byron had failed to do for herself—to vindicate her before the eyes of the world as a wronged, helpless woman trapped in a silence created by her husband's unspeakable crime.[67] She also attacked in her book the worst features of sentimental marriage: woman's silent, passive dependence and man's unquestioned authority and sexual license. She wanted not only to speak for the "child-like" and "artless" Lady Byron, but also for all "helpless" wives "cowering" before their "maddening" and "imbruted" husbands like "dogs . . . beaten, kicked, starved, and cuffed," whose "special grace and virtue" consists only in their "*utter deadness to the sense of justice*," and who swallow whole "what John Stuart Mill calls the literature of slavery for women." "I consider Lady Byron's story," Stowe wrote to Horace Greeley, "as a type of the old idea of womanhood, that is a creature to be crushed and trodden underfoot, when her fate and that of man comes into conflict." What might Lady Byron have been, Stowe asked Greeley, "if *she*, and not some man," had been given the opportunity to "control and guide the thought of England."[68]

In striking fiercely at Lord Byron, one of the great heroes of her youth, exposing and analyzing him as a tortured and twisted sexual fanatic, Stowe confronted, in a very different way from that of Antoinette Brown Blackwell, her own romantic desires. That fact gives *Lady Byron Vindicated* its special place in the history of feminism. It is one of the most disturbing indictments of romantic passion written in the nineteenth century. "He and he alone," Stowe wrote toward the end of her book, "is the cause of this revelation." Although she seeks to isolate the "real" Lord Byron, she is preoccupied with him as the mythological expression of the romantic impulse —a confusion that does not help Stowe as a biographer but frees her to display her own peculiar insights into the meaning of romantic love.

Stowe's Byron exhibits all the characteristics of the romantic lover. He had the enormous power to project himself either directly or indirectly into the lives of other people and to stimulate in others the capacity for fantasy. "Beautiful, dazzling, and possessed of magnetic powers of fascination," he transformed ordinary men and women into "blinded adorers who would swear that black was white; or white black," persuading them to "smile away their senses, or weep away their reason." He had a "most peculiar and fatal power over the moral sense of the women with whom he was brought into relation; and that love for him, in many women, became a sort of insanity by depriving them of the just use of their faculties." An originator, in Stowe's view, of the romantic literary tradition, Byron corrupted the literature of Europe and America, triggering, in dangerous ways, the fantasies of innocent men and women. Byron himself, as Stowe saw him, surrendered like a child to the attractions of fantasy by reproducing them in his poetry and by reconstructing the living objects of his desire into hateful demons or beautiful angels, or into both demons and angels at the same time.

The true romantic must also show a tormenting awareness of his own internal contradictions, of his capacity for both good and evil, of his worth and worthlessness. Finally, in order to exist at all, the romantic lover must have in his repertoire the passionate need for the intensification of his love. Such intensification entails, in turn, the yearning for obstacles to overcome, for new taboos to break, and for dangerous and forbidden things—a yearning, of course, that Byron supposedly had beyond measure. In Byron's poetry, Stowe writes, "the stimulus of crime is represented as intensifying love. Medora Gulnare, the Page in "Lara," Parisina, and the lost sister of Manfred, love the more intensely because the object of love is a criminal, outlawed by God and man. The next step beyond is *madness*."

Only a genuine romantic could have so heatedly depicted the essence of romantic love. Unlike Blackwell, who apparently still clung to a tempered version of romantic passion, Stowe chose to extirpate such passion from her experience. She did not analyze Byron's romanticism on moral or religious grounds. Like so many ex-Calvinists of her generation, she no longer treated antinomian heresy or personal hubris as a symptom of human depravity. She relied almost entirely on a new, increasingly fashionable branch of physiological inquiry known as the "science of brain affections," which dealt with deviance from a physiological perspective. "By all accounts," wrote Stowe, "it is made apparent that ancestral causes had sent him into the world with a perilous and exceptional sensitiveness of brain and nervous system." Byron's passions arose not only from an inherited pathology, but also from his secret act of incest. To bolster her argument on this point, Stowe quoted profusely from an article by Dr. Forbes Winslow, called "Anomalous and Masked Affections of the Brain," an article that examined the effects that hidden criminal acts, whether actually committed or imagined in dreams, had on the mind.

Stowe thus reduced romantic passion to an anomalous disease subject to treatment. She denied Byron his human identity by figuratively splitting him in half: one half was his normal, healthy human self still present in his feelings of "remorse;" the other half was his abnormal or inhuman self, the self of passion and contradiction, the self that mingles the need to destroy with need to worship, murder with idolatry. Stowe employed this clinical approach because she considered it the most compassionate approach to mental illness, yet she ended only in unveiling its underlying gruesomeness. In *The Island Neighbors*, Antoinette Brown Blackwell preserved the memory of romantic passion by relegating its victims to an isolated niche in the world. In *Lady Byron Vindicated*, Harriet Beecher Stowe consigned these victims to the hospital and to the humiliation of recovery.

Many feminists praised Stowe for her dissection of Byron, but the established male press for the most part defended him against his enemies as it tried to absolve Henry Ward Beecher from the consequences of marital indiscretions.[69] Elizabeth Cady Stanton compared *Lady Byron Vindicated* favorably to John Stuart Mill's *Subjection of Women*. Both books, she said, sought to free women from the scourge of old-fashioned marriage; she admonished women not to take part in the outcry against Stowe's attack. "Our present civilization," Stanton wrote for the *Independent*:

> is marked with as hideous outrages on the mothers of the race, in marriage and out of it, as have ever blackened the pages of history at any period of the world. . . . The true relation of the sexes is the momentous question at this stage of our civilization, and Mrs. Stowe has galvanized the world to its consideration. . . . Our low ideas of marriage, as set forth in our creeds and codes, making man master, woman slave, one to command and one obey, are demoralizing all our most sacred sentiments and affections, and making the most holy relation in Nature one of antagonism and aversion. . . . Before women who wield strong pens join in this hounding of Mrs. Stowe . . . let them analyze the real position of women today.[70]

In the *Revolution* the English correspondent Rebecca More concurred with Stowe's position on Byron, calling him a "malign antinomian," and Mary Livermore declared for the *Woman's Journal* that Stowe had fashioned her book from the "very noblest of motives" and had "placed Lady Byron forever in the innermost hearts of the good and noble, in small compensation for the awful injustice, and the desolate and bitter lot which closed so early and darkly around her."[71]

Influential male feminists, including George William Curtis, editor of *Harper's New Monthly Magazine*, and the progressive free-thinker Moncure Conway, also publicly joined hands with Stowe. Conway's thought graphically exemplifies the splitting off of reform thought from its romantic roots in the antebellum period. In *Earthward Pilgrimage*, a novel published in 1870, he wrote that "the ages of egotism reach their final flowering in Byron and perish."[72] The son of Virginia slaveholders, Conway left the South and Methodism in the late 1840s to become an abolitionist, radical Unitarian minister in the North. He owed much of his early intellectual development to the influence of Emersonian transcendentalism. By the 1870s, however, Conway had cast off his early romanticism and his Unitarian ministry for positivistic rationalism, and he had chosen new heroes: Ludwig Feuerbach, David Strauss, and Ernst Mach.[73] Byron and the romanticism he represented became for Conway the incarnation of the irrational. The spirit of Byron hovers over Conway's most important work, *Demonology and Devil-Lore*, a spirit that will find a grave in Conway's rationalist, scientific analysis of dreams and fantasies. Conway sought to demystify the "hellfire" dogmas of the past and to render powerless through analysis the dreams that "still haunt all the regions of our intellectual twilight, the borderland of mystery, where rises the source of the occult and mystical which environs our lives. The daily terrors of barbarous life avail to haunt the nerves of civilized people, now many generations after they have passed away."[74] Like Stowe, Conway wanted to uncover the latent irrational content hidden in history, to strip away what he called the infantile "costumes . . . masks . . . and sentimental glamour" of the "discredited deities and demons of the past," and to discover the "real," rational, and unambiguous core of human life.[75] For the early Emerson, the existence of the imagination itself depended upon the antirational acceptance of reality as a veil and a deception that "mocks and instructs the soul." Images, Emerson believed, mediated between the seen and unseen, and by so doing established a "real and passionate relation between the thought and some material fact."[76] Conway destroyed the mediator. "Taught by Science," he wrote, "man may with a freedom the barbarian cannot feel exterminate the Serpent."[77]

Rational Love Based on Knowledge, Not Passion

HARRIET BEECHER STOWE'S attack on romantic individualism and her exposure of its ultimate expression in incest, underlined for mid-nineteenth-century feminists both the problem implicit in greater social and economic freedom for women and the solution to that problem. By the 1870s the gradual secularization and democratization of American social life, coupled with a rapid capitalist development that increasingly subverted the older sexual division of labor as well as the sentimentalism attached to it, created conditions favorable to the emergence of women into the public realm with men. After the Civil War thousands of women took part in social organizational work that impinged either directly or indirectly on the lives of men: in temperance, social science, and moral education; in the reform over a forty-year period 1840-80 of the marriage laws that feminists helped to institute and that legally permitted women to transact their own business, keep their own separate earnings, and retain ownership of their separate estates; in the reform of many state laws during the 1870s that sanctioned the right of women, whether married or single, to employment in the professions; and in the growing employment of large numbers of women in the industrial sector of the economy and, in fewer but no less significant numbers, in the professions, especially medicine, journalism, and education.[78]

These changes represented noteworthy advances for women yet forced feminists to attack the economic individualism that produced them. By depicting the dangers of unrestricted, formless freedom in her book, Stowe captured the fears of this generation of feminists, who thought that further unadministered growth would topple the social and economic system into a chaos whose symptoms they already perceived around them in the proliferation of hotels, boarding houses, and apartment buildings; in the glaring number of divorces, wife beatings, patricides, and vagrant children; in the widespread occurrence of female diseases and nervous disorders common to both sexes—hysteria, depression, insomnia, headache, and epilepsy. Antoinette Brown Blackwell wrote in 1872, a year after she finished *The Island Neighbors*, "forces grope blindly and without aid in the shape of unregulated impulses" and "are devastating the world."[79] Feminists had no desire to endanger the sexual relation with further devastation; they were not anarchists nor radical individualists. They wanted, in fact, to reconstruct the sexual relation on a firmer, more "impregnable" basis by exchanging a sentimental system of sexual order for an egalitarian one.[80] Stowe therefore spoke for this intention as well as to the fear. She intended to slay forever the attractive potency of Byron's kind of love, an unrestrained, impulsive, and fantastic love that flourished on inequalities and deceptions, radically obscured and even violated the boundaries between the sexes, and sought only to invade and overwhelm. She suggested in her attack a love quite different from sentimental and romantic love, a love based on the rationalization of the male and female sexual impulse.

What Stowe merely suggested other feminists developed at great length. Rational love did not emerge from ignorance, subterfuge, and momentary dalliance, but only after an extended period of mutual self-scrutiny and revelation, and equal development of both sexes. And when feminists said "extended period," they meant what they said. Rational love depended upon a foundation built in childhood with both sexes becoming accustomed to the same diet, the same hygiene, and the same functional clothing. This process deepened during adolescence, with both sexes learning the same physiological and sexual truths, both becoming acquainted with the scientific facts of life, both equally developed in all their faculties, and, most important, receiving the same education in the same classrooms. "Restricting the social intercourse of the sexes," Stanton warned in 1868, "adds undue zest to their meeting. . . . If the sexes were educated together we should have the healthy moral and intellectual stimulus of sex ever quickening and refining all the faculties, without the undue excitement of sense that results from the novelty in the present system of isolation."[81] In a coeducational setting young men and women would have the chance to meet each other as "actual companions" and not as "gods and goddesses" as "natural enemies." "If the true lover comes," wrote the editor of the *New Century* in 1876, "he is tested by clear eyes, that have learned to estimate somewhat the uses of their seeing."[82] "It is better that the sexes see each other daily than dream of each other," Abigail Scott Duniway argued a year later in her newspaper, the *New Northwest*.[83]

Feminists advocated sexual integration from childhood onward and in most institutions because it promoted what they variously called "educated," "intelligent," and "organized" love between the sexes.[84] Men and women "fell" in love, then, only after, in Sarah Hackett Stevenson's words, "knowledge" had replaced "instinct" in the choice of companions and only after both sexes had fully developed all their faculties.[85] Why did feminists embrace a love that seemed, on the face of it, so lackluster? The answer lay partly in the fact that the sensuous distance so necessary for the cultivation of fantasy and passion did not exist when the sexes mixed in public and private. Fantasy and passion, feminists believed, tended to subvert the sexual bond with disequilibrium, conflict, and disease, and to throw women especially into dangerously vulnerable and dependent positions. Rational love, on the other hand, fostered harmonious sexual relations based on equivalent experience and growth; it provided constant opportunity for observation and scientific knowledge. It mirrored the rationalization occurring in other areas of American life, a process designed to bring cooperative harmony and order into a volatile economic and social system.[86] "Where there is a mutuality of interests," Paulina Wright Davis wrote her friend Anna Parsons as early as 1850, "such interlinking of life, there can be no real antagonisms."[87] True love, according to Faith Rochester in the *Revolution* in 1870, yields a "mutual reciprocity where each loves too truly to hold the other in bondage."[88] "Where true love exists," Victoria Woodhull said three years later, "no tie forged by the art of man can in any way add to its efficiency."[89] Rational love achieved more than the forging of stronger sexual bonds; by reducing the importance of love in the lives of women, it freed them for a more active and constructive part in the industrial society at large. In such a society romantic love became an anomaly: obedient only to the laws of sensuous delight, it not only squandered time but abolished it altogether. Rational love remembered time and carefully marked the path it must pursue.

Rational love reflected the legacy of early Puritan Arminianism, with its accents on institutional order and public morality just as romantic love reflected Puritan antinomianism with its allegiance to the rights of individual conscience, and as sentimental love reflected the Puritan sexual division of labor. Like the Puritan rationalists, mid-century feminists eschewed passion and placed great emphasis on rational self-control, harmony, organic balance, and moderation. They encouraged the sexes to calculate rationally their better interests before contemplating marriage, and they focused upon the roles men and women must play in contributing to and serving the community life around them. Yet the divergences between the early Puritans and the liberal feminists were as marked as the similarities. Both groups sought the same ends but by different means. The classic Puritans had adhered closely and openly to hierarchical social principles and advised the sexes to choose their partners rationally on the basis of property, social station, church membership, and birth: they paid almost no attention to love as a factor in marrying.[90] Puritans achieved order in sexual behavior, as in all other matters, by clearly locating moral and social authority in the hands of a powerful ministerial elite. The responsibility for maintaining sexual order resided in the communal elite and not in the private individual. In ths nondemocratic, authoritarian period the public sector remained the principal focus of administrative interest; the private lives of men and women mattered only to the extent that they did not violate communal norms and proscriptions. Men and women suffered less in their private lives from sexual constraints because they were expected to sin more. Finally, the Puritans preached moderation in love not only because community order required it, but also because God, not man or woman, was the true object of human love. Rational love, therefore, found a profound justification in a cosmic fact: love of men and women was a step toward love of God.[91]

Feminists, of course, repudiated traditional hierarchical social principles and patriarchal authoritarianism. For them personal love was the determinant factor in marrying. The problem for them was how to transform sexual love into an egalitarian relation while at the same time preserving social order and community. Thus feminist rational love suppressed those pre-Oedipal tensions that had their source in the breakdown of older forms of authority and that constituted the basis for romantic fantasy and passion. Based in part on deception, on the mythological reconstruction of the love object, romantic love liberated the sensuous imagination and gave it productive direction. By giving men and women complete control over the means of imaginative production, it allowed them, in effect, to "produce," "make," and possess freely and permanently their own love objects from both the materials of fantasy and the "real" materials of the objects of love. But in the interest of truth and equality rational love attempted to separate the means of imaginative production from the act of love, to control the fantasy process that transmuted the love object into an imaginative construction in order to possess it. Where the Puritans gave some vent to the expression of intense antinomian feelings in their devotion to God (Jonathan Edwards placed the affections at the center of the self), feminists permitted such feelings little or no expression.[92] Feminists sought to secure the existence of rational love by building a new, authoritative, bourgeois foundation for it or by surrounding it with requirements that equaled in their rigor and exclusiveness traditional criteria for marriage. They exchanged property, social status, church membership, and birth for scientific knowledge, shared and similar experience and development, and mutual surveillance. They demanded a deeper rationalization of the inner lives of men and women, of the private sphere, than Puritan rationalism had required.

Feminist rational love, therefore, supported a new stage in the emancipation of bourgeois women, but it immediately disclosed contradictions of its own. By demystifying romantic love and unveiling the source of its productive energy in fantasy, feminists believed that they had erected a new, more enduring basis for sexual intimacy in the actual reality of things. Yet this intimacy had a very special character: it began very early and it deepened between the sexes only through familiarity and the influence of a common history, never through the experience of intense passion and conflict. Unchallenged or mediated by the autonomous, productive power of the imagination, such love did not necessarily deepen at all, although, for the same reason, it both delayed marriage and discouraged divorce. Romantic love, for its part, began with a great distance and ended in a deep, private intimacy fraught with the potential for pain, despair, and betrayal. Like an intense conversion experience it transfigured time as well as the object of love. Unmediated rational love, on the other hand, exposed both sexes to the full weight of history and to the full weight of reality. Romantic love acknowledged, even fed upon, social and political obstacles to its realization—class barriers and oppression, public opinion, social traditions, social inequities and customs. It challenged these evils, the inner struggles of the self and the outer conflicts of the world, as the price paid for the passionate, always uncertain, possession of a specific object of love. Through such experience the romantic learned to "derive meaning from all experience" and to understand, even more deeply than before, the full complexity and contradictions of life in general and of capitalist society in particular. "The universality of beauty," Theodore Adorno wrote recently in *Minima Moralia*, "can communicate itself to the subject in no other way than in obsession with the particular. No gaze attains beauty that is not accompanied by indifference, indeed almost by contempt, for all that lies outside the object contemplated. And it is only infatuation, the unjust disregard for the claims of every existing thing, that does justice to what exists."[93]

Unrelenting nondialectical rationalism denied passion, conflict, and "evil" their right to exist. It stole the power from love and looked at the world without compassion. What D. H. Lawrence said of Benjamin Franklin applies with equal truth to American reformers: they took away the "illimitable background" of the self and thus the true foundation for freedom.[94] Rational love, and the symmetrical wholeness it was supposed to sustain, contradicted the meaning of individual growth and development. On the one hand, the advocates of rational love shared the realistic faith of the romantics—that is, they purportedly faced the "facts of life" with unflinching rigor and precision—but, unlike the romantics, they labored to rationalize conflict and to replace "evil" with a perfectly balanced and democratic system in which the feelings played an equal but never determinant role.

William F. Channing wrote to his friend Elizabeth Cady Stanton in 1899, repeating a position he had held since the 1850s: "The natural form of marriage can only be established and exist in a freely and perfectly organized social body in which all the parts work together."[95] Channing's use of the "body" as a metaphor for social and marital happiness was predictable and apt, expressing a favorite and central metaphor of feminists, indeed of all bourgeois reformers in the nineteenth century: the acquisition of a body "perfectly constructed and well adjusted in all its parts," promised to extend the lives of men and women beyond ordinary limits.[96] In principle, too, it promised to prolong marriage, but for a dear price. The fact is, rational love did away with the dynamic element necessary for genuine organic growth and development and implied the interlocked, fully compartmentalized human being.[97] Even more darkly, by attempting to muzzle sexual myths, it left the "faculties of the imagination" to the "incoherence of purposeless activity."[98] To the extent that they adhered to this rationalist model, to the extent that they cast off the best in the romantic tradition, or robbed it of all its complex and dialectical power, feminists found themselves thrown willy-nilly into an ironic alliance with the sentimental vision of the world.

Part 2

FEMINISM IN THE PUBLIC REALM

CHAPTER 6

FROM THE PRIVATE TO THE PUBLIC REALM: INDIVIDUAL FREEDOM AND SOCIAL HARMONY

IN A SPEECH delivered in 1882 at the Concord School of Philosophy, one of the most highbrow summer schools of the late nineteenth century, Julia Ward Howe reiterated a position she had held for a decade: "One thing each one of us should be, and that is an individual. And yet I have named this word with fear and trembling, for there is no monster like an exaggerated individual."[1] Many feminists joined Howe in yielding up the primacy of individualism to organization, equilibrium, community, and centralization. Distressed not only by sexual oppression but by urban crowding, labor unrest, filthy streets, crime, disease, and class conflict, they looked for ways of adjusting individual freedom to the pressing imperatives of social order and harmony.

To carry through this adjustment feminists had to break down the wall separating the private sphere, which trapped and isolated women from a rich experience of the world, from the public sphere, in which male power reigned. At the same time the public sphere itself had to be reformed, a necessity requiring more than the egalitarian integration of the sexes. Male interests, feminists believed, tied to an unregulated, competitive market, were destroying the welfare of the community. Women, however, would introduce new and needed elements. Armed with a new philosophy of cooperation and harmony, they would, in league with men sympathetic to their cause, humanize the community life.

Retreat from Individualism

SELF-RELIANCE, self-development, self-culture, and other conceptions of "individual freedom," did not disappear from feminist consciousness in the postbellum period. Still, the individual came to be viewed with different eyes. Many important feminists distanced themselves from their earlier individualist careers. Bronson Alcott, along with other male feminists like Franklin Sanborn and Thomas Higginson, left much of his individualist past behind him. His experiment in "consociate family" living at Fruitlands in the 1840s had prepared him for such a departure.[2] In the 1860s he had fallen beneath the spell of what was called St. Louis Hegelianism, a rationalist philosophy that expounded the virtues of institutional order, and by the late 1870s he had completely severed his bond with transcendentalism. "The individual sees only himself like Narcissus in the pool," Alcott wrote then. "Individualism brings men into opposition to the family, institutions of learning and the state," and "only as it is broken down is there harmony."[3]

Parker Pillsbury never went as far as Alcott, but in 1870 he wrote for the *Revolution* that the consideration of "personal freedom" was no longer one of the "grave problems" confronting Americans. The country was confused and divided, vexed by political corruption and bitter animosities, and needed a "readjustment of the whole political, religious, and social system" with a new "statesmanship broad enough to comprehend and harmonize the whole round of human needs and duties."[4] Pillsbury looked to the emergence, through the implementation of equal rights, of a rational and apolitical elite that might redirect the course of the nation intelligently. He urged that refined, educated, and civilized women enter the schools, the churches, and the state to mend what men had broken. "What she cannot refine and purify," he wrote, "through womanly instinct will be cast out like unclean spirits. The earth will be tilled for men and not for brutes. And then men will not be, nor beget brutes."[5]

In the 1830s, however, Pillsbury had been a fiercely independent Calvinist who, only months after ordination in 1839, left the ministry and bitterly attacked the intemperance, corruption, and proslavery attitudes of his denomination's officials. His mother "mourned him as one lost forever."[6] A Garrisonian and "no organizationist" in the 1840s, a self-reliant "soldier covered with scars," as he described himself, "with garments soiled and torn, determined to win or die," he hurled thunderbolts at all religious authority in the 1850s.[7] By the 1860s he was a feminist who argued for woman's right to marry and divorce according to her own best interests, or to remain single if she so wished, and to unite career with marriage whenever possible.[8] His libertarianism set him apart from the more moderate reformers, and by this time he espoused every principle in the reform lexicon. No one found greater charms than he in physical health, preventive medicine, the doctrine of no secrets, and in the welter of contemporary fads from Turkish baths to weight-lifting cures. He scoffed at fastidious people who refused to examine physiological truths and who ran "scandalized from the sight of children's naked bodies," or, for that matter, from the sight of their own naked bodies. Even in the 1850s, on tour as an abolitionist, Pillsbury had the habit of rising very early in the morning and "*stripping himself naked . . . and walking about . . . in a perfect nude state*" in the presence of other reformers of both sexes.[9] "Wash and be clean!" he exclaimed in 1868, praising the marvelous powers of water. "Wash and be healed, wash and avoid disease, is really the evangel to all people who have ears to hear."[10] He later took up the cause of sexual continence and social eugenics, and by 1880 he had joined Loring Moody's Institute of Heredity, an advanced agent in the field as Pillsbury had been so often for other reforms. For Parker Pillsbury the rationalization of the internal workings of the body and the subjugation of sexual fantasy and conflict represented "the greatest questions of the day."[11]

Like all leading feminists, Pillsbury wrote articles for such papers as the *Index* and lectured everywhere on Free Religion. He too clutched at the religion of science as the means to make straight the "crooked" complexities of life and to demystify the mysterious, irrational legacies of the past. "No miracles, no mystery, nothing supernatural, nothing *unscientific* will exist, when Science her whole lesson shall have studied and learned. . . . We know nothing of mind apart from matter. . . . I have no confidence in any faith which is not capable of a scientific basis." Let science work and man would triumph over all the "subhuman . . . undergrowths" of history, over the tumultuous struggles of the race, and, perhaps, over even death itself.[12] Love of health, obsession with bodily processes, adoration of science—all these things were symptoms of the rational frame of mind that marked the movement away from the excessive individualism of the antebellum period.

Most other mid-nineteenth-century feminists also adhered to a rationalist theory of harmonious development of all the faculties, including rational love and rational decision making based on the principle of no secrets and on scientific knowledge. They drew many of their principal ideas from positivism, which, more than any other contemporary system of thought, attacked romantic introspection, passion, and imagination, and sought to subject the individual to a new, harmonious, equilibrated social system under "collective guidance and control."[13] The English positivist Frederic Harrison, for example, said that reformers pursued a "three-fold work"—"to give unity to individual powers" (that is, to organize the individual symmetrically from within); "to bind up individuals into harmonious action;" and to keep that action true and permanent—unity, association and discipline."[14]

Positivist themes of "unity, association and discipline" were integrated in varying degrees of purity into a multitude of feminist publications in New York, Boston, Philadelphia, and elsewhere.[15] In New York City the Octavius Frothingham School of Free Religion, with its heavy positivist orientation, influenced the ideological character of the Free Religion press. Frothingham, a leading liberal minister and scholar, had read and absorbed a good deal of Comte in the 1850s. So had Julia Ward Howe, Elizabeth Peabody, and Theodore Parker.[16] Much of the feminist leadership was Free Religionist, espousing a new Religion of Humanity.[17] Indeed, the alliance between Free Religion and feminist leadership was so close that the offices of the *Woman's Journal* and the organ of Free Religion, the *Index*, were side by side on Trement Place in Boston.[18] Free Religion affirmed, according to its most vocal defender, the positivist editor of the *Index* Francis Abbott, "the Unity of Nature and Mankind . . . the supremacy of benevolence in matters of personal and social relations, and the supremacy of science in all matters of belief. It put the Church on the level of all other institutions, the Bible on the level of all other books and Christ on the level of all other men."[19]

Feminist Free Religionists, congregating in radical clubs in many major eastern cities, discussed the most advanced thought of the day on "physical, ethical and social science."[20] Thomas Higginson described these clubs in 1872 as "homes" for progressive men and women, "organizations to which they might join themselves without those mental reservations and exceptions which Christian organizations impose upon them."[21] The Boston Radical Club, founded in 1870, began as a public club that met to debate theological questions in the house of the tirelessly provocative liberal Erastus Bartol.[22] By 1872 it had spawned a network of clubs in New York, Philadelphia, Boston, Rochester, Buffalo, Syracuse, and other cities. In 1876, the year of the centennial celebration in Philadelphia, the radical clubs metamorphosed into the National Liberal League, with a broader organizational network that included an ideological range from extreme radicalism to conservatism.[23] Progressive spiritualists who had struggled to develop their own organizations in the late 1860s and early 1870s joined the radical clubs and liberal leagues. Spiritualists and Free Religionists often wrote for and read the same periodicals—the *Index*, *New Age*, *Banner of Light*, and *Radical*.[24]

In the early 1870s such clubs often displaced suffrage societies as places where feminists could gather to discuss such controversial subjects as eugenics, sexual hygiene, abortion, and marriage. "A Radical Club in Syracuse," Matilda Gage wrote to Martha Wright, president of the National Woman's Suffrage Association, "seems to have taken the place of the city suffrage association. It discusses all reforms. I occasionally attend."[25] Woodhull spoke at the Syracuse club in 1871 without a fee,[26] and Stanton addressed the Radical Club in Philadelphia. "It is a degradation of the religious element in women," she said, "to use it exclusively, as men now do, to mitigate by indirect influence the crimes of their making, instead of, by direct power, preventing them."[27] In 1873 feminist Mary S. Hibbard gave a lecture in tribute to Mary Wollstonecraft before the Rochester Radical Club, which met in the home of the progressive spiritualists Isaac and Amy Post.[28]

Positivists in the 1870s, including the most important American disciple of Comte, Henry Edgar, wrote columns for *Woodhull and Claflin's Weekly*. The Chicago feminist and labor reformer Elizabeth Osgood Willard wrote a neopositivist book called *Sexology* in 1867. In it she tried to "improve" or update the "theory of equal rights" by ordering astronomy, physics, and Spencerian biology into a systematic, synthesizing feminist statement.[29] *Sexology* argued at length that the law of sex runs through all forms of life, that life depends on movement and strives to reach an "equilibrium" or "balance of power" between "centrifugal" male forces and "centripetal" female forces. The book also claimed that "the organization of humanity is woman's work in the world." Willard wrote Caroline Dall in 1868, "You will find that the laws of nature as explained in *Sexology*, solve the problem of woman's position in a civilized harmonious condition of society, and also, that it settles the question between the laborer and the capitalist by a reconstruction or new organization of society in conformity with the natural laws of organization in the human system."[30] Her book influenced many feminists, including Elizabeth Cady Stanton, and represented perhaps the first exercise of its kind by a woman in this country. It expressed in a specific way the general theoretical tendency of mid-nineteenth-century feminism that appeared in the work of such feminists as Antoinette Brown Blackwell, Hester Pendleton, and Sarah Hackett Stevenson.[31]

Jane Croly and her husband, David, were both deeply influenced by Comte in the 1860s. Managing editor of the *New York World*, David Croly also edited the overtly positivist journal *Modern Thinker*, whose first issue was printed in 1870.[32] Jane Cunningham Croly had come to the United States from England with her father, the radical Unitarian John Cunningham, who opened free schools for factory workers and was apparently hounded out of Britain for his "infidelisms." His daughter adored him. "He taught me all the ignorance of men," she wrote for the *Woman's Journal* in 1870. "He wrapt his little daughter in his large man's doublet, careless did it fit or no. . . . While he lived there was always one who without explanation understood and believed in me."[33] In the 1850s Jane Cunningham attended Central College, in Geneva, New York, one of the first genuinely egalitarian, nonsectarian, coeducational institutions in the country, supported and funded by abolitionists and feminists. It had three women on its faculty and "utterly repudiated . . . the doctrine of exclusiveness and caste."[34] Jane Croly began her public advocacy of reform by joining Andrew Jackson Davis in his attacks on the "supernatural theology" of Horace Bushnell. She worked as a journalist, moved to New York City, married David Croly, and briefly became a member of Stephen Pearl Andrews's Club, where the "best" male and female "minds in New York City" discussed philosophical and scientific questions.[35] She also bore five children (one of them the Progressive Herbert Croly); edited a famous fashion magazine, *Demorest's Monthly*, for twenty-seven years; cofounded the first woman's club; and organized the Woman's Parliament in 1869.

Jane Croly and her husband each wrote a book on marriage and the sexual relation—respectively, *For Better or Worse* (1876) and *The Truth About Love* (1870)—making heavy use of positivist ideas. Both, of course, wrote pieces for the *Modern Thinker*. "Whatever the weakness and strength of its founder," Jane Croly wrote in an article called "The Love Life of Auguste Comte," "there is little doubt that the 'Religion of Humanity' will live and continue to attract, as heretofore, the attention of the wisest and the best among us."[36] The Crolys joined the New York City Positivist Society, which convened in the private homes of wealthy members in Harlem and in Brooklyn (that hub of so much feminist activity) and elsewhere during the early 1870s.[37]

Another important female reformer drawn to the moral certainties and clear institutional theory of positivist thought was Augusta Cooper Bristol, gifted member of that remarkable community of progressive men and women in Vineland, New Jersey. She read Comte, Spencer, and Henry Carey, presided over the Ladies Social Science Class in Vineland, became an important figure in the National Woman's Suffrage Association, and wrote articles for the *Revolution*. Along with Lita Barney Sayles, Helen Campbell, and Imogene Fales, Cooper also joined the New York City Social Science Association and the New York City Positivist Society. In 1880 she gave a series of lectures at the Positivist Society called "The Evolution of Character," and followed it with a series before the Social Science Association. Cooper, like Croly, Sayles, and others, was influential in the Association for the Advancement of Women, one of the most important feminist organizations of the 1870s. She was one of its directors and vice-presidents during the 1870s.[38]

Dr. Mary Putnam Jacobi was perhaps the most enthusiastic positivist of her generation. Daughter of George Palmer Putnam, the powerful publisher of *Putnam's Monthly*, she too was her father's favorite—"the great interest of his life."[39] She was ambitious and precocious, free from sentiment and fashionable frippery, scornful of all theologies, and independent-minded almost from the moment she could walk. Both her parents warmly supported her desire to become a trained, first-class physician. In 1866 she went to Paris to study at the *Ecole de Medicine*, and five years later became not only its first woman graduate but a graduate with highest honors. While in Paris she read Rousseau and Comte and participated in the political ferment of the times. She often ate at the same table with "radical republicans and socialists" and relished their company, learning to "identify herself with the oppressed millions."[40] The passionate republican anarchist Elisée Réclus was her closest and dearest companion in Paris. At the moment of the outbreak of the Franco-Prussian War she fell in love with a French medical student who fought on the front against the Germans, an affair that threw her into "infinitely painful" conflict, made her miserable and happy by turns, and "revolutionized" her emotional life.[41] After the war she lived through the exhilaration and horrors of the Paris Commune and suffered with her friends the anguish of its failure. "My interest is immense," she wrote her mother, "in the events that are passing, especially since the Republic, and as far as I myself am concerned, I feel really quite ready to die in its defense."[42] She acutely saw the politician Jules Favre as a fool and Thiers, the butcher of the Communards, as an "arch-mischief maker."[43]

In 1871 Putnam returned to America with her degree but without her lover and without the political fire she had experienced during the most intense moments of the Commune. "Too analytic," "too soul-searching," too "rational," she refused to surrender to romantic impulses and sacrifice allegiance to her own country for the prospect of an anomalous, uncertain life as a doctor married to a doctor in France.[44] She married instead the appropriately cool and stable pediatrician, Abraham Jacobi, a German émigré and state socialist, the antithesis of her friend Réclus and of her French lover. Politically she seemed to have forgotten the heat and agony of the Commune, or perhaps she never fully understood its political significance. When she joined the New York City Positivist Society she deepened her knowledge of Comte, absorbed the Comtian rule to "live for others," discoursed on the "new moral safeguard, live in the open air," and celebrated the coming of spring in the positivist way with recitations from pagan literature on the seasons.[45] Seven years later she wrote the most thoroughly positivist book ever written by a woman in the nineteenth century, *The Value of Life, A Reply to Mr. Mallock's Essays "Is Life Worth Living?"* Her book was clear evidence of how completely she had settled her debts with Paris. Her mind had fixed itself upon a stable point, never to be shaken again.

Jacobi tore into Mallock's Catholic arguments against Comtian positivism brilliantly and sarcastically.[46] She adhered to the "true positive method" that substitutes "the study of the invariable *laws* of phenomena for that of their *causes*, immediate or final." She conceived of society as an "organism whose several parts are necessarily subordinated to one another," and of unquestioning individual devotion and service to society as noble, truly human, and above all "scientific," since all individual functions and roles are "socially interdependent" and "inseparably united;" thus "real isolation of elements" is theoretically "impossible." In her view, "The social functions of human beings are directed, first, to the maintenance of the existing order of society, and only afterwards to its progress towards any new order." Progress, if it comes, can be effected only "by the elite of humanity constantly engaged in the discovery of new laws" and in observing certain "sequences which will be uniform with those of the past." She believed that social and individual "happiness" results "when all the powers" are "in equilibrium," with a modicum of "tension . . . necessary to social vigor." She hurled at her Catholic adversaries what she considered the final indictment against the Catholic Church: it based its truth and laws on myths and figments of the "imagination," while positivism established them in the incontrovertible "realities of existence . . . in the general conditions of existence in society . . . in the relation of things to one another, to their personal or social antecedents and consequences, to physical conditions which may coincide with them"—in things "as they really are," and not as the mind "imagines" them to be.

The feminist leadership of the National Woman's Suffrage Association similarly borrowed many ideas from positivist thought, and applied some major unifying concepts from Comte's *Positivist Politics* to the cause of woman's advancement. One example comes from Martha Wright, Lucretia Mott's sister and the wife of a wealthy lawyer, David Wright. In 1870, a few years before she became president of the National Woman's Suffrage Association, she wrote to Stanton, "We must organize from the present disorder, systems embracing the interests of all classes. O! how I see the want of regulation in national affairs."[47] Elizabeth Smith Miller, NWSA member, translator of French texts for the *Revolution*, and daughter of Gerrit Smith—one of the wealthiest, as well as most consistently individualist, feminist reformers of the antebellum period—also actively turned to social science in the late 1860s. She and her husband, Charles, were both philosophical materialists, devotees of Huxley, Buckle, Spencer, and Comte, and friends of American Comtians. They believed, according to Elizabeth Cady Stanton, in "a government of immutable laws for the race: wise, beneficent for those who knowingly bring themselves into line with eternal principles; merciless on the ignorant masses, who, through the centuries, have been ground to powder by the forces, that in obedience to law, might have secured them highest growth and development."[48] In the 1870s Elizabeth Miller joined the American Association for the Advancement of Women, lectured, and wrote a book on the science of rational cooking and training of cooks.[49]

The early and brilliant organ of the NWSA, *Revolution*, published long excerpts from *Positive Politics* that clearly conveyed the nature of the feminist interest in Comte. The quotations reflected Comte's concept of the three stages of history—theological, metaphysical, and positive. Mankind, Comte believed, had evolved into a positive stage, which entailed the abandonment of the ideological remnants of the two previous stages and which demanded the "organization of society upon a new and purely scientific basis."[50] Historical developments—specifically, the secularization of society—had at last made it possible for man in a genuine sense to balance the three components of his "moral constitution—reason, activity and affection"—and had liberated the "affective" component ("Women," "Altruism," the "Love Element") in particular, to play a principal role in securing and deepening this harmony and in freeing the world from the tyranny of the "selfish instincts of mankind." "As Priestess of Humanity," Comte wrote, "woman's office consists in cultivating and developing the affective principle of human unity, which is our principal safeguard against the immense social disturbance resulting from existing intellectual anarchy" and is "destined to modify the spontaneous reign of material force."[51]

Elizabeth Cady Stanton and the Positivist World View

THE LEADING PHILOSOPHER of the feminist movement, Elizabeth Cady Stanton, made these positivist concepts her own. Like other feminists, she displayed the same shift in ideological consciousness from a strong emphasis on individualism in the antebellum period to an attempt after the Civil War to combine individualism with structure, organization, and centralization. In the same breath in which she heralded an age in which both men and women would be "absolutely free," she sought to control this freedom and harmonize it with the demands of social order and community.

Historians of feminism have tended to describe Stanton as a great nineteenth-century exponent of individual freedom. Looked at from one angle, her life and her thought on major social questions from the 1850s onward do suggest this interpretation. She was the most brilliant, witty, sarcastic, and dynamic of feminists, and she had a romantic side. Some of her favorite writers were Emerson, Carlyle, Ruskin, and Charlotte Brontë. She exhorted others to love and live to the fullest, gloried in "sunrises and sunsets," and enjoyed the pleasures of self-reflecting introspection.[52] "How we mortals cheat ourselves out of our birthright," she wrote her cousin, Elizabeth Smith Miller, in 1853. "How few ever taste the blessedness of loving nobly, generously, passionately. . . . Have you, dear Liz, lived long enough to enjoy solitude . . . to look for it and long for it as an epicure does for his dinner?"[53] Mother of seven children, daughter of a prosperous judge and a mother whose ancestors had fought in the Revolution, Stanton was a woman who felt expansively and relished open, natural spaces and uninhibited movement. She wrote back to her husband from a Kansas speaking tour: "I would love to live here . . . so that our children could ride and breathe and learn to do big things. I cannot endure the thought of living again that contracted eastern existence. Here the boys would rise. . . . You would feel like a new being here."[54]

"*Laissez-faire* with all my heart," Stanton said in 1868.[55] Like every good nineteenth-century liberal, Stanton defended the principles of free trade and free labor—open competition in open markets—and attacked all political and economic concentrations of power that prevented both men and women from "naturally" seeking the "prizes of life."[56] She brought her liberalism into the home, applying it to married life, an application that shows most clearly in her lifelong attachment to liberal divorce and to the free-love principle of individual sovereignty.[57]

As in work and production so in matters of divorce and separation. According to Stanton, the state had no right to enforce or coerce contracts freely entered into by men and women. Individuals alone determined the nature and extent of their contracts. In an address to the New York State legislature in 1854, Stanton said, "If you take the highest views of marriage, as a Divine relation, which love alone can constitute and sanctify, then of course human legislation can only recognize it. Men can neither bind nor lose its ties, for that prerogative belongs to God alone, who makes man and woman, and the laws of attraction by which they are united."[58] At a woman's rights convention in New York City she described marriage as a civil contract founded on "friendship" and "love;" and in a long review of Caroline Dall's *College, Market and Court* for the *Radical*, she observed that the "sacred affections and delicate sensibilities of the Marriage Relation cannot be measured by the clumsy machinery of law. We only ask of it to act impartially by both sexes and to leave moral and spiritual relations to their own remedies and compensations."[59] She gave the most sensational rhetorical rendering of this position in a speech on marriage and divorce in 1870. "We are all free lovers at heart," she declared, "although we may not have thought so. . . . We all believe in the good time coming . . . when men and women will be . . . a law unto themselves and when, therefore, the external law of compulsion will be no longer needed. . . . The element of legal compulsion," she added, "is all that distinguishes marriage from those natural and free adjustments which the sentiment of love would spontaneously organize for itself."[60]

This kind of talk angered such feminists as Wendell Phillips and Thomas Higginson, who apparently thought of marriage as something unique and sacred and not as a civil contract terminable like any other. In the late 1860s and early 1870s, however, many non-free-love feminists basically agreed with Stanton either openly or in secret.[61] Even the paper that initially objected to such discussion, the staid *Woman's Journal*, yielded to the intensity of the discourse. Suffrage did not dominate its pages in the 1870s; education, sex and health reform, professional work, wages, and even facets of the marriage question often held sway. The *Woman's Journal* enlisted the Christian radical Jesse Jones, a man widely respected in Boston feminist circles, to write a long series of articles on marriage, culminating in two articles on divorce. "Divorce," Jones wrote, "so far as women take the initiative in it, is one phase of the revolt of women against the harem idea," and however "corrupting" it may appear to some, "it is yet a movement which seeks for good, for it is a movement to escape out of tyranny into freedom." Like many other feminists, Jones blamed the lingering power of the "harem idea" in the minds of men on the "church," "the bulwark of the supreme paganism of this day."[62]

Stanton's laissez faire views and her pastoral romanticism notwithstanding, neither Stanton nor most of the feminists who held some of these views can be called a radical individualist. Nor can she be lumped with many of the free-love anarchists who hated the growing centralization, the industrialization, and the complexity of American society; they aspired to an untrammeled pastoral existence and turned their backs on all political activity.[63] Stanton loved the West, the open country, its expansiveness, its heavy potential for fulfillment, but she lived willingly in the East, near or in New York City, for much of her adult life. She also confronted but did not repudiate the capitalist system in its increasingly industrial form. Referring to woman's inefficient "business habits," she wrote that "the strict discipline of the factory is a hard but needed training school."[64] She shared the ideas of several contemporary economists that "in the present condition of civilization, no change in labor can take place without some individuals suffering thereby. The power loom and the sewing machine throw many out of employment, they end in producing the necessaries of life in greater abundance, so that all are enriched thereby." Jane Croly took the same position and so, later on, did Charlotte Perkins Gilman.[65] Furthermore, Stanton did not oppose organized political activity or political parties. Her very devotion to constitutional republicanism (symbolized by her admiration for Jules Favre, Giuseppe Mazzini, and John Stuart Mill), coupled with her efforts to build third parties, removed her completely from the anarchist-individualist camp. In 1868 she worked to create a new political coalition among skilled workers, labor reformers, and suffragists; later she sought to enlarge this coalition to include temperance reformers, spiritualists, and even members of the First International.[66] Like most feminists in the 1870s, and like many Republicans, some of whom tried unsuccessfully to organize a new political party, Stanton condemned the existing party system and conventional politics for partisan conflicts, corrupt patronage, and general blindness to the institutional needs of the country. She wrote that she would have followed the breakaway banner in 1872 if "the Liberal Republicans had put a woman's suffrage plank on their platform and nominated Benjamin Wade and George Julian for President and Vice-President."[67] Like other reformers, Stanton vigorously proposed a scientific politics that transcended the limitations of the party system and the character of ordinary political life.

More important, she attacked laissez faire individualism itself. Retaining faith in liberalism, she championed cooperative industry and cooperative unions, and maintained that "charity must come in to relieve individual cases which have suffered from changes for general good." She wrote in 1868, "We must be tender to those who fall back, stragglers from the great army of progress and with a compassion that does not degrade them, help them to start again on their toilsome way."[68] She hoped to transform mental asylums and prisons from authoritarian, punitive enclaves into "moral seminaries" run by women, "where all that is good and noble in these unfortunate ones might be awakened into life."[69] Stanton reacted deeply to the social and economic problems caused by urbanization. She worried about the thousands of immigrants who poured yearly, even monthly, into already congested cities plagued by bad sanitation, bad housing, bad educational facilities, and corrupt politics. She feared the domination of "selfish" capitalists who monopolized wealth and oppressed the working classes, men whose positions depended on the liberal capitalism she admired but who promised to destroy completely the sobering and harmonizing influence of the "better" and more educated classes. As she observed such changes, Stanton's older romantic individualism was challenged and transformed. Romanticism had not directed her inward; instead it moved her to confront the inequities embedded in political, social, and economic institutions.

The distress and misery that wracked economic and social life, Stanton believed, were the consequence of male dominance. Men were unfit to assume those tasks requiring "unceasing love, patience, and mercy."[70] Unregulated male excess, unrestricted male "brute force," male "indulgences, appetites, and vices" encompassed the worst aspects of modern competitive individualism, subverting social "equilibrium." The "male element," she wrote, setting the direction of feminist complaint on the subject, "has held high carnival thus far, it has fairly run riot from the beginning, overpowering the feminine element everywhere, crushing out all the diviner elements of human nature, until we know little of true manhood and womanhood."[71] In the 1850s Stanton had sneered at the "mysterious twaddle, the sentimental talk at all our conventions about the male and female element!" Later she continued to ridicule "sentimentalism," but she also developed an unwavering belief in the basic "human" differences between the sexes, a belief shaped by her reading of Comte.[72]

"When Comte asserts that recognition of woman's thought is primal to the reconstruction of the state, the church, and the home, he grants all we have ever asked. He disposes of every question he discusses on the basis of social harmony."[73] Stanton turned time and again to the advice and wisdom of positivist philosophers—to J. H. R. Wilcox, for example, the eugenicist who wrote for *Modern Thinker*, became chairman of the Committee on Arrangements and Credentials for the NWSA in 1870, and remained secretary of the Executive Committee of the New York State Woman's Suffrage Society into the 1880s. She also turned to Albert Brisbane, Stephen Pearl

Andrews (whom she called "Pearlo"), William F. Channing, and Rutger B. Miller, a devoted Comtian who pleased Stanton greatly when he promised to dedicate his book on the "reorganization of Society" to the Woman's Committee of the NWSA.[74]

In 1868 Stanton's cousin, Gerrit Smith, recommended that she read Miller's translation of Comte. He challenged her, tongue in cheek, to grapple with Comte—"or will you be constrained," he asked, "to admit that woman's only empire is the heart?"[75] Stanton felt no constraints and met the challenge by coming back to Comte's affective concept in editorial after editorial for the *Revolution*:

> We insist . . . that Comte's principles, logically carried out, make women the governing power in the world . . . Comte makes man a personal, selfish, concentrating, reasoning force. He makes woman impersonal, unselfish, diffusive, intuitive, a moral love power. He divides society into three classes, Intellect, Affection and Activity. He says intellect and activity, capital and labor, ruler and ruled, can only be harmonized through affection, which is the feminine element in woman, and this he exalts above the intellect and activity, conception and execution. "Love for others," he says, is the great law on which society . . . is to be reorganized. This can only be done by the cultivation of the unselfish, the moral, the diffusive, the woman; and thus we actually reverse the present order of things.
>
> In the restoration of the love element, which is woman, capital and labor will be reconciled, intelligence and activity welded together, forming a trinity that shall usher in the golden age.
>
> We have thus far lived under the dynasty of force, which is the male element, hence war, violence, discord, debauchery. From this we can only be redeemed by the recognition and restoration of the love element which is woman.[76]

From her editorial days on the *Revolution* to the first volume of the *History of Woman's Suffrage*, Stanton described refined, educated womanhood as the regulating, "centripetal force" of civilization, the instrument of social "equilibrium," "harmony," and, above all, "organization." She began her autobiography with: "Social science affirms that woman's place in society marks the level of civilization."[77] As a Comtian she believed that the secular "reorganization of society" required the "restoration of the love element" consistent with the needs and problems created by a new stage in human evolution and based on "immutable" physical or material laws "governing not only the solar system, the vegetable, mineral and animal world, but the human family, all moving in beautiful harmony together."[78] Along with Matilda Joslyn Gage, a physician's daughter who was coauthor of the *History of Woman's Suffrage* and editor of the *Ballot Box*, and Paulina Wright Davis, editor of *Una* in the 1850s, a Rhode Island feminist and wife of a merchant, Stanton introduced these ideas in slightly modified form into the National Woman's Suffrage Association platform of resolutions for 1872. They openly asserted what many feminists, by this time, shared only privately with one another.

Be it resolved, declared Gage,

> that the movement for the enfranchisement of women is the movement of universal humanity . . . that man, representing force, would continue . . . to settle all questions by war; but woman, representing affection, would, in her true development, harmonize intellect and action, and would weld together all the interests of the human family—in other words would help to organize the science of social, religious and political life.[79]

The emphasis Stanton placed on refined womanhood as a principal "organizing force" points to a larger commitment after the Civil War to a general, rational "reorganization" of the public and private life, a commitment that divorced her further from those diminishing numbers of reformers who spoke and wrote only in terms of human spontaneity and human freedom. First, apropos of this new perspective, she appealed to feminist women to organize on a national level in order to get their rights. In the 1850s most feminists had objected to the organization of a national society on explicitly individualistic grounds. In a letter to the 1853 Syracuse convention, Angeline Grimke Weld explained this position:

> Organizations do not protect the sacredness of the individual; their tendency is to sink the individual in the mass, to sacrifice his rights and to immolate him on an altar of some fancied good. . . . We are bound together by the natural ties of spiritual affinity. . . . We need no external bonds to bind us together, no cumbrous machinery to keep our minds and hearts in unity of purpose and effect.[80]

Under the leadership of Stanton and others, feminists reversed this focus after the war. Where previously organization had by its very nature manacled individual freedom, now it supposedly paved the path to its fullest realization. The unencumbered work of spiritual affinity could no longer be relied upon; it needed support, structure, organization.

Second, Stanton believed that the complex demands of public and private life in industrial society required an organization that was rational and centralized, with intelligent, educated administration. "Because politicians are ignorant of the laws of political economy," Stanton wrote for the *Revolution*, "and leave everything as they say to regulate itself, is no reason why the people shall wreck their own interests by following their example. Do not let us longer confound the designs of Providence with the legitimate results of human legislation."[81] In contradistinction to her formal laissez faire economics, therefore, Stanton increasingly came to maintain that the best "interests" of the people depended not on an unfettered market, but on national centralization.

Like her close associates Gage and Anthony, and like the other social science reformers who also paid close attention to these matters, Stanton saw in the concentration of national power the basic prerequisite for the expansion of human freedom. "Our age has annihilated space; danger lies in darkness and distance. . . . The dangerous is that which has grown to such dimensions in the various states, multiplying legislation and regulating each petty concern within its borders."[82] She therefore looked from the atomistic state governments to the national state to legislate and administer on such key social questions—as health, education, and the sexual relation. She advocated, for example, that the federal government institute a uniform divorce law, that it pass legislation on suffrage and equal rights, and that it impose compulsory education.[83] In 1870 she and Elizabeth Smith Miller became the principal American representatives of the International Woman's Association, an organization founded by Maria Goegg of Switzerland in 1868 to foster *solidarité* among women throughout the world. Many of its members believed, as Stanton did, that in regard to education the "abrogation of the family tie is absolutely needed in the highest interest of the new social order." Society, not parents, had "the best right to the education and training of children."[84] Impressed by what many considered the efficiency of the new German "eugenics" laws, she proposed as early as 1870 that the state should prevent the marriage of the morally, physically, and mentally unfit.[85] She also advocated a "scientific government" guided by the most intelligent men and women, "gifted with the genius of coordination or the power to harmonize, organize and direct. . . . If the average ability were raised a grade or two," she declared in 1882, "a new class of statesmen would conduct our complex affairs at home and abroad, as easily as our best business men now do their own private trades and professions. The needs of centralization, communication and culture call for more brains and mental stamina than the average of our race possesses."[86] It was for this reason, in part, that Stanton favored educated suffrage, minority representation, and civil service reform.[87]

Finally, and most germane to the question of married life, Stanton supported a rational reconstruction of marriage and the sexual relation—or, more generally, the private life—so that they would contribute to social harmony and social order. She wanted marriage to serve the best interests of the state. It is true, of course, that she warmly endorsed individual sovereignty in matters of divorce and separation. It is also true that she applied this principle only to divorce and separation. "I advocate individual sovereignty," she wrote in her diary in 1898, "only in the matter of separation, when the parties find themselves wholly incompatible and antagonistic. Here they should have complete freedom of choice."[88] Married life, as well as the act of marrying, demanded more caution, more deference to the needs of society, and considerably less exercise of the individual will. In her controversial and widely distributed 1870 speech on marriage and divorce, Stanton asked for greater freedom of divorce, but she also denounced the custom of common-law marriage that put no constraints on individual choice and action and that, by its example, gave moral support to "polygamy," "bigamy," and "adultery." In this speech Stanton focused at length on the history of marriage and on the laws that tried and failed to regulate it, a focus that distinguished this document from her earlier utterances on divorce and made clear her equally strong interest in the character of married life. "In many of the states of the Union," she said,

> a legal marriage may be contracted between a boy of fourteen and a girl of twelve, without the consent of parents or guardians, without

> publications of banns, without even the signatures of the parties, the presence of a priest, court justice or any officer of the State. Such absence of all form and dignity in the marriage contract is unknown in any other civilized nation.[89]

Stanton's critique of common-law marriage and her accompanying proposal that "homogeneous legislation" be passed to eliminate it implied a deeper attack on the careless and dangerous freedom with which men and women entered the marriage relation. It was a critique that was elaborated throughout the rest of the century by reformers eager to raise the age of consent and to rationalize the marriage laws.[90] It appeared, for example, at the end of the century in the work of the Progressive historian of matrimonial institutions, George O. Howard, who deplored common-law marriage as the most invidious form of "unrestrained individualism" and who agreed with Stanton that "the highest individual liberty can be secured only when it is subordinated to the highest social good."[91] However much Stanton defended divorce, she sought to create the conditions that would remove its necessity. She advised young men and women thinking of marriage to attend coeducational institutions, to study scientific principles and facts, to avoid biologically inferior and psychologically dissimilar partners, and to dodge the pitfalls of romantic love. Stanton felt a strong intellectual hostility toward "heterogeneous unions" that were based on "every possible inequality of condition and development"—disparities of age, beauty, education, grace, goodness, and intelligence. Such disparities ignited the passions and quickly wrecked marriages. Only rational, egalitarian unions could flourish and survive.[92] In 1868 she wrote, referring to woman's domestic life, that "instinct is a great matter, and serves Indians and bees admirably, but in the world of civilization and railroads reason and education are as important;" in 1870 that "science must be called in to investigate every part of the subject;" and in 1871 that "the fundamental conditions of life" must be "based on science."[93] The sexual relation foundered on impulse, stupidity, and instinct. Shaped and governed by rational, appropriately positivist laws, it yielded untold blessings for both the individual and society.

Feminists Revise Positivism

IT CANNOT be claimed that Stanton and other feminists swallowed positivism whole. Like the southern sociologists of the Civil War period, who, as Louis Hartz has observed, countered Lockean liberalism with Comtian theories of association and organization, these northern feminists could not escape, nor did they altogether want to, "the Burkean power of the liberal tradition."[94] For one thing, many of them—Julia Ward Howe, Thomas Higginson, Anna Brackett, Ednah Cheney, for example—found Hegelian idealism more attractive than positivism from a moral point of view, although it can be argued that both these philosophical systems, as they took shape in the minds of many feminists, tended ultimately to lead to the same intellectual consequences. Howe read Comte and Hegel but never wholly renounced her Christian faith for rationalism: according to the freethinking Chicago feminist Kate Doggett, Howe served as a "connecting link between the radicals and conservatives, softening and toning down the somewhat startling words of Rationalism to suit the taste of the latter. She knows how to prepare the meat for those who are not yet strong men, but are ready to abandon the cradle of childhood."[95] This much could also be said of the *Woman's Journal*, which Howe helped to edit. More willing than other periodicals to play down the antireligious character of feminism, it nevertheless claimed in one form or another in countless articles that "the leading principles of Science must be deeply implanted in the general mind."[96]

Comte outlined in his philosophy an extremely hierarchical, organized, and antidemocratic statism with a "scientific priesthood" that would be in the strongest possible position to shape and direct human affairs; most American reformers, on the other hand, had sought—although often unsuccessfully—to blend the need for social order, for system, and for enlightened government with democracy and laissez faire economics. Comte replaced all religion based on supernatural beliefs with an atheistic humanism, but most reformers—in spite of private convictions and frequent attacks on institutions of the church, and in spite of the profound transformation of the religious mind that made Comte's views attractive—could not, or refused to, espouse atheistic ideas publicly. And, finally, feminists drew implications from

Comte's position that they knew Comte himself never intended; he insisted that women remain deified within the domestic sphere, that they depend for their livelihood on men and never yield the "infantile" qualities which set them apart from men and which represented their greatest charm.[97] Most feminists found this extreme formulation regrettable, if not—Lester Ward's description—"execrable."[98] When it came to the "woman question," even the positivist contributors to the *Modern Thinker* peddled such a decidedly diluted Comte that their English and French counterparts rebuked them for paying too much attention to the "dogma of the right of private judgement," and especially for having fallen into the "heresy" of assigning women an "undue prominence" and being "tainted by the woman's rights ideas prevalent in the United States."[99]

Jane Croly feared the effects of conventional political life on female sensibility. She doubted whether marriage, motherhood, and professional independence could be managed easily at the same time. "Home is woman's domain," she said. Nevertheless, she demanded for woman "equal pay for equal work;" she labored for the professional advancement of single women, widows, and married women (she herself led two careers); and she could conceive of "no reason why women should not vote."[100] At a time when her life spun in a multitude of directions, she wrote the feminist leader Anna Dickinson to plead that the two should publish and edit a new paper "that would attract the attention of the whole country" and would "become a political, as well as social power. . . . I feel myself better equipped today . . . than I ever was, and my relations are such as almost to command success, but I do not feel like going it alone, and you are the one person above all others to help, and unite your forces with mine." She added winningly, "We should have a mighty good time and secure fame and fortune."[101] Positivist Albert Brisbane deplored "social inequalities, the pride of caste, and the subordination of women," and Comte's chief American apostle, Henry Edgar, thought "it quite certain that we need—absolutely and urgently need—that women should interest themselves in public affairs both social and political."[102] Mary Putnam Jacobi became an ardent supporter of woman's suffrage and agreed with the French utopian Fourier that "one third of the total number of girls thoroughly hate the restrictions and limitations imposed upon their sex—and crave the energetic turbulence of boys."[103] In 1876 she wrote *The Question of Rest for Women During Menstruation*, a brilliant medical attack on the myth that menstruation debilitated women and the most impressive single physiological statement in favor of coeducation written by a feminist. The book won the Harvard College Boylston Prize for the best scholarly manuscript of the year. In the early 1870s Putnam Jacobi devoted her considerable powers to the organization of feminist social science and to the widening of woman's professional opportunities.

Positivism, especially the positivism espoused by Comte, was committed to the development of a new social system that would eliminate the disruptive and anarchic tendencies caused by individualism. Fully accepting developing industrial capitalism and the social relations produced under it, positivists sought to remove its conflicts and excesses by reconstructing it on a more efficient and scientific basis. Positivists wanted to reconstruct society on what they perceived to be the objective facts of social and economic life. They applied the techniques of the natural sciences to society. They condemned all speculative, imaginative, and introspective thought—all thought that came from the dreams of individual men and women and had no root in the existing "tendencies and developments actually inherent in society." These tendencies and developments, of course, were those of industrial capitalism.

The discovery of scientific social laws made possible the realization of the positivist goal: the unified and centralized organization of the social order. This organization would be obtained by reliance on a new class of scientific experts and by the moral regeneration of the people. Scientific experts were supposed to discover the laws of society in an apolitical, nonpartisan way, while it was the government's purpose to implement these laws. Other intellectuals prepared the people for the moral acceptance of this organized society. For positivists, moral reform had to precede other material changes. Before the introduction of other reforms, people had to learn to sacrifice their individual, private interests to the interests of the community.

Many mid-nineteenth-century feminists, as well as later feminists like Charlotte Perkins Gilman, adopted these doctrines, but in their attempt to create a more cooperative, egalitarian public realm, they did not or could not embrace them completely. Rather, they tried to revise positivism. They tried to integrate it into an older democratic humanitarianism.

The spirit of positivism, however, even if modified, pervaded the thinking of American feminists in both specific and general ways from the Reconstruction years into the late nineteenth century. The feminist affection for George Eliot alone attested to the strength of this spirit. At the time of Eliot's death, the authors of the first volume of the *History of Woman's Suffrage* called her "the greatest novelist of the nineteenth century. . . . Dignified and transfigured" by the "tenets of Comte," her "genial altruism" supplied the "place of creeds" and her "scientific knowledge of the natural sequence of law" superseded "a blind faith in an irresponsible arbitrary something." In the words of the popular feminist and sex reformer, Eliza Bisbee Duffey, Eliot "inculcated the purest morality and the broadest humanitarianism" and "made selfishness stand out in all its deformity."[104] Feminists regularly quoted from Eliot's works in their journals, books, and newspapers. Croly, for example, chose these lines from Eliot as a motto for her book *For Better or Worse*: "If the past is not to bind us, where can duty lie? We should have no law but the inclinations of the moment."[105] The feminist *New Age* quoted this line in 1876: "You cannot isolate yourself, and say that the evil which is in you shall not spread. Men's lives are as thoroughly blended with each other as the air they breathe: evil spreads as necessarily as disease."[106] The rise of women's social science associations in the 1870s was also influenced by this ideological tradition. As one positivist observed in 1873, it was

> a remarkable sign of progress of the age that moral and social questions should be in so many and various quarters discussed from the standpoint of Positive Science. . . . The possibility of a Positive Science is now generally admitted, or at least, tacitly assumed. Social Science Congresses are attended by people who would be loath to incur the suspicion of heterodoxy.[107]

Like Comte, feminists attacked "selfishness," "materialism," and, in the words of the positivist feminist Augusta Cooper Bristol, "the crushing and disintegrating spirit of individualism." They heralded the "coming woman" as the vehicle for the creation of a truly altruistic, symmetrical, and rational society based on the "equilibration of forces." Both Bristol and Stanton also used the five historical categories or departments of Herbert Spencer, another widely read positivist and favorite intellectual source for many reformers. Spencer's categories depicted the evolution of society from "activities ministering to self-preservation" to "activities . . . devoted to the gratification of tastes and feelings" or to leisure, but both Bristol and Stanton emphasized the fourth stage, of "activities . . . involved in the maintenance of proper social and political relations," and agreed, in the positivist manner, that the "fourth department of activities can never be adequately realized except through the moralization of wealth; the consecration of capital to social welfare; and the scientific adjustment of society."[108]

Feminists also believed, as one put it, that "permanent political changes cannot be effected without previous social and moral changes," and they did gradually come to maintain that all truth originates in physiological, "anatomical," or material law, and that humankind can "predict and thereby prepare for and modify events and results." They were convinced that reform must rest on "observed facts" and nonpartisan science.[109] The spirit of Comte, coupled with the more immediate and diffused impact of Spencerian positivism, seems to have contributed more to the shaping of postbellum reform thought than any other single influence. Positivism gave a language to indigenous trends already manifest in American reform movements. By their adherence to individual and social symmetry and to the cult of no secrets, by their devotion to scientific knowledge, and by their concerted efforts to reconstruct the educational system along more symmetrical lines, feminists underlined the reform attempt to combine individualism with a more structured, rationalized, centralized, and harmonious social system. They hoped to bring private life more closely in line with public welfare.

CHAPTER 7

ECONOMIC AUTONOMY FOR ALL WOMEN

"WORK, AND PAY FOR WORK!" Jane Croly exclaimed. Feminists glorified work for all women, an ideological position that followed the impetus of bourgeois liberalism. To them female economic dependence was "social fiction."[1] In an essay for the *Radical* called "Parasites" the feminist free-thinker Marie Brown wrote:

> Dependence is the condition that insures enslavement. It craves the firmness and solidity of weight of a stronger and greater nature, and is crushed by these preponderating elements because it lacks the corresponding forces that resist and balance. Its weakness invites oppression, its servility tempts tyranny, its helplessness provokes contempt and invites cruelty.

Women, the archetypal parasites both by necessity and by choice, had "clung in the most abject helplessness to anything that promised support."[2] Stanton put this in a slightly different way in her speech before the American Equal Rights Association in 1869. Dependence resulted in an exaggeration of sexual difference not to be found under natural conditions. "The strong, natural characteristics of womanhood," she said, "are repressed and ignored in dependence, for so long as man feeds woman she will try to please the giver and adapt herself to his condition."[3]

Ardent feminist expression of these liberal ideas did not take place in a historical vacuum. Before and after the Civil War, thousands of women worked in the factories of the Northeast and elsewhere. More important for middle-class interests, women also increasingly entered such professional fields as medicine, journalism, and education. The gradual reform of the marriage laws and laws that affected working women also buttressed feminist appeals. These changing conditions created a concrete foundation from which feminists could level attacks against woman's dependence on man. It was within this context that feminists could demand greater economic autonomy for all women, greater activity in the public realm.

These concrete historical conditions also established the bases upon which feminists could organize in behalf of working women, and, above all, in behalf of women like themselves who yearned for success through professional independence.

The Virtues of Independent Labor

IF DEPENDENCE stood for much that was bad in woman's condition, than independence, "courageous individuality," "heterogeneous" self-culture, "self-activity," "self-formation," represented much that was good.[4] Anna Dickinson angrily declared in 1870, in the middle of the uproar over the McFarland-Richardson scandal, "I would make every woman understand that she was born for herself and not for another. . . . Woman was *made for herself*—to round out herself. Let her live to the full and make a complete woman of her in every respect."[5]

Such a firmly held conviction unraveled into a multitude of implications. It entailed, for instance, an attack on the concept of woman's sphere, and certainly on the "pedestal and goddess theory," with its related ideological hardware—piety, submissiveness, self-sacrifice, and sentimental ornamentalism—and it embraced the centrality of womanhood, not of wifehood and motherhood, in woman's experience.[6] "We do not hesitate to proclaim," the *Revolution* observed in 1871, "that womanhood is greater than wifehood or motherhood, includes them as the greater includes the less," a proclamation Stanton fixed in history when she stated in her famous speech on marriage and divorce "that womanhood is the great fact, wifehood and motherhood its incidents."[7] This individualist stance produced an important series of corollaries—among them the nearly unanimous feminist demand that all women should be educated, on principle, for trades, businesses, and professions and thus become fit to make an honorable living. To protect themselves from the disabling consequences of economic depression, they should learn the value of money, how to spend, save, and invest it, and they should never be forced to marry for support or to marry at all if marriage promised only the degradation of an economically dependent status. No longer should marriage be regarded as the goal of a woman's life. The *Woman's Journal* called "ridiculous" the "idea that the sole aim of a woman's life should be to be married." The *Woman's Advocate* pointed out that marriage was "not the only aim of a woman's life." She should also have an "occupation, trade or profession."[8]

Feminists tried to make these arguments convincing to other women by resorting to evolutionist theory, but they lapsed into cliché in their efforts to prove the irrelevance of the "empire of muscle" or male brute strength to the work of the modern age, that is, that the modern age had created employments both men and women could perform with equal success.[9] They also attacked the weakening religious traditions that had once powerfully sustained woman's subjection. In 1869 a writer for the *Woman's Advocate* asked, "Why should a few men who lived some eighteen hundred years ago, and were full of the prejudices of their age and country, be referred to as authority, direct or indirect, in the great practical question of the present day? Living in an age when the world has attained greater Liberty than in that of Paul . . . is not John S. Mill a better authority upon the subject?"[10]

Feminists appealed to the present and past experience of women to prove that, in spite of impossible odds, women were not and had never been completely passive or powerless to take command over their own lives. The feminists of the 1860s and 1870s celebrated every demonstration of female capacity and bravery they could find: the exceptional exploits of such contemporary women as Ida Lewis, who had the fortunate habit of saving the lives of shipwrecked men off the coast of Rhode Island; the extraordinary women of the Civil War period who worked tirelessly for the Sanitary Commission; and the other, more ordinary women who struggled against poverty during the war, caring for their children, serving as "mowing machines," haying and harvesting, and later supporting their families while tending crippled, invalid husbands. They praised the widows who inherited the burden of household responsibilities, who went into real estate, business, the professions; and all those other women who worked at more grueling and arduous tasks than men in the factories and fields of England, France, and the United States.[11] When Thomas Higginson made the mistake of claiming in the *Woman's Journal*, in the fiatlike and patronizing way so characteristic of him, that women were too delicate to endure the stresses of protracted work, Lydia Maria Child retorted with an angry article proving precisely the opposite. She triggered a burst of similarly angry correctives.[12] As one feminist wrote, after reading Caroline Dall's *College, Market and Court*, "There has been, and can be, nothing from the hardest manual labor upwards, which we women have not done, are not doing, and cannot do."[13]

The experience of women, past and present, yielded a wealth of data to substantiate the feminist case for the capabilities of women. Feminists plunged into the past not only to trace a dismal history of oppression, but also to bring to light, for polemical and genuinely historical reasons, those propitious, often heroic, and not too infrequent moments when women shaped the course of history and the character of civilization. Celia Burleigh resurrected Xantippe, wife of Socrates, from the calumny heaped on her by classical scholars; Parker Pillsbury rescued the Amazons and Thomas Higginson Sappho.[14] Higginson, in fact, believed that the modern "discovery of women" had made it necessary to "re-write all history."[15] Caroline Dall wrote articles and two books in the 1860s on ancient, Renaissance, and Enlightenment women, including some sketches of such women as the Countess de Cinchon, who, Dall wrote, invented quinine in 1661; Madame Ducondras of Rome, who in 1712 was the first physician to use a manikin while lecturing; and Lady Mary Wortley

Montagu, who supposedly introduced inoculation to Europe in 1771.[16] Dall berated scholars who sniggered at Aspasia, Hypatia, Mary Wollstonecraft, and even Lucretia Borgia. To Dall these were great human beings who managed to rise to considerable heights in spite of enormous odds.[17] "It required courage," Stanton wrote in 1868, "in any woman to vindicate the fair fame of Mary Wollstonecraft. Mrs. Dall has done it ably and we trust that her verdict will stand against all objectors."[18] Elizabeth Stanton herself did brilliant feminist biblical scholarship, first in articles on biblical heroines for the *Revolution* and later in the form of the *Woman's Bible*, published at the end of the century.[19]

Matilda Joslyn Gage, perhaps the most important of all nineteenth-century feminist historians, ceaselessly mined the past for material on gifted women. She maintained that a careful "archeological" study of mythology would unearth much about the early creative and inventive genius of women. In the 1860s she began her investigation of women inventors from the ancient to the modern period, published it in article form in the *Revolution*, and later made it into a book. According to her, women invented embroidery, bread, the cotton gin, the manikin, pillow lace, the straw bonnet, and the science of medicine.[20] In 1879 Gage wrote her famous essay, "Woman, Church and State," and shared in the writing of the *History of Woman's Suffrage*, a book bursting with historical data on great women, living and dead, and on contemporary feminist women. It was because Gage knew so much about these women, in fact, that ignorance on the part of her contemporaries surprised and occasionally angered her. In 1871 she wrote to Martha Wright after reading an article by Horace Greeley in the *Herald Tribune* that suggested that women had "no maintaining or resisting power." "What an ignorance of History this shows," said Gage, and then went on to list for Wright the names of Catherine of Russia, Elizabeth of England, Maria Theresa, Joan of Arc, Madame Roland, Isabella of Castile. "Artemisia, Philla, Semisanis, and a myriad of other names come up, and as I read, I grow more angry."[21] In 1878 a Missouri feminist, Rebecca Hunnicott, asked Gage if any information existed at all "regarding woman's work or progress in arts, science and literature, as showing equal capacity with men for such developments and pursuits." Gage responded in the *Ballot Box* with an array of achievements that showed the depth of her reading and must have staggered Hunnicott. "Our friend," she wrote,

> has placed an Herculean task upon our shoulders; merely to catalogue the books giving such information would be a task for years and require more room than all the Suffrage papers in the land could give until the end of time. We should have to run through the library of women authors alone, once owned by Peter Paul Ribero, which numbered thirty-two thousand volumes, and which now could be multiplied tenfold.

Gage suggested that Hunnicott begin with the Bible, then work up through "Herodotus, Champollian, Lepsius, Williamson, Bryant" and many others, "for merely an insight into Egyptian women's work and progress in art, science and literature. . . . We could not even catalogue women's names that have been eminent in art, science and literature."[22]

Women of the Working Classes

BY THE post–Civil War period, historical conditions had made it possible for feminists to claim the legacy of individualism for women on more than theoretical grounds. From the late 1850s to the 1870s, several factors converged to form a strong, though unstable and vulnerable, base on which feminists might approach the economics of marriage in new ways. The collapse of woman's traditional domestic industries and the accompanying rise of the new factory system that replaced them, absorbing more and more female labor, was the first of these conditions. As Mary Putnam Jacobi wrote in 1873,

> Today, when a new industrial system of aggregate and factory labor has supplanted the old domestic disseminated industry, women continue to throng factories, printing houses, binderies, slop houses, etc. But they no longer direct handmaidens, nor cooperate with fathers, husbands and brothers, but they slave in the service of machines, and do the bidding of overseers.[23]

By 1870 more than 350,000 mostly single women were employed nationwide as industrial manual laborers, four-fifths of them in the clothing industry.[24] They endured the worst conditions of any workers and got starvation wages as hoop-skirt workers, seamstresses, glove makers, hat makers, feather and ornament workers, lace makers, and so on. Other women, mostly immigrant, found employment in the equally oppressive cigar-making industry.[25] A smaller, sometimes more fortunate group worked as printers, telegraph operators, and department-store saleswomen.[26]

Their presence was visible and acknowledged. A leading feminist, Anna Dickinson, told a lecture audience in 1870, "Twenty years ago Society said *nothing* on the conditions of laboring women. Nobody in this day is ignorant of the fact that multitudes of women are toiling in garrets and cellars . . . for a pittance of bread, for scanty garments and insufficient fire."[27] Few sought to remove them from industrial employment in hopes of reversing the clock. Instead, there were institutional efforts throughout the East and elsewhere to find industrial work for single women, to train them for trades, and to improve their working conditions. "There are great benevolent societies," the *Springfield Republican* observed in 1874, "in New York City and in nearly every large city which attempt to find work for unemployed women."[28] Innumerable charities run by women, devoted to the welfare, education, and employment of working-class women, appeared from the middle of the nineteenth century on. According to a catalogue of charities compiled in 1876, women operated more than two hundred charities before 1860, thirty-three of which tried to attend to the needs of working-class women. After 1860 more than three hundred new charities conducted by women came into existence, fifty-four of them conceived explicitly with the interests of working women in mind. Most acted as employment agencies, but they paid some attention to industrial education and working-class uplift. Four of them—the Day Nursery in Lawrence, Massachusetts (1875), the Union for Home Work in Hartford, Connecticut (1871), the Day Nursery for children in Philadelphia (1873), and the Ladies Relief Association in Baltimore (1873)—offered primitive day-care facilities for the children of working-class women.[29]

In 1863 several wealthy New York City merchants, in response to the growing crime rate, vagrancy, illiteracy, and poverty in the city—symbolized by the draft riots of that year—founded a nonsectarian

Working Women's Protective Union. Superintended by upper-middle-class women, it was a clear attempt on the part of the city's capitalist elite to introduce order into the chaotic lives of working-class women. A letter to *Woodhull and Claflin's Weekly* called it "the one real Woman's Institution in New York."[30] An employment agency for single working-class women not engaged in domestic service, it included a library and educational programs; equally important, it collected claims for women swindled in their wages by corrupt employers.[31] By 1873 it had helped more than 27,000 women find jobs and had collected more than $10,000 in claims. By the end of that year applications for aid had multiplied so fast that the union could not accommodate the demand.[32]

In 1870 Jennie Collins, former Lowell Mills girl, labor organizer, tailor, and a woman impressed by the feminism of Susan B. Anthony and Elizabeth Cady Stanton, set up an organization similar in some respects to the WWPU, although less influential. She converted a spacious hall on Washington Street in downtown Boston into Buffin's Bower, a home for needy women new to the city. "Girls coming to the city without money or friends have been taken in and given a home until work was found for them," she wrote in her Sixth Annual Report. "So destitute have they been, in some instances, that it was found necessary to supply them with articles of clothing. No estimate can be put upon the value of such a place to girls who come from cold, cheerless lodging-rooms, and suffering from the effects of low diet."[33] With the cooperation of her own employer and with financial assistance from both Boston merchants and the Massachusetts prison authorities, Collins provided food, clothing, shelter, and employment services for shopgirls, restaurant girls, and others, and enlarged her establishment to include a workshop and a reading room.[34] Like the WWPU, Buffin's Bower relied in part on capitalist wealth. It too acted as a nonsectarian employment agency and, above all, grappled with the new industrial conditions that had altered the nature of woman's work. By 1876, according to the feminist paper *New Century*, there were 30,000 young women in Boston, doing work that ranged from work in restaurants to the manufacture of artificial flowers; they represented 10 percent of the entire city population.[35]

Like these employment and relief agencies, the industrial education for working-class women that began after the Civil War indicated how far the American bourgeoisie acknowledged and tried to shape the new conditions. Industrial education existed before the Civil War, but in a primitive form, designed to teach women such domestic tasks as sewing, cooking, and housework.[36] In the early 1870s schools like Cooper Union in New York City and the Packard Institute in Brooklyn expanded their curricula to include newer fields of specialization for women. One of the first schools in the country to provide instruction to women in art, Cooper Union by 1869 was offering classes in wood engraving, telegraphy, wood carving, and painting. S. S. Packard of the Packard Institute made a public proposal in 1871 to educate, at his own expense, fifty women for business clerkships.[37] Of greater importance, however, were the industrial schools that sprang up from Illinois to Maine, teaching women in both single-sex and coeducational settings such crafts as telegraphy, printing, machine-shop work, farming, and horticulture, as well as domestic skills.[38]

Of these schools, Mrs. Bachelder's Industrial School, in Boston, was one of the most interesting and innovative. Like so many educational reformers of the day, Bachelder bemoaned the absence of decent education and decent remuneration for women in the industrial and domestic-related trades. Claiming to speak for ten thousand working women in New England, she told the 1869 Equal Rights Convention in New York City that "industrial schools, not almshouses, should be built, that the talents of women should be far more recognized and that they be accorded the same compensation for their labor that is given to men."[39] She attacked the existing division of labor for reducing male and female workers to machines, confining them to the production of single parts and not "whole" commodities. She proposed to remedy this by teaching girls every facet of clothes design and construction from plain sewing to crocheting. Students would be prepared to make complete products, and even though they might be forced to perform repetitive tasks, they would have a command of all other tasks and "feel the spirit of the place."[40] A hundred women enrolled in Bachelder's school in the summer of 1869, and the Boston Board of Education absorbed her plan into the city school system, putting the industrial school on a par with other public schools. Bachelder's paper before the Boston Social Science Association extolling the virtues of her plan was warmly greeted. By the mid-seventies the American Social Science Association itself championed industrial education along the lines of her original conception.[41]

Professional Working Women: Marital Ideology Strengthened

AS THE economic system assimilated more and more female labor, an important change was taking place not only in the lives of working-class women but in the lives of middle-class women, a change that more than anything else determined the volatile conditions within which a feminist marital ideology might more clearly appear. Women in the industrial work force were insufficient to support the emergence of a new ideology. Alone, their existence would probably have subverted feminism by establishing stronger distinctions between working-class and middle-class women: immigrant, lower-class women worked, but refined women did not. To prevent the hardening of the sexual division of labor along these new lines, feminists were forced to argue in the interests of both classes of women, with support from the new historical conditions.

Reformers pointed to the end of domestic industries, to the passing of older caretaking functions, and to the failure of the domestic sphere to satisfy the productive aspirations of many ambitious and resourceful women. "Previous to the discoveries which have revolutionized labor," Octavius Frothingham wrote in 1867 for the *Herald of Health*,

> the home was the workshop, the manufactory, the mill, and the superintendents and operatives were the women of the households. . . . The daughters of even the wealthiest houses took on themselves the manufacture of a hundred things which now are made by steam or machines. . . . Woman's occupations have been taken away; and no others have been devised to supply the place of them.[42]

Reformers also pointed to the expansion of universal education, to the "opening of the vast regions of the West for settlement and enterprise," and to the growth of democratic institutions that raised the expectations of both men and women. Conversely, they described the unpredictable downswings of the economic cycle that threw husbands out of work and pushed both single and married women into deeper poverty, and to the rigid sexual division of labor that favored the intellectual and social advancement of men.[43] Some, like Francis Walker, saw a contradiction between "artificial restrictions," "tender illusions," and "older prejudices," and an expansionist economy that could easily support more women in the work market.[44] Reformers also observed the impact of such change on the lives of single and married middle-class women: chronic illness, boredom, humiliating economic dependence; absorption of intelligent labor in "primitive directions;" emotional, physical, and intellectual estrangement between husbands and wives; and widening geographical distances between single, active young men and an expanding pool of single women constricted in their movements by social conventions.[45] As Stanton said in 1868, "The advance of opinion on this subject is so great and so constant that we feel it hardly necessary to argue it further."[46] Even Horace Bushnell, in his otherwise shallow book on woman's suffrage, conceded that all the trades and professions should be opened to this class of women.[47]

All these changes, then, caused confusion in the economic lives of women that needed to be confronted either individually or generally if the social system was to reproduce itself without conflict, and after the Civil War it was apparent that a major readjustment was indeed underway in the work arrangements of middle-class women. The voluntary charities that aided working-class women employed middle-class women. Protestant churches too enlarged their organizations to accommodate more female labor in both the foreign and domestic missions. In 1870, for example, the Methodist Church granted its Women's Foreign Missionary Society the authority to dispatch female missionaries to foreign lands.[48] The following year the General Convention of the Episcopal Church created sisterhoods to care for the sick and needy urban poor.[49] To some extent these Protestant undertakings were responding to the incipient Catholic challenge in America as well as to the necessity of finding new employments for "superfluous" middle-class women; they implied respect for the historically effective Catholic organization of female labor. As the *New York World* put it in 1870:

> In the Catholic Church, a woman who deserves well . . . is very sure of being put in a position to make her merit effective and to make it known. She has a choice of enlistment in any one of those corps, some of them centuries old, by which the church succeeds in supplementing the labors of the clergy. And she has the example of canonized and worshipped women to stimulate her to her labors.[50]

Even feminists admired the organizational brilliance of the Catholic Church, its sisterhoods, its hierarchies of subordination and superordination. The feminist social scientist Ella Giles Ruddy wrote in her novel, *Maiden Rachel*, that "a grand work has been done the last fifteen centuries by the Catholic sisterhood; a work which is worthy of imitation by unmarried Protestants."[51] Even Stanton expressed awe at the "wonderful organization of the Catholic Church. . . . In these convents and sisterhoods," she wrote in her diary in 1881, "it realizes in a measure the principles of cooperation. My dream of the future is cooperation."[52]

The growth in opportunities for secular professional advancement also reflected changes in women's work. The expansion of the textile industries fostered a demand for managerial female labor. In the 1860s and 1870s a small group of women managed and owned millinery and fashion stores in major cities and some did so on a large scale. The demand for female milliners penetrated the newspaper want ads for females, hitherto confined mostly to general housework, cooking, ironing, and sewing.[53] Milliners had been and continued to be the "elite among needlewomen."[54] In Springfield, Massachusetts, in 1875, for example, nine of the eleven millinery stores were operated by women; a number of women owned fashionable shoe stores in New York City in the mid-seventies; and by 1877 Pittsburgh was reputed to have the most successful fashion businesses owned by women in the country.[55]

Among the few flourishing businesswomen of fashion, Madame Demorest of New York City was certainly the best known. A married woman with children, she not only owned a large fashion establishment but also published a popular magazine. She contributed to the emergence of the American dress-pattern business and organized one of the first large-scale importing and wholesale firms in this country entirely operated and managed by women.[56] Founded as the Woman's Tea Company, Demorest's importing concern attracted the notice of both feminists and nonfeminists. In 1877 several affluent New York women who had donated considerable capital to the enterprise celebrated the founding of the company with a sail up the Hudson River in a ship named for Madame Demorest. City bankers and merchants were there and the keynote speaker, a Reverend Mr. Deems, observed without a trace of irony that "a ship bought by women, the money earned by women, and sent out by women on a voyage round the world for commercial purposes, would be a sight that prophets and kings ought have been glad to see. . . . Wherever it went, it would be the evangel to preach a new gospel, not the gospel of woman's rights, but of woman's act."[57]

Women who succeeded in business caught the eyes of feminists anxious to prove that women could compete with men in an open market. Like Caroline Dall, many longed to make capable women "the heads of firms, the movers of great undertakings, the contractors of supplies, persons conversant with large interests."[58] The evidence that both single and married women could perform well and consistently in the world of "profitable and respectable" professional labor as well as in business also began to appear after the Civil War, when women pioneers entered careers in journalism, medicine, education, and government.

Social historians have time and again pointed to journalism as one of the earliest and most fertile fields in which women demonstrated their capacity to succeed and persist. Widespread literacy, a vigorous free press, the desire of men to give the reins of "culture" to women, the economic duress of educated women, and, above all, the stunning growth and technological transformation of the publishing and newspaper business after 1860, opened up a wide area (but by no means an equally competitive one) for the employment of women at all levels in writing, publishing, and editing. "Hardly another single profession," according to the *Springfield Republican* in 1875, "has within the last twenty-five years undergone such radical change and rapid development as journalism," an advancement, moreover, that was dependent upon "mechanical invention."[59]

Many married women had already become editors, writers, translators, and publishers for varying lengths of time—among them Elizabeth Oakes Smith, Lydia Maria Child, Sarah Hale, Caroline

Dall, and Amelia Bloomer.[60] With the advent of the new machines, however, many more women took up careers as newspaper and magazine writers. A few, like Mrs. Frank Leslie and Mary Mapes Dodge, rose to top positions with yearly salaries of five to fifteen thousand dollars; others, including Susan B. Anthony, Abigail Scott Duniway, and Jane Cunningham Croly, worked for or owned newspapers operated and managed for the most part by women.[61] Several women in San Francisco incorporated the Woman's Pacific Publishing Company with a capital of twenty-five thousand dollars, just the amount necessary to purchase a modern press. Every facet of the company, from management and superintendence to business agents and typesetting, was handled by women.[62] In 1884 Julia Ward Howe compiled a list of almost a hundred woman newspaper editors, correspondents, and reporters.[63]

The transformation of medicine after the Civil War also created a climate more favorable to women. Three factors, in particular, helped bring this change. The first was the unprecedented and widespread appearance in industrialized countries of chronic female sexual disorders of and diseases of the nervous system that affected both sexes, and the simultaneous growth of a new science of preventive health and therapeutics to treat these diseases considered "peculiar to modern civilization." The second was the reconstruction of medicine into specialized programs within organized institutional structures, a new professional basis intended to order and satisfy the ambitions of middle-class participants. The third was the powerful tradition of female nurture that strengthened the desire of women to become doctors.[64]

Many women (mostly trained in "irregular" medical schools) who became experienced in the study of woman's diseases did so on the grounds that women, not men, should treat other women suffering from sexual and nervous diseases. Female doctors also felt a special affinity for preventive medicine, believing that traditional medical practices had contributed to woman's physical degradation more than to her physical well-being. They needed an opportunity to pursue regular practice. The early phases of professionalization let some women translate desire into action; it produced a liberal climate conducive to the entrance of women into medical schools and it offered the occasion for women to free themselves from their stigmatized, primitive, and disorganized place within the "irregular" areas of medicine.[65]

Before the Civil War only a handful of women had received regular medical training. The rest had studied at irregular and less respected schools. Almost four hundred women had graduated from regular institutions by 1880, and well over a thousand by 1900.[66] Many had successful practices and a few drew yearly salaries ranging from three to twenty thousand dollars.[67] Some cities even appointed a woman as city physician. In 1870 Grand Rapids, Michigan, not only had a woman physician in the post; there were also women serving in the ministry, and a female city librarian and city historian.[68] In Springfield, Vermont, in 1872, one woman acted as city physician and another was superintendent of schools.[69] Indianapolis added a female doctor to the staff of physicians in the City Dispensary in 1880.[70] More than any other group, women doctors provided strong models for other women wishing to enter the field of medicine or, for that matter, any other profession. "I know of no women," declared a writer for the *New Century* in 1876, "who have it in their power to do more for their sex."[71] These women appeared at a time when the medical profession itself had become more respectable and within a context, according to Caroline Dall, "when the heroes of most books are physicians"—an observation confirmed by the publication during the 1860s and 1870s of such novels as Marie Howland's *Papa's Own Girl*, Henry Ward Beecher's *Norwood*, Caroline Corbin's *A Woman's Secret*, Mary Clemmer Ames's *Eirene; or A Woman's Right*, Diana Murdock's *Life for a Life*, May Braddon's *The Doctor's Wife*, Lillie D. Blake's *Fettered for Life*, and Andrew Jackson Davis's *Tale of a Physician*.[72]

Most important of all, and more to the point of this chapter, a good percentage of these women married before or after entering medical school. The female medical students at the University of Michigan who became the wives of physicians in the mid-1870s had decided to become doctors at the suggestion of their husbands or with their complete sympathy; others chose to marry their male classmates without sacrificing career to marriage.[73] According to the *Ballot Box*, 20 percent of all regular women doctors in 1881 were married, and another 6 percent were widows.[74] Women doctors married more frequently in the rest of the century than any other group of college-trained women.[75]

Post–Civil War legislative reform tended to enhance woman's status as a worker in important, if limited, ways; and it did so not only in publishing and medicine but in government, law, and education. During the Civil War the Treasury Department hired on a provisional basis female clerks, copyists, and counters of greenbacks, women soon known as "government girls."[76] Soon afterward, legislative reform fixed women's right to these employments, and by 1875 government at all levels employed them as stenographers, receptionists, and bookkeepers.[77] State and local governments appointed some women to profitable political clerkships, several women became notaries public, census enumerators, and postmistresses. In 1874 alone, 10 percent of all post office positions in Massachusetts were held by women.[78] In addition, legislative reform in many states sanctioned for the first time woman's right to employment in the professions. Illinois led the way in 1870 by formally admitting women to the bar, and the state legislature established a new Committee on Social and Domestic Relations intended to place the state "in advance of all others on the question of protection for woman." Michigan, Iowa, and Missouri also officially approved women's right to practice law; Minnesota and California passed comparable laws.[79] In 1879 Congress passed the Lockwood Bill, permitting any fully accredited woman to practice law before the United States Supreme Court.[80] Only twenty-six women were practicing law in the United States in that year, compared to almost four hundred practicing regular women physicians. Margaret Fuller's contention in the 1840s that no woman would ever practice law had been refuted,[81] but women did not have the traditions or the historical reserves of motivation to challenge the sexual prejudice and competition from men that met them at every point. Notoriously and lucratively related to capitalist enterprise, law was still in the hands of men.[82] Medicine, journalism, and education were not, because they had a more oblique relationship to male capitalist production.

By the 1870s woman's place as teacher and administrator within the educational system had also improved. Men had partly relinquished their monopoly on the teaching profession and women had already constructed for themselves an educational ideology based on the sexual division of labor.[83] In 1860 woman teachers numbered only 20 percent of the work force at the elementary and secondary school levels; by 1880 this number had increased to 60 percent and, in the cities, to 90 percent.[84] Women began to assume positions of authority in the educational hierarchy. Legal reforms in many states had much to do with this change. In 1872, for example, Illinois was in the van again, with a law that permitted women, married or single, to run for school offices; thirteen other states and two territories followed suit a few years later. Women responded to these reforms with immediate enthusiasm. Kansas elected its first female superintendent of schools in 1872; in 1874, Massachusetts elected five; Iowa elected five in 1873, ten in 1876.[85] In 1873 thirty-four women ran for the office of county superintendent in thirty Illinois counties. By 1874 eleven counties had women as county superintendents of schools.[86]

As Sheila Rothman has recently pointed out, the promise of these early triumphs was not fulfilled. After 1880 no other state granted women the right to run for school office, and by 1900 the number of women superintendents had declined.[87] In 1880 several boards of education reacted against married women teaching. Chicago followed a pattern set earlier by Washington, D.C., maintaining that "when a lady teacher married, the Board would consider it equivalent to her resignation." The *Cincinnati Commercial*, a paper with feminist sympathies, was indignant when Cincinnati considered imposing a similar restriction. Widows, it observed, could teach, "but the quiet, domestic wife, she who lives with her husband, and takes care of his buttons and his dinner, thus setting a model example to all the daughters of Eve—no, she shan't have a school. Husbands are an ornament not to be allowed to women who work."[88]

As important as the laws that expanded womens' employment was the continuous reform of the marriage laws in nearly every state and territory from the 1840s well into the latter part of the century. This reform reflected attempts by the bourgeoisie to order married life and the sexual relation—the whole private sphere—according to the shifting priorities of public life under capitalism. It helped significantly to create conditions favorable to the emergence of a feminist ideology of married life.

In the antebellum period several northern and western states adopted what one writer called the "separate property system."[89] Maine led the way, followed by Massachusetts, Pennsylvania, Connecticut, New York, Illinois, California, Wisconsin, Kentucky, Iowa, and Ohio. They exempted the real and personal property of married women from attachment to their husbands' debts, permitted married women to own real and personal property and to make separate contracts with people other than their husbands, allowed them to make wills and to dispose of property in the same manner as men, and to some extent freed husbands from liability for their wives' prenuptial debts. Kentucky, New York, and California covered the most ground. Kentucky, in its revised statutes of 1852, also permitted married women to keep the earnings of their personal labor and the profits of any trade or business.[90] The New York law gave women the power to trade and to make contracts alone, to accumulate property through real estate and commercial ventures, to receive legacies and to bequeath them in their own names, and to earn their own wages and spend them as they saw fit. *Woodhull and Claflin's Weekly* commented on these early New York reforms: "There are now few rights which woman does not possess equally with man and there are even many privileges possessed by her of which he is deprived."[91] Under a California law of 1852 married women were also granted the right to do business under their own names to keep their own earnings, a reform that resulted, according to one observer, in newspapers teeming with notices written by women advertising stock raising, farming, blacksmithing, carpentry, and bricklaying.[92]

After the Civil War marriage reform proceeded with speed, provoking Joel Bishop, a popular nineteenth-century historian of the law, to write that "the cheek of the idol [of women's subjection] has fallen to the ground. The huge idol will sooner or later be broken into pieces." By this time thirty states had separate property systems for both sexes.[93] During the Reconstruction years and under Radical Republican regimes, several southern states—notably Georgia, South Carolina, and North Carolina—instituted legal changes that followed to some extent the pattern already established in the North. These laws exempted the property of women from liability for their husbands' debts.[94] In the northern and western states the provisions of the marriage laws expanded, giving women an increasing number of rights. Even feminists had little to quibble over. "In this country,"

Eliza Duffey wrote in 1876, "women possess an independence with accruing advantages, not to be found in all its depth and breadth anywhere else in the world."[95] In 1871 the governor of Massachusetts made recommendations to the state legislature about women's marital rights that "marked an era in the history of our executive department."[96] In 1874 Massachusetts granted women the right to contract, to make promissory notes or mortgages, and to sell real estate on the same grounds as men; this strengthened provisions that already allowed women the right to their own earnings and to sell personal property sole.[97] Of these sweeping reforms Thomas Higginson wrote in 1874 that they

> did palpable injustice to men. . . . It is unjust that the wife should be free as the husband to transact separate business and yet shall have the power to make him responsible for her debts, when she is not responsible for his. It is unjust that she should carry his credit with her, when he does not carry her credit with him. It is unjust that, however unsuccessful he is, he and not she, should be legally liable for the support of the family.[98]

In 1878 the Massachusetts House of Representatives narrowly defeated a bill that legalized contracts between husbands and wives.[99] Oregon, Illinois, Nebraska, New Hampshire, and Connecticut passed laws entitling women to separate estates and separate earnings.[100] By 1891 Joel Bishop could write that

> almost everywhere . . . statutes have preserved to the wife her antenuptial property, and entitled her generally or in special circumstances to her earnings, and imparted to her some power of contract. Through a not unnatural inadvertence, this class of legislation has in some localities been carried so far as practically to reverse the parties, making the wife the master and the husband the servant.[101]

California established the most complete legal equality between the sexes in marriage by freeing them to contract with one another and to "owe each other the same as if they had never been married;" and in 1873 a Nebraska lawyer (wrongly) told the National Woman's Suffrage Convention in Omaha that Nebraska had been the first of all the states to uphold the power of the wife to make contracts with her husband and to enforce them against him in her own name.[102]

In varying degrees these marriage reforms subverted the older common-law system of paternalistic and protective law based on fixed precedents, as well as the older system of equity law based on principles of natural justice and abstract principle. They show the progressive and pervasive invasion of the expansionist market and of its legal handmaiden, the free contract, into the social sphere.[103] Thus the authors of the first volume of *The History of Woman's Suffrage* declared in 1880 that "the change from the old Common Law of England in regard to the civil rights of women, from 1848 to the advanced legislation in most of the Northern States in 1880, marks an era both in the status of woman as a citizen and in our system of jurisprudence."[104] The modern Marxist historian of the law, Morton Horwitz, has written of the rise to hegemony of the contract in the legal thought of the nineteenth century. "Only in the nineteenth century," he observes, "did judges and jurists finally reject the long-standing belief that the justification of contractual obligation is derived from the inherent justice or fairness of an exchange. In its place, they asserted for the first time that the source of the obligation of contract is the convergence of the wills of the contracting parties."[105] These reforms had strong feminist implications, exemplified in some states by the legalization of contracts between husband and wife, by the widening of the grounds for divorce, and, above all, by the granting to married women the right to do business in their own names and to keep their earnings.[106]

It would be a mistake to assume that most state legislatures, especially those in the South and in some northern and western states, reformed the marriage laws with feminist intentions or that the older patriarchal common law had ceased to operate in the area of the sexual relation as it had in other areas of economic and social life. Marriage and divorce reform in many states may have been one of the earliest state attempts to regulate as well as to liberate the expansionist tendencies of the market and thus establish an ordered context for capitalist venture and expectation.[107]

Judges, in particular, who had unprecedented and immense power in the nineteenth century and who tended to quash the feminist orientation of legislative reform, placed great emphasis on the protective rather than the enabling aspects of the married women's property acts. "Their object," an observer familiar with judicial decisions wrote in the 1850s, "is to *protect* married women and to secure their property from improvident and dissolute husbands who might squander or appropriate it to the payment of their own debts."[108] Stanton, Gage, and Anthony recognized this object in 1880 when they said of the early laws that fathers who bequeathed their estates to daughters "could see the advantages of securing to woman certain rights that might limit the legal powers of profligate husbands" and "husbands in extensive business operations could see the advantages of allowing the wife the right to hold property that ought not to be seized for his debts."[109] Many nineteenth-century judges—such as the influential Joseph Story, who contributed to the liberation of the private sphere from the constraints of public welfare—insisted upon distinguishing marriage from all other contracts; as the "basis of the whole fabric of civilized society" it could not, unlike other contracts, be dissolved by mutual consent.[110] Regardless of the instrumentalist bias of the law, many judges persisted in applying patriarchal principles of common law to the status of women within marriage. Chief Justice Lowrie of Pennsylvania, for example, paid no attention whatsoever to the phrase "to secure the rights of married women" as it appeared in the language of the 1848 statute; he not only read the act as "protection" but also claimed that a married woman did not hold her property as a person unmarried, but only as if "it were settled to her use as a *femme couverte*. Accordingly, the common law is not removed."[111] Thus the *New Century* could observe of the Pennsylvania reforms: "The laws concerning separate estates of married women have reached some degree of enlightenment in Pennsylvania, so far as the letter of the law is concerned—but the judicial decisions, heretofore, have so construed the statute as to consider her husband as her trustee."[112]

Although the laws permitted married women to earn their own living and to keep their own earnings during marriage, they did little or nothing to alter the sexual division of labor within marriage. In 1873 Mary Williams reminded readers of the *Woman's Journal*—in an article called "No Equal Rights For Wives Anywhere"—that the law still failed to acknowledge the money value of woman's household labor. The wife has a legal right to subsistence from her husband, but she "is not expected to receive wages for what she does in the house of her husband."[113] Of related and greater significance, the marriage laws conflicted with a much broader sexual division of labor that made, in Isabella Beecher Hooker's words, "every woman a consumer and every man a producer."[114] On the one hand the progressive features of the market, captured to some extent in the marriage laws, enabled women to work outside the home; on the other hand the demands of market, coupled with the weight of cultural stereotypes, tended to fix women in the status of consumers and to deepen their attachment to the domestic sphere.

These qualifications aside, the marriage laws taken as a whole had real feminist content. Contradiction compelled feminists to find ways to resolve it. Feminism was not antithetical to marriage reform; it expressed both the individualistic and enabling tendencies of reform as well as the protective and restrictive ones. Free-love feminists and pro-divorce feminists replicated the line of thought embedded in legislative reform: they emphasized equal rights and tended to define marriage as a contract like any other contract or as one entered into and potentially dissolvable through the exercise of the individual wills of both parties. At the same time, however, feminism objected to the atomistic, individualistic circumstances fostered by socioeconomic development and sought to reduce the conflict between the sexes induced by these circumstances and exacerbated, to some extent, by the marriage laws themselves. To put it another way, bourgeois feminism can be viewed as protective and restrictive in its own right. Feminists therefore attacked the institution of common-law marriage that permitted men and women to establish conjugal unions without constraint or without formal intervention of the state; some feminists worked for a homogeneous divorce law that would do equal justice to both sexes while simultaneously removing the law's inefficiency and contradictions; and most feminists sought to eliminate marital risk by legally stopping the physically, morally, and intellectually unfit from marrying and by creating egalitarian, institutional settings wherein young men and women might more rationally and intelligently make marital choices. Feminists wanted to construct a scientific, egalitarian communal life that gave new expression to the protective, harmonious spirit of the older common law.

Feminists Organize for Reform

FEMINISM was not a fixed ideology of equal rights, but a multi-faceted egalitarian ideology determined and limited by its historical matrix. It sought to give shape to these historical conditions. Feminists told other women about marriage reform and they also helped to introduce it. By 1861 nineteen states had passed new laws, inducing Caroline Dall to write that "the credit for this change should certainly rest with the women and men of this reform, for in every state, its sympathizing friends helped to frame the laws."[115] After the war the *Woman's Journal* observed that "we have effected a general reformation in the laws that concern Woman as wife, mother and widow."[116] Feminists continuously memorialized the state legislatures, and in 1873 they addressed memorials to every legislature that had failed to introduce any reform.[117] They took their case directly to the legislatures and even drafted some of the laws.

In 1869 several influential members of the Illinois Woman's Suffrage Association—including Mary Livermore, Kate Doggett, James and Myra Bradwell, Judge Charles A. Waite, and Rebecca Mott —organized a committee that traveled from Chicago to Springfield to plead before the legislature that the earnings of women be "protected from the rapacity and brutality of drunken husbands." "In ten days," Mary Livermore reported, "the older law was stricken from the statute books, and, today, the married women of Illinois can control every penny of their property."[118] When Ann H. Connelly, of Rahway, New Jersey, was prevented from keeping her only child after divorcing her husband, she persuaded the New Jersey legislature to pass a law equalizing parental ownership of children.[119] The laws of California, Massachusetts, and Connecticut also displayed the signs of feminist politics. Under the guiding touch of his feminist wife Sarah, William Knox, a state senator in California, drafted and pushed through passage an 1872 bill granting married women the right to dispose of their separate property by will.[120] Governor Claflin of Massachusetts recommended marriage reforms to the legislature, reflecting the influence of two powerful feminist journalists, William Robinson and Samuel Bowles, and in 1874 the laws were significantly altered. Governor Hubbard of Connecticut supported reform in his first message to the General Assembly in 1877; by the end of the year Connecticut had a new law securing equal property rights to married women.[121]

In the late 1860s and early 1870s many feminists, especially those of the National Woman's Suffrage Association who, like Stanton, rejected the sectarian timidity of the Boston reformers associated with the *Woman's Journal*, could be found in the vanguard of support for single working-class women.[122] They backed the construction of industrial and trade schools and played an important role in organizing working women.[123] Several women printers, under the leadership of Anthony and Stanton, established the Working Women's Association, designed to "act for the interests of its members, in the same manner as the association of working men now regulate wages of those belonging to them."[124] The Working Women's Association attracted —indeed, was intended to attract—mainly skilled workers, and soon several hundred young single women aged fifteen to twenty-five were attending regular monthly meetings. Anthony and others also supervised the formation of two unions for women—the Sewing Machine Operators' Union and the Woman's Typographical Union. Anthony said, "You must take this matter at least seriously now, for you have established a union and for the first time in woman's history, you are placed, by your efforts, on a level with men, as far as possible."[125] Stanton and Anthony labored to bring women into the ideologically sympathetic National Labor Union, the most advanced union of skilled workers in the country, and to forge a political alliance with it.

These feminists formally adopted as their own the platform of the National Labor Union, stressing the duty of the state to grant and secure the rights of the whole people; they demanded a system of paper currency, equal distribution of property, and the end to landed monopolies; they demanded the eight-hour day and the creation of working men's trade institutes, lyceums, and reading rooms "as a means of advancing their intellectual and social improvement."[126]

In the early 1870s the NWSA feminists' pursuit of a political alliance with labor, and with other elements unrepresented in the established political parties, continued, although this attempt marked a movement away from the explicit goal of improving the lot of working-class women. In 1872 feminists of New York City sought to unite under the banner of the "People's Party" with labor reformers, members of the International Working Men's Association, prohibitionists, and spiritualists. Feminists, Stanton wrote to Isabella Beecher Hooker in 1872,

> have culture, refinement, social influence but no political power. The Labor and International meetings have the votes and the political power if organized: but lack all we have: now put the two together and see what a combination we have. . . . Our platform will combine the Labor Reformer's demand for paper currency, "land for all" and woman's suffrage: these will be our three grand planks.[127]

The People's Party's published call attacked the major political parties, charging the government with "financial and military despotism," oppression of "millions of people condemned to continuous servitude and want," landgrabbing, and "wholesale robbery of the people's wealth."[128] Moreover, as to the alliance itself, Stanton even suggested, in a consciously sensational maneuver to seduce allegiances from the established parties, that it might give women the power to rise up in the spirit of Charlotte Corday if they were not legally granted their political rights. "It may be," she wrote to the *Golden Age*, "that this step in progress is to be achieved not by argument but by blood, and *that* bolder hands shall hew short, straight paths for womanhood."[129]

These feminist ventures, however, and especially the specific efforts to combine forces with working-class women, ended in dismal failure. Historians have assigned various reasons for it, but clearly one of them was the conflict between bourgeois feminist consciousness and the existing social and economic conditions of working-class women and men. In the post–Civil War period the composition of the working class was changing; it consisted increasingly of uneducated immigrant groups and less of native men and women with whom feminists could more easily identify. Apart from the fact that Stanton, Anthony, and others knew that real power lay in the hands of the Democratic and Republican parties and would have allied themselves with either one had it embraced woman's suffrage, and apart from the fact that they did respond compassionately to the exploitation of the working class and did tend to see the status of women in relation to the general material conditions of life, these feminists failed to consider objectively the interests of women they intended to defend.

As Ellen Dubois has shown, most feminists did not understand the particular character of wage labor. More often than not they described all nondomestic labor, "even the veriest drudgery," as "enobling"—a polemical point, since feminists wanted to get women of "culture and refinement" into the professions, not into the factories.[130] Furthermore, feminists focused not only on professional work but also on marriage and suffrage and other matters that, in the minds of many women, bore little or no relationship to the concrete economic problems of working-class women. Looked at from another angle, however, these feminists ventures into labor politics produced positive results. Through these unstable and abortive alliances, feminists learned much from working-class women about the meaning of work in the same way that they learned about "self-ownership" from the free-love anarchists who represented a continuously troubling but necessary left-wing element in the forging of feminist consciousness. Feminists gained greater insight into the economic plight of bourgeois women—the pervasiveness of an unequal wage scale, the effects of long hours on health, the lack of remuneration for domestic labor, and the limited number of employments open to women. It was in relation to women different from themselves that they developed a deeper understanding of their own position in society and of their own need to unify among themselves for social, economic, and political action.[131]

In the midst of their short-lived quest for allies in the working class, feminists began to organize middle-class women into sisterhoods committed to the professional interests of women. Although since the 1850s feminist leadership had worked for the expansion of employment opportunities, condemned the unequal wage scale, and encouraged women to enter the professions, only after the war did feminists organize.[132] The Working Women's Association fizzled as an organization for working-class women but prospered as one for the middle class, attracting professionally active women.[133] It paved the way for the formal establishment of professional woman's clubs in major cities like New York, Boston, Chicago, and Washington.

Founded in March 1868 by Jane Croly, Charlotte Wilbour, and

Alice Cary and originally called the "Blue Stocking," the first of such clubs, the New York City Sorosis, attracted many New York women active in some kind of professional labor, among them some of the most important feminist free-thinkers and spiritualists of the day.[134] By 1870 it boasted thirty-eight writers, six editors, twelve poets, six musicians, four professors, two artists, one historian, nine teachers, ten lecturers, three philanthropists, and two physicians.[135] Sorosis almost immediately dissociated itself from the Working Women's Association and described its purpose as the advancement of women's professional interests.[136] It served a function for professional women unmatched by any other male or female group of the time: it praised the accomplishments of individual members at almost every meeting, congratulated members when they married, proffered condolences when they became ill or when family members died, reprobated gossip and slander, and generally cultivated a spirit of friendship, fidelity, and unity among its members.[137] William Alger's *The Friendships of Women*, in fact, represented the club's principal text on that subject.[138]

In a speech to the club Jane Croly gave the strongest explanation for the existence of Sorosis: impetus for the "employment and incorporation" of women into state affairs that the state itself did not provide, by

> creating a system of rewards, of recognition of merit in women. . . . Every boy born in America looks forward to the possibility of personal distinction. . . . But women have nothing of this sort to anticipate. The household is their only acknowledged sphere and in it there is neither reward nor promotion. The more strictly and conscientiously a woman fulfills her duty, as laid down, the more narrow and contracted her life becomes. Marriage does not alter this condition of things. It is the refuge of some, but it has been the grave of many clever women, especially those who have married clever men.

Croly later explored this position at greater length in her popular book *For Better or Worse*, describing in painful detail the dreariest features of contemporary marriage: girls forced by circumstances to marry for bed and board and economic security, or worse, brilliant and resourceful women condemned by marriage to dependence, seclusion, intellectual sterility, and even madness. The picture is so bleak that marriage, even on feminist terms, emerges as a risk too dangerous to be taken.[139]

The Sorosis took only the first step in the feminist organization of professional middle-class women. The next step, and a faltering one, came in 1869, when Croly, Fanny Fern, Mary Davis, Laura Curtis Bullard, and "other ladies of high social standing" organized the Woman's Parliament in New York City. Croly gave the opening address and, like many feminists at this volatile moment in the history of feminism, denied the central importance of suffrage for women. "The ballot," she said, "is at best only one agency; it cannot do everything. It does not do everything for men." What we need now, she urged, is the "combined efforts of intelligent and active women in their own organizations to work for humanity."[140] The parliament set up committees on education, household reform, health and social reform, newspaper and magazine work for women, and so on. Although it lasted only a year, it was resurrected again in a more sturdy incarnation in 1873, with the establishment of the Association for the Advancement of Women.

Occasionally labeled the "Ladies' Social Science Association" and convened as the Woman's Congress during the 1870s, 1880s, and later, the AAW was organized on the initiative, again, of the New York City Sorosis. Paulina Wright Davis had the idea for a Woman's Congress in 1869. "This congress would be a school," she wrote the *Revolution*, "to prepare women to act with men in the future" and, being a body made up entirely of women, it would encourage its members "to dare to utter their bravest thoughts."[141] It was not until 1873, however, that the first congress was held.

In that year Charlotte Wilbour, president of Sorosis and a member of the NWSA, sent out a call to women to come to the first meeting in New York City and join the AAW—"all women," she emphasized, "who by voice and pen or practical work, have conquered an honorable place in any of the professions or leading reforms of the day."[142] We must "combine," exclaimed one congress woman, and not squander "our strength . . . in solitary struggle!" How else could we expect to "influence masses of women in different centres . . . in our large cities?"[143] Julia Ward Howe insisted that "the great segregation in which women have hitherto lived" must end. Women must exchange "ignorance" imposed on them by the "opposite sex" with knowledge "of the great body of womanhood."[144]

The response to the first call was overwhelming. It drew an overflow crowd—four hundred women from eighteen states—that spilled into the halls and stairs outside the densely packed convention room.[145] Equal numbers of active middle-class, upper-middle class, and upper-class women poured into later congresses.[146] Two of the wealthiest women, Howe and Kate Doggett, frequently served as presidents. Doggett married one of the leading Chicago merchants, who left her a sizable fortune after his death in 1876. In 1869 the NWSA Women's Bureau financed her trip to Germany as one of its delegates to the Woman's Industrial Congress in Berlin.[147] In the same year she joined the executive committee of the Union Franchise Association, an ill-fated organization founded by Theodore Tilton to reunite the American Woman's Suffrage Association and NWSA. Doggett also created the first Chicago Woman's Club in 1873, and joined and supported the Chicago Philosophical Society, a progressive-scientific study group that first met in the home of the wealthy feminist Fernando Jones, in 1874. Sarah Hackett Stevenson, Helen Shedd, and Professor Emily Chapin (all feminist social scientists) came to the first meeting, and so did several female students from the University of Chicago. Howe and Nancy Swisshelm managed to find time to visit this society.[148] A powerful figure in Chicago reform circles and a freethinking woman, Doggett habitually sent expensive gifts to her reform friends during vacations in Europe. She mailed Susan B. Anthony an "exquisite gold and ruby pearl broach" from Rome.[149]

Daughter of a rich banker and wife of the humanitarian reformer Samuel Gridley Howe, Julia Ward Howe acquired a passion for social science in the late 1860s and early 1870s, attending the meetings of the Boston and American Social Science Associations. She blossomed as a feminist at the same time. In 1873 she signed the call for the First Woman's Congress, but did not "hope for much unless this should be organized with much care and order."[150] For nearly a year she continued to doubt the usefulness of the association, but once enlisted into its executive ranks, she put her doubts aside. In 1875 she wrote the call for the congress, and on a European tour two years later she regretted that she was not at home organizing for the professional advancement of women.[151] In 1882, at a Woman's Congress in Portland, Maine, she spoke with pride of the Association for the Advancement of Women. "At first the cry was: why do you try to do anything? It is now: why don't you do more?"[152]

The Association for the Advancement of Women encouraged women to enter the professions and specialized scientific fields and to do so through the front doors of respectable institutions of learning untainted by quackery. In 1876 the AAW's journal, *New Century*, declared that every sensible woman knew "the value of special lines of work" and did not pretend "to grasp the entire circle of the sciences." The sensible woman also knew "that there are enough quacks and frauds in the world already, and every day brings us more conclusive proof that not only *training but proof of it*, is requisite, before one may speak as with authority, in any profession, or branch of trade."[153] In Mary Putnam Jacobi's words, not only would "the fellowship of common membership in a profession afford a model of association far better suited to modern ideas," it would also demonstrate to mankind that women had the brains to take their places in the political governance of society.[154] Unfailingly, therefore, every Woman's Congress during the 1870s and 1880s made a special place on its program for papers dealing with professional careers for women in architecture, law, journalism, medicine, museum work, statecraft, banking, music, and other areas.[155]

At the 1876 Centennial Exposition, in Philadelphia, feminist women set up a Woman's Pavilion that gave concrete illustration, perhaps for the first time in the United States, of the productive abilities of women. Historians have not noticed that feminist social science women built and supervised this pavilion and published their own newspaper, the *New Century*, for it. Published for only a year, the *New Century* was also in effect the organ of the Third Woman's Congress, which convened in Philadelphia the same year. Such women from the AAW as Elizabeth Churchill, Antoinette Brown Blackwell, Anna Garlin, Kate Doggett, and Jane Croly wrote articles for this paper; Sarah Hallowell, wife of a wealthy Quaker stockbroker and a member of the Pennsylvania Woman's Suffrage Association, coedited it; many other AAW women helped with the exhibits.[156] Both the pavilion and the paper gave graphic testimony to the existing weaknesses and strengths of woman's work. If the Woman's Pavilion

> has permitted the entrance of many petty prettinesses, it is for the good of the senders thereof that they may measure these trifles with the dignified and substantial exhibits of others. And as even the wax-fruit and embroidery stand for the living of these women, their means of honest support, we fail to see the objection to it as an exhibit of industry, save as it sets forth, in unmistakable speech, by how few methods hitherto has a woman been permitted to make a living at all.[157]

The Pavilion did more than display "petty prettinesses;" it exhibited the inventions of women—a fountain griddle greaser, a patent stove, a hand attachment for a sewing machine, and other related appliances. As such it represented only one display in succession of displays of women's industrial work that were typical of the industrial expositions of American cities in the 1880s and, earlier, of Paris, Berlin, and Florence—the Industrial Fair in Florence being suggested by the Italian feminist Salvatore Morelli "to learn what the actual condition of woman's work was in Italy."[158] In subsequent issues the *New Century* presented article after article on new work opportunities for women and on the women who pursued them, in the trades and handicrafts, in artistic work, in the scientific and learned professions, and in education.

AAW women confirmed indisputably Elizabeth Stanton's contention in 1870 that one could distinguish "strong-minded" women from all other women by their endemic commitment to the reform of "education, asylums, hospitals, and sanitary arrangements."[159] Such women promoted the inclusion of professionally trained women as managers and administrators in the bureaucracies of every reform institution. At the Syracuse Congress in 1875, for example, the wealthy Rhode Island Quaker Elizabeth Chace, a feminist who, according to Caroline Dall, owned "the whole town of Valley Falls and the whole water-power of the Blackstone River" in Rhode Island, advocated the election of women as equals with men to the boards of state charities, boards of inspectors, and boards of trustees of reformatories.[160] And in her report to the 1876 Congress, Ellen Mitchell wrote that women's prisons, reform schools, and children's aid societies should be staffed and run by women. The president of that congress, Kate Doggett, proclaimed on another occasion that women "should be found on all State Boards of Education, in the Management of Asylums for the Insane, in Reformatory Institutions," and

Dr. Julia Holmes Smith of Chicago urged that women share equally in the supervision of hospitals.[161] Even more startling was the Centennial Committee of Women's proposal that the committee itself be converted, after the demise of the exposition, into a sanitary body with expansive city-wide powers. The committee had spent three years in Philadelphia preparing for the exposition; it had examined all twenty-six wards of the city. Its members had traversed every street and peered into every neighborhood. "In matters pertaining to the health of the cities," the committee inquired, "what body is so competent to watch over the hygienic conditions, to report nuisances, and insist on the observances of sanitary laws, as the Centennial Committee of Women?"[162]

Some women justified these demands for integration on the grounds that women were more temperamentally suited for bureaucratic work than men. Caroline Dall quoted approvingly the words of the French feminist Ernest Legouvé: "Why should not the immense variety of bureaucratic and administrative employments be given to women? . . . For all this woman is far better prepared than man. Her eye is quick; her common sense ready; she sees the consequences of a course and does not need to debate long-disputed points."[163] More often, however, social science women took the position—one they would cling to tenaciously for the remainder of the century—that women's older caretaking and educational functions had been plucked from the home by the state, that social institutions as well as the state had acquired a domestic complexion, and that women, therefore, should follow their functions out of the home. As May Wright Sewall, one of the directors of the AAW in the late 1870s and an active member of the NWSA, wrote in 1881, "The state has assumed a domestic meaning" —homes once served the state, now the state obeys the imperatives of the "Home." Moreover, "whereas at the outset, most laws were prohibitory, most laws are now permissive," dealing "tenderly" with "all the interests of humanity."[164] Sewall drew the obvious connections. "Life" itself had determined the entrance of women into the public institutional life of the nation.

CHAPTER 8

THE SEXUAL DIVISION OF LABOR AND ORGANIZED COOPERATION FOR PUBLIC LIFE

"THE INEQUALITY of women finds its origins in marriage," said Sarah Norton, a middle-class feminist and member of the Working Women's Association. "To make political equality possible to her, social equality of the sexes must precede it; and as marriage is the backbone of social life, the backbone of social life must be broken."[1] The editor of the *Revolution*, Laura Curtis Bullard, wrote:

> We hold that the only means to prevent a floodtide of license from sweeping through society is to elevate marriage. The highest aim of our movement is to fit man and woman to stand as equals in this relation. . . . The ballot is only one of the agencies for her elevation. Although it can, and will, accomplish great and important results, it cannot secure to women all the social freedom which she demands.[2]

Paulina Wright Davis, a leader and a historian of the women's rights movement, read a resolution at Apollo Hall in New York City that represented the views of a large minority of feminists:

> The evils, sufferings and disabilities of women, as well as of men, are social still more than they are political, and a statement of woman's rights which ignores the right of self-ownership as the first of all rights is insufficient to enlist the enthusiasm and even the common interest of the most intelligent portion of the community.[3]

For many middle-class women, speaking at conventions in the 1870s and earlier, writing articles and books and joining organizations of overlapping memberships, the central question to resolve was the inequality and instability of married life.

In the mid–1870s the members of the Association for the Advancement of Women pursued this course in a muted voice, within the shifting limits of permissible discourse, but steadily all the same. The first congress dealt with such questions as "equitable monetary division between husband and wife" and cooperative households. At later congresses members discussed the organization of domestic labor and the concept of two careers for married women—marriage and motherhood combined with a public career. "Respectable" feminists began to consider more deeply the reconstruction of marriage and the domestic life, paving the path for the theoretical work of such feminists as Charlotte Perkins Gilman at the end of the century.

This consideration joined individualism with cooperation and organization. On the one hand feminists emphasized women's right to an economically autonomous position within marriage, even to the point of married women taking on two careers. On the other, feminists sought to organize women's working experience, both within and without marriage, according to cooperative principles. Like many industrial and agricultural reformers and activists of the 1870s, leading feminists did not seek to liberate the worker simply into an unregulated, competitive market. In place of individual competition, they increasingly put cooperative industry, cooperative sharing, and national organization.

The Backbone of Social Life

FOR THOSE WOMEN "who surrender" themselves "wholly to domestic life," as Elizabeth Cady Stanton put it in 1880, "in order to make it possible for the husband to have a home and family," decent, egalitarian compensation had to be given.[4] Nothing distressed feminists more than traditional exercise of paternal authority in the home. To break its grip (already weak and more readily challenged), male and female feminists demanded absolute equality within marriage.

"Equality with me," Henry Blackwell wrote to his future wife, Lucy Stone, "is a passion. I dislike equally to assume or to endure authority;" and: "Marriage is degraded by our statutes from the noble and permanent partnership of Nature with reciprocal rights and duties, into an unnatural mercenary relation between Superior and Dependent."[5] "The only mastery in love," observed Samuel Blackwell, husband of Antoinette Brown Blackwell, "inheres alike and equally in both the allied hearts. The insulting idea of headship should be exorcised as an alien thought."[6] "There is no need for mastership on either side," Jane Croly wrote in *For Better or Worse*, "in fact it cannot exist with equality in marriage."[7]

As historians of feminism have recently shown, feminists strengthened their case for equal "mastership" by appealing first to the slave analogy in its plantation form and later, after the Civil War, to its proletarian form.[8] "Beyond the question of black slavery," wrote Matilda Joslyn Gage in 1870, "lay . . . a deeper moral question in the 'subjection of woman,' for it spreads over a wider surface, sinks deeper into the national life, and upon it is built not only the physical well-being of the whole race of man, but the intellectual progress of humanity."[9] Feminists saw in the slavemaster, and later in the capitalist, the figure of the economically rapacious patriarchal husband who robbed his wife of her just remuneration, threw her into extremes of "parasitical" degradation, bound her to narrow patterns of behavior, and prevented the "growth, maturity, and health" so typical, they thought, of egalitarian relationships. Of the rapacious patriarch in his capitalist form, the Third Decade Meeting of the NWSA agreed unanimously in 1878: "Man, standing to woman in the position of capitalist, has robbed her through the ages of the results of her toil."[10]

To achieve egalitarian marriages, feminists looked in part to industrial principles and technologies. They imagined efficient, well-ordered homes based on domestic science and the use of new household appliances; the application of physiological principles to home life and the exclusion of delinquent, irascible, and "half-barbarian" servants;

varied domestic architectural forms such as the apartment, flat, or boarding house; and the rational use of time. All these were intended to remove the burden of drudgery and to shorten labor time, permitting the symmetrical development so necessary for health. One of the most articulate spokeswomen, Antoinette Brown Blackwell, put it this way: "Any woman of average health and ability" can and should "pursue systematically together . . . as many as three hobbies," while still remaining a "good mother, good housekeeper and good student." She told the First Woman's Congress that every married woman should have "three to six hours of systematic leisure . . . at her disposal."[11]

Economic autonomy required that married women be given complete authority to administer the domestic sphere ("It is her world, her domain"); it required that their work be considered an "inestimable . . . productive industry" on a par with the work of their husbands. "Until this woman's work," wrote Stanton, "is recognized as a great, and the most important of human work, it will not come into true relation with the other callings."[12] "The theory that a wife who . . . bears her fair share of the joint burdens," Blackwell wrote in 1869, "is yet 'supported' by her husband, has been the bane of all society. It has made women feel that it is their right to be dependents and nonproducers. Thus, for women, all work has until lately come to be regarded as degradation."[13] Feminists argued that married women should receive pay as other "industrial workers" did for their own labor. They believed that unremunerated labor or the symbolic absence of a wage, coupled with the myth of male "protection," produced nonproductive dependency and idleness.

"All work," wrote the editor of the *New Northwest*, Abigail Scott Duniway, "becomes oppressive that is not remunerative."[14] In the early capitalist period, when the economic lives of men and women were not yet completely separate from or invisible to one another and when labor had yet to become a fungible commodity, money did not have the power to represent the status of labor or the difference between the sexes. The concept of *femme couverte* had not shown itself for the contradiction it would become. By mid-century, however, and especially after the Civil War, money made most male labor respectable. It rendered invisible woman's work in the private sphere and, by blinding both men and women to the value of domestic work, sustained the pretext of male "protection." To make woman's work "visible" once again, feminists had to attach money value to female labor in the home, and thus to demystify the "myth of protection." "The estimated worth of a thing," Helen Jenkins told readers of the *New Northwest*, "is its money value."[15] Wives without an independent source of income were vulnerable to beggary and manipulation.

"Of all the little foxes," wrote Lucy Stone for the *Woman's Journal*, "that help to destroy the domestic vines . . . the worst is that which makes it necessary for the wife to *ask* her husband for money to supply the daily recurring family necessities and her own," thus feeding her "sense of humiliation, degradation and separation."[16]

In the 1850s the argument for compensation had become widespread. Writing for *Una*, Paulina Wright Davis advised beleaguered housewives and their husbands to "share everything. Extend to one another the same material and pecuniary rights." Elizabeth Oakes Smith wrote in her book, *Woman and Her Needs*, of the "hundreds of women who would rather go without money than ask for it . . . They feel mean and childish to have it doled out to them in little sums." Sarah Ernst, an abolitionist from Ohio, declared in the *Anti-Slavery Bugle* in 1853 that "the wife should have on half the profit of the joint stock company of marriage. Let every woman have a separate purse."[17]

After the war several feminists said they would no longer ask for a separate purse. They described husbands as "wage robbers." They considered the possibility of woman's banking institutions in which married women might make money from interest as well as protect themselves from philandering husbands. Most important, they demanded that the state legalize a married woman's right to a wage. In 1878 the women of the National Women's Suffrage Convention formally resolved that the "right of woman to the proceeds of her labor in the family" must be recognized by law. These women appealed again, in the first volume of the *History of Womans Suffrage*, for a "law securing to the wife the absolute right to one half the joint earnings" and denied the myth that women "are destitute of the natural human desire to accumulate, possess and control the results of their labor."[18] A few women, including Elizabeth Osgood Willard, even proposed that the state secure legally a home for every woman, married or single, to fix woman's independence and to encourage men, not women, to marry for domestic security.[19]

Of greatest significance to the discussion that follows, feminists began to repudiate the idea that married women should rely at all on their husbands for an independent source of income. "We don't want the purse of our husbands, brothers, parents," one feminist wrote for the *Woman's Journal* in 1875. "That is just what independent minded women rebel against; we want an independent way of earning honestly our own supply. . . . We want every woman to be pecuniarily independent."[20] The concept of two careers had begun to emerge.

Two Careers

THE NEXT important feature of the feminist economic critique of married life was its attempt to combine marriage and motherhood with a public career. It has often been claimed, on the basis of considerable data, that when it came to the matter of two careers, feminists failed to challenge the sexual division of labor. Thomas Higginson, one of the editors of *Woman's Journal* in the 1870s and often a president of the American Woman's Suffrage Association, thought he spoke for all feminists when he declared in 1871 "that we all recognize" that no woman should "be self-supporting at all during her career of motherhood," even if such a career stretched into decades; and ten years later he said that the "advocates of Woman's Suffrage" consistently assigned the outdoor work of life to one sex and the indoor work to the other.[21] Hardly a feminist of this generation, in fact, denied that mothers should not work during the years of childbearing and child rearing.

There was, however, no unanimous feminist capitulation to the sentimental sexual division of labor. A feminist from Virginia, Mrs. S. L. F. Smith, might say that, yes, while maternity made her "helpless" a wife did have a "valid claim" to her husband's support, but "the period of maternity is not the whole of a woman's life, nor is woman simply a propagator of human stock. She is a being . . . with intellect, with talent, with ability to stand side by side with man, and again and again win the laurels with man in fair competition in innumerable spheres of industry."[22] Even Higginson recognized the need for some kind of plan "by which the independence in business matters shall be combined with equal responsibility; and by which the children of a marriage shall be provided for, and the support of a wife shall be guaranteed while rearing her children. . . . This is not so easy," he wrote, but a "partnership system should be devised, whereby the present inconsistencies would be done away with; Women recognized as an individual as Man is recognized, both held responsible for debts contracted which pertain to the family, and neither called to account for the personal liabilities of the other."[23] Without ever sacrificing wifely and maternal duties, feminists sought to devise a plan that let married women become mothers and still, at some point in their married lives, professionally independent women.

What may be the earliest feminist statement about two careers was also one of the fullest. In 1853 Henry Blackwell wrote a series of letters to Lucy Stone, assuring her that he would follow the highest egalitarian standard in their marriage. "My idea of the relation," he wrote,

> involves no sacrifice of individuality but its perfection, no limitation of the career of one, or both, but its extension. I would not have my wife a drudge. . . . I would not even consent that my wife should stay at home and rock the baby when she ought to be off addressing a meeting or organizing a Society. Perfect *equality* in this relation I would have but it should be the equality of Progress, of Development, not of Decay. If both parties cannot study more, think more, feel more, talk and work more than they would alone, I will remain an old bachelor and adopt a Newfoundland dog or a terrier as an object of affection. . . . Let a woman prove that she can speak, write, preach, edit newspapers, practice medicine, law and surgery, carry on business and do every other human thing. And if possible let her prove too that she can do each and all of these and be a true woman in all other relations also. If it be true that a woman cannot be a wife and mother consistently with the exercise of a profession, it justifies to a great extent the argument of our antagonists who say that very thing. For myself I protest against this doctrine. . . . Herein I think is the legitimate function of Reason to so organize and construct our circumstances and duties and be true to our *whole* nature and live a symmetrical rational life true to all our faculties.[24]

Such a clear position did not find public expression until after the Civil War, when it began to appear frequently in feminist novels and periodicals and in public debate.

Take, for instance, the novels of Lillie Devereux Blake, Epes Sargent, and Mary Clemmer Ames. We remember that Blake's heroine in *Fettered for Life* refuses to marry her lover unless they live "equal in all things" and he "permits her to follow out her own career in life."[25] In his blank-verse novel, *The Woman Who Dared*, Sargent not only has his heroine propose marriage to the hero but also makes her a successful wife, mother, and professional artist at the same time. "Take all your rights," the hero urges.[26] Although Ames's *Eirene; or A Woman's Right* stops just short of reaching the same conclusion, it is a more probing and more historically sophisticated novel than the other two. The novel traces the career of Eirene Vale, eldest daughter of Aubrey Vale, from early youth to her marriage in the late 1860s. After leaving her home town for factory work in Busyville, she falls in love with the handsome, patronizing, and dandified son of the wealthy factory owner, transgressing class barriers. When his mother intervenes to end the relationship, he fails to propose marriage, and she departs for a bigger city. "Now, for the first time," writes Ames, "she had found herself face to face with the great and unresolved problems of the daily life, as seen in the vast city with its inequities, its miseries, its temptations, its uncertain rewards! The problem of labor, of sex, of condition, confronted her." Eirene enrolls in a business college, takes a degree, and, once she finds employment, insists upon and gets a wage commensurate with her education and equal to a man's wage in the same position. During the Civil War she volunteers to serve as a nurse in army field hospitals and experiences the worst of the war—the waste, the mutilation, the death—losing the remaining sentimental and romantic "illusions" that city life itself had failed to destroy. "What could such a love as Paul Mullane's be to her today? Nothing." Eirene meets a "woman's rights" physician during the war, a rational man who treats her as an equal and sneers at the notion of "female distinctions." He believes that a "woman has a right to earn her own living as long as she chooses." After they marry, Eirene decides to abandon her business career. "What I owe my husband and child," she says, "is the first and deepest obligation of my life, but not the only one. If I had no husband and child, if I were a solitary force in the world, I deny the right of any man to set a limit to my advancement, as I deny mine to set a limit on his." But, still unsatisfied, she uses her own money to establish a "large house of rest and help for women."

In a bold speech on woman's work in 1870, Anna Dickinson championed the "exceptional" women who had professional careers and also remained faithful to "home duties." "I know a woman in my own city," she declared, "whose husband is capable of earning eight thousand dollars a year. She might stay at home, economize, and with the help of one servant and contrivances get the work done to make both ends meet at the end of the year. But this woman has a genius for medicine. She studied it, she practices it. She earns two hundred dollars a year by it. She has good servants, an admirable housekeeper. Let the ninety-nine women gratify themselves and stay at home, let the one hundredth gratify her nature and go abroad."[27] Elizabeth Stanton struck the same chord in 1868 when she asserted in the *Radical* that women's domestic work "need not and often does not prevent careers in other professions. Artists, physicians, actors, singers, have been also good wives and mothers;" and she added in the *Revolution* that "once and forever . . . motherhood is compatible with voting and with holding office."[28] "The Coming Housewife," an article in Theodore Tilton's paper, the *Golden Age*, in 1871, was probably the most advanced statement of its time to appear in a "respectable" journal:

> The Girl of the Period does not aspire to be merely the keeper of the house. She would make her house, it is true, a pleasant one; but this is not the Alpha and Omega of her ambition. Besides this, she would be independent—would have a profession, or trade, would earn her own living and carry her own purse. . . . Marriage with her will be an incident in her life—a grand experience, a solemn relation truly—but it will not be the purpose of her existence. . . . The coming woman will be the keeper of the home. She will do all needful work and have time left to study law and medicine, to keep books and take photographs, to edit papers.[29]

The high-water mark of this discussion of two careers seems to have come in the middle and late 1870s. Feminists debated the question either directly or indirectly in their clubs, congresses, and social science associations. The New York City Sorosis, for example, always seemed on the verge of meeting the matter and often did, in subtle ways, by publicly congratulating professionally trained women in its membership when they married and still chose to pursue their careers. It acknowledged one physician on her marriage: "Resolved that Sorosis congratulates Alice DeBawn Benedict, M.D., upon the event of her marriage and desires that as the twain are of one mind in their professional life they may realize the double joy of one heart in their conjugal life."[30] More important, it was the women of the Sorosis who organized the first Woman's Congress in 1873, at which the subject of two careers entered fully into public feminist discourse. At the congress Putnam Jacobi denied with clear, rational thoroughness that childbearing and child rearing necessarily consumed all the energies of women. They did not, she said, during the period of "primitive segregated industry," when most women bore several children, and they should not in the middle of the nineteenth century, when the "modern form of aggregate industry" had superseded the older system of production. Under both productive systems women expended "about the same amount of labor." Career and marriage were essentially compatible. She underlined the "intermittent character" of childbearing and the "limitation increasingly imposed on it by . . . rapid increase of population" and by "our high standard of comfort," resulting in "the general and marked diminution in fecundity of marriages." With the lives of married female physicians as illustration she showed that the exigencies of child care need not extend beyond a short period—at the least two months, and at the most nine years.[31] Antoinette Brown Blackwell focused on the concept of symmetry: "It is utter desolation for any human being to be compressed wholly in any merely domestic world." All "laborers," male or female, needed "two complementary occupations, each of which would be a relief against the other," and if women found it difficult to pursue two occupations then they should apply the "mutual exchange principle." "Two poor neighbors," she suggested, "might help each other, one superintending the children of both in the morning and the other in the afternoon."[32]

Blackwell, the most remarkable and creative figure in the making of this new marital program, had pondered and studied the matter since the 1850s, casting off older views, fumbling her way to the mature formulations of the 1870s. At the New York City Woman's Rights Convention in 1860 she insisted for the first time that "you cannot have true marriages" unless women have work "outside the family," although she appeared to concede that those with young children should not work. Women's hopes lay rather in delaying marriage until twenty-five or thirty, having children in their thirties, and preparing for late middle age, or "just where our great men are in the very prime of life!"[33] Repeating these ideas for the *Woman's Advocate*, she deepened them by criticizing the American cult of youth. Americans, she said, should pay more respect, give more power, to "*older* men" and to "*older*, mature women."[34] Her brilliant series of articles for the *Woman's Journal*, later collected into a book called *Sexes Throughout Nature*, included an essay, "Work in Relation to Home," in which she hailed "symmetrical action" or "nature's regimen" as the future basis for all female labor and as the escape from "woman's kingdom of irrational forces."[35]

In a later version of the same essay, delivered to the New England Woman's Club in Boston, Blackwell questioned the Darwinian notion of nature as a state of "permanent warfare" and claimed, instead, that nature promotes the "cooperation" of the "domestic instincts and wholesome outside activities" in women. She lauded Margaret Fuller as an ideal model, a woman, Blackwell said, who "would have died" not only for her husband and her child, but for "a free Italy with equal cheerfulness!" And like other feminists, she singled out the married female physician as an example of what women could do. Such a woman might have a "dozen children" and still practice her profession.[36]

"Womanly methods of working," Blackwell believed, "mean something more than working in lady-like refined ways. They mean adopting the work to her woman's nature and to her special and personal relations whatever these may be."[37] In subsequent papers she examined thoughtfully the subjects that haunted nineteenth-century feminist debate. She recommended "heterogeneous household labor" for women or a "systematic" command of "three hobbies" to sustain the tone and vitality of life. She said that only "historical" conditions had limited woman's mind and that only objectively obtained "physical data" could ever determine sexual differences. At the Syracuse Congress in 1874, she sloughed off her earlier confidence in the efficacy of delayed marriage and condemned it as "a most mischievious concession." She now gave unequivocal support to two careers. "It is time that we utterly repudiate the pernicious dogma that marriage and a practical life-work are incompatible."[38]

Other women argued the same position in the congresses and social science associations. Take, for instance, the paper of young Anna Garlin, "The Organization of Household Labor," given at the fourth congress in Philadelphia. Daughter of abolitionist parents, she was only twenty-five when she appeared before the congress. Her intellectual development resembled that of other women in social science. In the year of the congress she abandoned Congregationalism, the faith of her youth, and joined the Free Religious Association.[39] Two years later she was to marry a liberal minister, William Spencer, a sensitive man prone to invalidism who often allowed his wife to preach in his stead and even backed her successful effort to become a nondenominational liberal minister. In her paper Garlin asserted that modern women who had learned to appreciate "moral, mental and physical culture" anticipated, like men, the pleasure of both self-support and "homemaking" at the same time. Homemaking need not mean drudgery or confinement, the very things that educated women sought to escape. Unlike other congress women, Garlin distinguished between homemaking and housekeeping. To her, homemaking symbolized the "spiritual" or emotional realm, one equally desired by men and women, separate from the world of work. Housekeeping, on the other hand, "is a collection of industrial pursuits which lie nearest the home-life, and are therefore dependent on the homemaker's direction; but which are susceptible, like all other industries, of organization into an orderly process of business." She believed that "organization" would liberate the home from housework without subverting it, and at the same time create work for women in the public realm.[40] Garlin was a member of the National Woman's Suffrage Association in the 1870s and 1880s, and became a leader of the American Purity Alliance and the American Social Hygiene Association, an organization dedicated to sexual enlightenment.

At the Syracuse Congress in 1875, Jane Croly allowed her audience a glimpse into the private workings of her own two-career life. She described the long hours she spent as a journalist, often working in her office from seven in the morning to twelve at night. "The care of a home and the rearing of children during this time," she declared, "have been recreations, and sickness has been kept at arm's length by actual want of time to pay it proper attention. This is the way men work; this is the way women work who accomplish anything." Her intention was not to glorify this regimen to other women interested in newspaper work, but to make clear to them that professional success depended upon the acquisition of the very best professional education.[41]

At the Illinois Social Science Association women fiercely debated the two-career question. On the one side stood the Chicago progressive, old-time abolitionist, and former member of the Working Women's Association, Helen Starret, and a Chicago physician, Mrs. Dr. Severance; on the other was a Mrs. Willard, a woman with unflinchingly traditional views. Starret said that married women had "been crowded out of the home circle" by industrialization and now "must take their places with men." My daughter, Willard responded, will never "study out. . . . Domestic work is best for girls."[42] This question reached an international level in 1878, when the International Congress of Public Morality, in Geneva, Switzerland, declared that the successful reconciliation of two careers—of woman's "practice of a profession . . . with her position within the family"—depended "upon the profession and upon the individual position of each woman."[43]

Organization and Cooperation: Keys to Emancipation

PROMINENT feminist leaders believed that marriage and "practical life work" could be reconciled by cooperation and organization. Like the "human body," Julia Ward Howe said, "the working bodies of human society are organizations."[44] The ideology of cooperation and organization that permeated feminist discourse came to the feminist through the utopian communitarian movements of the antebellum period. Many feminists, from Albert Brisbane and Jane Croly to Elizabeth Cady Stanton and Paulina Wright Davis, were influenced by the socialism of Charles Fourier.[45] "Fourier," Mary Putnam Jacobi said, "in many respects, was the most sagacious of modern Socialists."[46] Paulina Wright Davis wrote for *Una* in 1855:

> We appeal to Fourier's principle of the "oneness of the race" and human brotherhood as the grand superstructure. . . . Fourier saw society throughout, organized upon that fundamental principle, every member interlinked, for all purposes and in all activities, with every other member. . . . We want associative, attractive and incorporated labor. . . . There is a broken balance between the sexes that must be restored.[47]

After the Civil War the ideal of organized cooperation took an even stronger hold on the feminist mind, and in every major feminist periodical articles appeared on cooperative industries, cooperative houses for working-class girls, and cooperative associations. Feminists as diverse in some respects as Thomas Higginson, Anna Dickinson, Louisa May Alcott, Elizabeth Stanton, and Wendell Phillips looked to cooperation as the wave of the future. "That there is to be a gradual reorganization of labor," Higginson wrote for the *Woman's Journal* in 1872, "on the system of cooperation, is as sure, in some form, as that the world moves."[48]

In 1867 many progressive spiritualists gathered at a Convention of the Friends of Human Progress to discuss a proposal made by a New Jersey land improvement company to build a "cooperative manufactory" combining the capital of the company with the labor of reformers. The company promised to share the profits and emoluments of the business on equal terms with the reformers. The reformers accepted the plan, thinking it symbolized the "transition from the isolated to the cooperative order by means of the Friendly Neighborhood, the Cooperative Village and the Unitary Home."[49] The members of the New England Woman's Club heard a paper by a female factory worker on a woman's "cooperative manufactory" in Troy, New York, and a speech by the Reverend William Henry Channing, cousin of William Francis Channing, on the general subject of cooperation.[50] In 1872 the noted feminist philanthropist Elizabeth Thompson incorporated the Humanitarian League out of a pressing need, she wrote, to "alleviate suffering" and to create cooperative and humane "neighborhoods and communities."[51]

Throughout the 1870s and early 1880s feminist interest persisted in forms of industrial cooperation, and especially in the cooperative enterprise of Jean Baptiste Godin in Guise, France. Inspired by Fourier, the industrialist Godin was famous for his Familistère, or workingmen's palace, an integrated communal manufactory that took nearly twenty years to build and that provided cooperative stores, nurseries, kindergartens, elementary education, a system of profit sharing, and a system of old-age and health benefits for the workers.[52] Elizabeth Stanton's niece, Kate Stanton, a marriage reformer and a member of American Labor Reform League, and Augusta Cooper Bristol, a social scientist, both returned from France convinced that, in Stanton's words, Godin "had made the best and most hopeful experiment in the solution of the labor question and that his example should be known by all men and women, especially in America."[53] Women of the AAW organized societies to spread the principles of cooperation according to Godin.

In 1880 a leading AAW woman and historian of that organization, Lita Barney Sayles, founded the New York City Social Science Association with Mrs. A. H. Whipple, hoping to popularize Godin's ideas. Sayles edited *Cooperative News* and the S S A published her book, *Rules for a Retail Cooperative Store*.[54] Sayles was doubtless influenced by the idealistic cooperative ideology of the Knights of Labor, the most progressive labor union of its day, the one most critical of the capitalist system, and the only union to admit women.[55] In 1882 Sayles, Helen Campbell, and Imogene Fales organized the Sociologic Society of America. Like other associations of its kind, it stressed industrial cooperation, putting "harmony" and "sharing" in place of individualism and the "competitive" ethos.

"It is because all are parts of one great whole," Helen Campbell believed, "that real progress will be impossible 'till the new ideal reaches out and enfolds all; 'till competition and the baseness born of it retreat to the shadows from whence they came and cooperation on its largest, noblest sense is the law of life."[56]

In the late 1880s, Sayles, Fales, and Campbell joined the feminist nationalist clubs of Edward Bellamy, the author of the enormously popular *Looking Backward* and a man who made socialism acceptable to many middle-class men and women. While denouncing the idea of class struggle and retaining a strong respect for property, Bellamy defended the cooperative principle and the ownership by the state of all the means of distribution and production. Like the contemporary Fabian Socialists, he advocated the formation of a nonpartisan welfare state. Other feminists such as Mary Livermore, Julia Ward Howe, Florence Kelley, and Abby Morton Diaz also joined nationalist clubs in this period. Lucy Stone became a member, thus setting an ideological direction for her daughter, Alice Stone Blackwell, who became a Communist sympathizer in the 1920s.[57]

Marietta Stow, no feminist in her early years, had married an affluent California merchant and enjoyed the life of a fashionable woman, but her husband's lack of foresight left her without a penny at his death. That transformed her into a feminist, and she heatedly denounced the California probate laws in a popular book called *Probate Confiscation and Unjust Laws Which Govern Women*. She also went to Washington to plead for the passage of her own bill for the relief of widows. Belva Lockwood, the first practicing female lawyer in the District of Columbia, presented the legal arguments and the Boston feminist Caroline Dall led a delegation to lobby for the bill.[58] Unable to endure the penury resulting from an unfortunate marriage, Stow struggled for a measure of social influence by writing books and directing reforms. In 1881 she organized and was the first president of the San Francisco Social Science Association, which started a silk industry and worked to establish cooperative industrial colonies in California.[59]

How did the cooperative principle bear on the question of the reconciliation of marriage with a "practical life work" for women? The view of sexual differences that took shape in feminist ideology after the Civil War throws some light on this question. Feminists like Elizabeth Cady Stanton, Antoinette Brown Blackwell, Matilda Joslyn Gage, and a host of others viewed the sexual differences as relatively static. Womanhood implied love, cooperation, harmony, purity, and altruism. Manhood implied intellect, egoism, sexual passion, and strength. Historical conditions under masculine regimes had repressed and perverted these qualities and destroyed the "equilibrium" between the sexes. Female love was changed into "puling," dependent "passivity" and selfishness, and male strength into a driven, dangerous, and selfish individualism. "The world is oppressed by masculinity," Mary Hubbard told the Rochester Radical Club in

1873.[60] "The male element," Matilda Joslyn Gage said in 1881, repeating an old positivist formulation of Stanton's, "has thus far held high carnival, crushing out all the diviner elements of human nature."[61] This thinking seemed to clash with the fixity of sexual difference, observing as it did the power of history or environment to alter sexual behavior. Indeed, even as feminists wrote of "natural" sexual differences, they argued for the fluidity of sexual identity and increasingly adopted an evolutionary, environmentalist perspective. The hygienist Lydia Fuller, the phrenologist Nelson Sizer, and Daniel Schindler, the Christian radical who was president of the Pittsburgh Woman's Suffrage Association, maintained that, as Schindler put it, "no male but has the female element, no female but has the male element."[62] Thus Caroline Dall proposed in 1867 the balanced diffusion of masculine and feminine elements, lest woman's "tenderness grow into tawdry sentimentalism" and man's "strength grow into cunning, rapacity and tyranny."[63] Influenced by Darwin, Spencer, Buckle, and others, feminists believed that an unfettered evolutionary process, unencumbered by artificial checks on female and male development, would not only liberate the "harmonizing" power of women but would culminate in a new race of "cooperative" and "equilibrated" men and women who shared between them the best traits of both sexes. Blackwell expressed this position succinctly in 1875: "With the continual evolution of each [sex] in process of time there arise more and more points of mutual adjustment, of complex adaptation; then one or both sexes begin to develop the characters of the other."[64]

Egalitarian marriage, according to feminists, established the principal instrument, the master medium, of evolutionary progress toward sexual cooperation. In such a marriage the sexes would perform "equal duties," support "equal responsibilities," display the same "interest" in the life of the family, and prepare with equal thoroughness for "paternity" and "maternity."[65] Egalitarian marriage was the very model of cooperation. "A man and woman who are truly married," Frederick Hinckley wrote in an article called "Cooperation" in the *New Age*, "who makes life a study of how to develop each other and how to produce healthy and beautiful children, are the finest example we have of cooperation. It is an ideal state, some will say."[66] As a psychological process, moreover, marriage "humanized" the sexes. Laura Curtis Bullard wrote in the *Revolution* in 1871,

"Marriage, which is the culmination of the idea of sex, was designed to make man more womanly, and woman more manly." Eliza Duffey wrote in 1876 that marriage "means the union of all that is manly and womanly in a perfect whole" and "will develop the traits of the individual . . . as no other discipline can."[67]

This humanization did not mean the end to sexual differences. The logic of the evolutionary argument notwithstanding, few feminists wanted sexual homogeneity. They were attacking "one-sided" development. Biologically, egalitarian marriage meant the transmission of the best traits of both parents to the children, the girls inheriting male strength and the boys female tenderness; both sexes, in fact, beginning life with their "excessive" tendencies tempered by rich genetic endowment. Most important, a "cooperative" marriage freed both sexes to engage in a breadth of activity from hobbies and reforms to professions and politics.

The relationship between cooperation and the expansion of woman's work beyond the home, however, rested on more than the unfolding of evolutionary forces and the promise of cooperative marriage. Noting the passing of domestic industries, many feminists—especially those of the Association for the Advancement of Women—found in the organized socialization of woman's domestic work the key to emancipation. No one extolled its virtues better than feminist Anna Brackett in a paper read at the Syracuse Woman's Congress in 1875.

Brackett had studied educational philosophy under the St. Louis Hegelians, William T. Harris and Henry Brokemeyer. In 1870 she had become an important figure in educational reform in St. Louis and headed her own normal school, with a salary of $2800 a year, more than any other female teacher in the nation at the time. In the mid-1870s she went to live in New York, where she moved comfortably within the most progressive reform circles and became the director of another school.[68] Feminists used this school as an illustration of what women could do.[69] A Free Religionist, Brackett wrote essays for Harris's *Journal of Speculative Philosophy* and for the *Radical*, including one on Margaret Fuller.[70] She evangelized everywhere for coeducation, and in 1874 edited a famous book on the subject, *Education for American Girls.*

Brackett's paper for the Syracuse Congress, "Organization as Related to Civilization," blended a Comtian faith in science with Hegelian concepts of duty and service. She began with a hymn to Grand Central Station, one of the colossal symbols of mid-century organization and an image that would appear often in feminist literature to the end of the nineteenth century.[71] At Grand Central, according to her, everyone had his place—the ticket cashiers, engineers, conductors, and so on—and everyone did his individual job while at the same time serving the station and the community. Summing up the central thesis of contemporary social science, she observed:

> Each man has his own functions to perform, and does it, and alone. . . . Each and all are controlled through numberless subordinates, by one superintending mind, which thus orders to the minutest detail with a view to the interest of the whole. . . . Of civilization, no better index. . . . Civilization means nothing more than individual caprice subjected to rational will; the individual to the general good.[72]

Mary Putnam Jacobi said precisely the same thing from the Comtian perspective. "Positivism," she wrote, "derives from the consciousness of social welfare, and from the cultivated habits of subordinating immediate personal impulses to higher sentiments, of honor, of sympathy. . . . Interdependence of functions . . . means the sacrifice of one part to another or to the welfare of the whole."[73] Should Brackett's feminist audience wonder about what had happened to the concept of individual symmetry or human wholeness, Brackett reassured them that symmetry had simply taken on a cosmic dimension. Organized division of labor, of course, required the "renunciation of a certain roundness and perfection in the individual" or "the attainment of many minor faculties;" but in return the individual gained enormously. "By yielding up his will to the good of the whole, he is served in turn by the will of every individual of the organized state." By making "himself voluntarily one-sided," he "thereby becomes many-sided on a higher plane."[74] To emphasize the feminist significance of her argument she unveiled another symbol, the "common window" in a commercial bank. In the old days, she reminded her audience, only one bank window existed and that was given over totally to male transactions. In the very recent past two windows had appeared, one for men and the other a "courtesy" window for the "minor" business of women. Now, however, she could see a time when the two windows would metamorphose "back again to one, but that one, common ground;" a time when men and women would be able to use, administer, and serve as equals all the most important "external organizations of society."[75]

The women of the AAW and other feminist social science affiliates shared Anna Brackett's faith in the powers of organization. These women never denied the pivotal importance of woman's domestic sphere, an importance many of them recognized as both the great strength and the bane of mid-nineteenth-century feminism. Their intentions were to raise woman's work, woman's propensities, woman's maternal functions to an "equivalent" social status. "Equivalence," indeed, was the hallmark of their ideological stance. As Lucinda Chandler observed on the third day of the First Woman's Congress in 1873, "The kind of equality which is needed is an accepted equivalence of the different powers and uses of man and woman. . . . Til she can lay her hands upon every instrumentality—educational and legislative—whereby she can make suitable conditions for her child before the cradle, and suitable provisions for its training, she cannot fulfill the law of motherhood."[76] May Wright Sewall, a director of the AAW in the late 1870s and an active member of the NWSA, made a similar point before the Boston NWSA Convention in 1881:

> In the ultimate analysis there are not separate interests in all humanity; absolutely all will be served when each is best served. But only in this ultimate sense is it true that woman's interests are identical with men's. The interests of women and men are correspondent, not identical, and their common interests will be best subserved as the correspondent interests are advanced.[77]

Sewall and other feminists had stuck to this position ever since Stanton first formulated it. "The best interests of the race," she announced in 1868, "demand that the equilibrium of sex be restored."[78] And they stuck to it because they believed that every facet of social existence had taken on a "domestic meaning." "Domestic life *is* our sphere," Sewall wrote, "but man has never measured for us the circumference of its orbit, and what woman has ever reached such height, depth and breadth of things as to have touched at any point the boundaries of her inherited domain?" They stuck to it, moreover, because they believed that women, not men (who had failed at this enterprise), would harmonize society and bring "reason" into the public realm. To them "maternal" meant reason, and reason in its most rationalist form. As Sewall put it, "The degree of civilization is measured by the degree of allegiance to reason."[79]

In this spirit female social scientists recommended that household industries—cleaning, cooking, and sewing—be "organized" outside the home in the manner of male work and according to the principle of specialization. Anna Garlin opened her 1874 Congress paper, "The Organization of Household Labor," with a reflection on the male division of labor:

> Organization as applied to labor, is that process of civilization which separates the industry of the world into parts, defines the details of each part, and sets each man to work upon that part and detail for which his taste and training suit him. . . . The method of man has changed, until now he has become a *specialist* in labor, fulfilling his father-office in whatever suits him best.

Woman's household labor, Garlin went on, had escaped the beneficent effects of this industrial process. Women still suffered from the drudgery of undifferentiated labor and, unlike men, who enjoyed both public and private lives, still endured the isolation of a completely private, primitive existence. This fact had inhibited young single women from seeking the "average condition of family life, which has in it little or no place for the individual mental development of woman who is its head." Specialization, Garlin hypothesized, would reverse this trend; it would give women both public and private lives by bringing the "mother-office" into the public realm. Organized into specialized occupations and means of self-support for women, household industries would thus depart from the home, leaving it an "emotional and spiritual" enclave for both men and women.

Zena Fay Peirce (like Brackett a devotee of Hegelianism), Anna Brackett, Mary Putnam Jacobi, and others agreed with Garlin that the complexities and "pressures of modern life," the high standard of living, and the "increasing public demands placed upon women" had made it difficult for housewives to do their "three-fold" task intelligently or hygienically.[80] Weighed down by the "general scope" of their work, they could do nothing thoroughly or well.

Women, Peirce observed at the same congress and in the same

vein, must "organize the great functions which God has given them—that of housewifery—on the fundamental principles of Capital and Labor."[81] Garlin and Peirce proposed that "private kitchens and industries" (Garlin, Putnam Jacobi, and Blackwell also included private "nursing" functions) be arranged into women-administered occupations; Peirce favored the creation of elaborate cooperative establishments and of Rochdale-style cooperative stores. Augusta Cooper Bristol held up the Oneida Community as a model of communal living, repudiating its sexual practices but praising it as "an enlarged inviolate household—a complex speciality. The weaned babe is in the arms of its own family, in the arms of unity. And I predict, that it is through enlightened motherhood that many more enlarged homes, that new and improved Oneidas, will become possible in the days that are coming."[82]

In 1873, before the First Woman's Congress, Stanton spoke eloquently to this point:

> . . . rest and peace for women, and protection for the children, and the best possible condition, is not in the isolated home but in cooperation. You put together the means of a dozen or so families and you can realize . . . the coveted hours of freedom for mothers of these homes, but as it is—I have been through it all myself, and hence I speak from experience—I have lived in an isolated home surrounded by children, and I know that the minds of mothers must be forever in contact with the needs of infancy, and we never can have it otherwise until we have cooperative homes.[83]

Caroline Dall presented an outline for industrial cooperation in *College, Market and Court.* Drawing on her experience in both the American and Boston social science associations, she argued the feasibility of government-sponsored, publicly operated, "self-supporting laundries" as well as "ready-made clothes rooms."[84] Stanton liked these proposals but did not see eye to eye with Dall on government intervention. "It is dangerous," she warned, "to let government meddle with labor."[85] Anna Garlin leaned away from such elaborate affairs toward less costly, more workable neighborhood organization. "Four women of character and thorough business ability," she suggested in her 1874 paper, "an experienced cook, a capable laundry supervisor, a trained nurse and practical artiste in clothesmaking—

could organize the household work of a neighborhood with no more risk or outlay than is necessary to the man who starts a new branch of business in a small way."[86] Frances Rose McKinley, a free-love feminist concerned with social science, summarized the optimistic basis for all these reform schemes:

> In times past, women have been in the necessity of the case, mothers, nurses, cooks, housekeepers, or drudges. As science gradually systematizes human exertion to its most economic applications, all this labor will be done by the aid of machinery or organized division of effort, in such a way as to be no longer a drag and a degradation, but an exercise and recreation.[87]

Free from the burden of "primitive, segregated industry," women would now have the chance to perform genuinely "individualizing" and productive labor while at the same time contributing to their communities. In Garlin's words, a reconstructed "domestic sphere would afford opportunities for special training and congenial life business, for definite and increasing compensation as one rose in its ranks, to those who chose it, whether married or single." Whether a woman was married or single, the idea had a certain imaginative brilliance to it, fusing as it did domestic with public feminism. These feminists found an important key to the emancipation of middle-class women in the organized and cooperative transfer of woman's work from the private to the public realm.

CHAPTER 9

THE BEE AND THE BUTTERFLY: FASHION AND THE DRESS REFORM CRITIQUE OF FASHION

FASHION was a powerful adversary to feminism. It competed with feminism for women's allegiance in both the private and public spheres. While fashion promised to tie middle- and upper-middle-class women to the sentimental sexual division of labor within the private sphere and to give them public definition as consumers, feminism sought to break the attachment to the home and to free these women into production and public life.

In the 1850s middle-class dress, male and female, struck the eye with its variety and color, but by the 1870s men had relinquished color and style for unrelieved drabness, while women had become subject to every wave and flutter of fashionable style. The period between these two decades witnessed the emergence of department stores, feminized monuments to the interests of sentimental womanhood. These "marble palaces" seemed to reinforce woman's isolation from the world, although they bore a direct relationship to male production.

Dress removed women from the time-bound, workaday society of men. Not only was dress the prism, as Nathaniel Hawthorne put it in his *Blithedale Romance*, through which the "splendid beams" of woman's love, tenderness, and beauty "shone forth," it had also become a symbol of both a private timelessness and a time gone by, protecting women from the pressure of present events and the weight of the future, The new stores institutionalized this protection. As one dress reformer plaintively remarked, fashion withdrew women from the "con-centered time in which generations exist," leaving them languishing in a "status quo" that "turned only backward."[1] Another wrote that fashionable women knew only "children's time," or "no time," a slow, unstructured, nonexistent time filled by the "pasting of pictures on old jugs and by the making of all the useless litter called fancy-work."[2] Men, on the other hand, made history and actively participated in public life.

Yet feminism did not seek to overthrow the traditional feminine virtues of selflessness, love, intuition, and personal purity that fashion claimed heir to. Many feminists themselves were fashionable; many were inextricably implicated in fashion-related industries; and others simply feared to berate a system of dress that appeared to offer the only visible sexual order in an otherwise disorderly universe. Feminists did, however, attempt to make rational what historical conditions had made irrational. In the 1850s feminist dress reformers were boldly individualistic and often stubbornly sectarian; twenty years later they were more restrained and pragmatic. Yet neither group was quite able to resolve the contradiction of perspectives. The dress reform movement struggled to break the sexual division of labor that found its strongest bastion in the fashionable world. In the process both feminism and fashion were transformed.

A Picnic *en Costume*

"LIFE," Frank Goodman observed in Melville's *The Confidence Man*, "is a picnic *en costume*. . . . To come in plain clothes, with a long face, as a wiseacre, only makes one a blot upon the scene, and a discomfort to himself."[3] Frank Goodman touched the pulse of the 1850s, a great decade in American fashion, when men and women—but especially women—dressed in new and often wildly extravagant styles. "Nothing to wear" became the nagging fear of male and female alike, and a poem with that title, published in *Harper's Weekly*, was the decade's most popular satirical spinoff.[4] The vagaries of fashion even fascinated Walt Whitman in his early years. In 1856 he wrote a long article on Broadway fashion that began: "Soldiers and militiamen are not the only people who wear uniforms. A uniform serves two purposes; first, to distinguish the wearers from others and, secondly, to assimilate them to each other." Whitman labored to "detect some people"—ministers, dandies, workmen, and prostitutes—"by their uniforms." And he described himself, as he often did in poetry and prose: a "tall, large, roughlooking man in a journeyman's carpenter's uniform . . . careless, lounging, sturdy, self-conscious, microcosmic."[5]

The middle-class interest in embellishments, in color and ornamentation, reached almost epidemic proportions. The journalist and aesthete Bayard Taylor wrote in 1856, "Just now the fashion runs to jewelry; we have ruby lips, and topaz light, and sapphire seas, and diamond air. . . . We have such a wealth of gorgeous color as never seen before."[6] Americans both at home and abroad appeared to have just discovered color. "The marvelous significance of *color*," wrote an American visiting Florence in 1854, "the inconceivable magnificence of *color* . . . colors too gorgeous for mortal vision were to me a perpetual joy, at all times and everywhere. . . . I believe, and still believe, that no single influence acting on the spirits through the senses, has the power that hues and tones have."[7] Henry Ward Beecher himself fell willing victim to this new wave. He collected stuffed humming birds and carried in his pockets amethysts and opals—he did not like diamonds—fingering and studying them, arranging and rearranging them for hours in solitary reverie.[8] "I have an opal ring (on view only)," he wrote a woman neighbor, "which, in many respects, is the finest I ever saw. . . . Perhaps your husband could invite you to come with him, and then I can let you see some of my treasures."[9] The famous editor of the *Home Journal*, Nathaniel Willis, also "relished a fine appeal to his senses." Willis, "a born shopper," had a "feminine eye for the niceties of upholstery, pottery and all kinds of purchasable knick knacks." He enjoyed being surrounded by living color as well. Peacocks—those "gorgeous creatures"—freely roamed his spacious grounds at Idlewild, New York.[10]

This male attraction to elegant, fashionable attire for its own sake, though not long-lived, echoed the excesses of the early nineteenth century, when gentlemen wrote tight-fitting pants and flashy surtouts, furbelowed shirts of worked cambric, and an occasional corset for that "pinched-in-waist" effect.[11] Urban middle-class male dress in the North recalled the dress of Yankee traders who worked on the southern levees in the 1820s wearing green coats, pink waistcoats, and yellow trousers.[12] Studs and scarf pins adorned men's clothes; tailors made suits of satin and velvet, often with a rich brocade or embroidery.[13] Hats had a vogue. When the Hungarian patriot Kossuth came to New York in 1851, his bold fur hat with a black feather in the brim ignited the hattery business, and for a moment the "Kossuth hat"—along with the Italian Alpine and the Mexican sombrero—threatened to weaken the growing dominance of the stovepipe. In this decade the clothing industry developed the technology to produce a broad range of male apparel—overcoats, linen dusters, summer suits, and so forth—that would not be surpassed until our own time. As one historian wrote in the 1890s, "The ten years ending 1860 will always be remembered as a period when the styles and fabrics for men's wear were of greater variety than ever before or since."[14]

Examples of sartorial display in the urban centers were easy to find. An important arbiter of taste in the 1840s, Willis admired the leisured life of the "gorgeous dandy," and despite some waning of his influence in the 1850s, he tried to stamp in the minds of gentlemen a model of masculine appearance that was elegant, carefree, "bohemian," and, most of all, "ornamental."[15] Such a model did not signify extravagance or gaudiness, but a studied simplicity of dress that might take hours to create but never drew attention to itself. The *Home Journal* informed its gentleman readers:

> The best possible impression that you can make by your dress is to make no separate impression at all; but so harmonize its material and shape to your personality, that it becomes a tributary in the general effect, and so exclusively tributary that people cannot tell after seeing

> you what kind of clothes you wear. They will only remember that you looked well, and somehow dress becomingly.[16]

Willis's plea for "ornamentality" met with a warm response from a cluster of dress-conscious "bohemians" who wore their hair long, grew beards in the "manner now so prevalent among officers and dandies," and strutted forth in the street with "walking sticks and canes."[17] William Prescott, the historian, wrote his friend George Tichnor from London, where he had been presented at court: "I wish you could see my new costume, gold lace coat, white inexpressibles, silk hose, gold buckled patent leather slippers."[18] American court costume during the antebellum days was wonderfully flamboyant, shaped as it was by southerners at European courts, who continued to dress in this fashion even after northerners began to emulate codes established by the military and President Franklin Pierce tried to complete a reform begun earlier by Andrew Jackson, recommending a simple homogeneity of court dress in the "democratic manner."[19]

These "dressers"—or the worst of them—drew continuous comment from the contemporary press. The "Easy Chair" of *Harper's Monthly*, for example, described the human "poodles" who paraded both in the city and in the summertime spas with their "velvet tunics and Honiton lace," "long glossy locks," and dyed mustachios. "We have a fear," the Chair declared, "that the velvety race is on the increase." Amelia Bloomer's *Lily* often commented on masculine dress, sometimes with a liberal tolerance and sometimes with glancing blows at the male "butterfly." Even Whitman had some harsh words for a "painted" Broadway male:

> Somebody in an open barouche, driving daintily. He looks like a doll; is it alive? We'll cross the street and so get close to him. Did you see? Fantastic hat, turned clear over in the rim above the ears; blue coat and tiny brass buttons; patent leather shoes; short frill; gold specs; bright red cheeks, and singularly jetty black eyebrows, moustache and imperial. . . . How straight he sits, and how he fingers the reins with a delicate finger stuck out, as if a mere touch is all.

So spoke the man who had only recently tossed off a dandy's dress, paid a lot of money for his carpenter's uniform, and still favored frills on his blouses "open at the throat."[20]

Still, male dress of the 1850s never approached the female in extravagance, style, and color. The movement in male fashion in the North, in fact, was toward subtlety and simplicity, and ultimately toward a basic homogeneity. The ideal fashionable male could dress, in good taste, in ten minutes while his female counterpart, according to the *Home Journal*, still toiled at her toilet for hours, though she was "not so artificial in her make-up."[21]

Female fashion was remarkable both for its unusual underclothes and for the various costumes worn over them. In 1853 and 1854 the hoopskirt and crinoline, both inspired by French fashion designers, appeared in New York shops. Originally a haircloth petticoat, the crinoline had evolved by now into a whalebone or metal frame, designed to support other petticoats and the visible overdress. The crinoline was "all the rage" and soon became the center of a "crinoline war," between those who united behind the crinoline's principal champion, Empress Eugénie of France, and those who condemned it as a criminal addition to female underclothes—French Bourbons and American dress reformers. "The fate of the crinoline is dubious," wrote the *Home Journal*'s fashion editor, Genio Scott, but as a warm admirer of the Empress he never doubted what the outcome of the war would be. He considered the crinoline an inspired and permanent accessory to female costume.[22] The crinoline also produced a new "crinoline etiquette." "There is a change in the laws of politeness," the *Home Journal* explained in 1857. "It is now decided that, in consequence of the new difficulties in female squeeze-past-ability, it is the imperative law of etiquette to pass to the head of a pew or omnibus, according to the *succession of arrival*."[23]

The existence of the hoop and the crinoline permitted women to wear more and more petticoats and, especially after the adoption of the sewing machine, led to a boom in the petticoat industry. In 1856 one Broadway manufacturer was turning out more than three thousand petticoats a day. Two years later, as mass production came into its own, firms such as Douglass and Sherwood of Broadway were manufacturing four thousand hoop skirts daily.[24] Designers were busy improving the petticoat to insure its marketability. As if reflecting the jewels and ornaments that fascinated the urban middle class, petticoats themselves appeared on the market in solferino, magenta, and mauve, colors created from new chemical dyes. The red petticoat flared into fame, first brought into use by Queen Victoria—"to re-awaken," according to the *Home Journal*, "the dormant conjugal susceptibility of Prince Albert"—and it was soon a triumph in Washington, New York, and Philadelphia.[25]

Designers created costumes for all ages, complexions, and occasions—for balls, promenades, ice skating, fishing, yachting, eating, and sleeping. They covered simple materials with vast ornamentation, decked them with countless flounces, used rich and striking colors. Long, trailing skirts were often remarked on as the great "street sweepers" of the day. Fashion magazines—*Godey's*, *Peterson's*, and the *Home Journal*—reveled in the most detailed descriptions. Genio Scott, perhaps the most original and the most curious of all fashion editors of the period, composed the following portrait of a ball robe for blondes. Sexually nuanced and even titillating it is a perfect representation of the genre of fashion writing:

> Now, in describing a ball dress for a blond, the "crème of the crème" is a sky-blue *taffeta* robe, with *jupe* divided equally into three full flounces. Over this blue robe, is one of white *tarletane*; the flounces of the *tarletane* disclosing three inches as a border of each blue one. The flounces are edged with a deep, plain hem—no ornamentation to detract from the fresh transparency of the lively *jupe*. The body—very *décolletée en cœur*—is pointed at the waist and rather long; the top is *orne d'une belle draperie* composed of blue and white plaits alternating, in the surplice waist genre, both across the back and stomach, uniting gracefully on the shoulders by *agrafes* of diamonds. The short sleeves are lightly composed of two full flounces . . . and of white tarletane, showing a little border of blue silk in keeping with the flowers on the skirt. The stomach is ornamented with a *bouquet agrafe*. The hair is parted over the center of the forehead, slightly crimped and brushed backward from the temple in full *bandeaux*, over which is worn a crown of forget-me-nots mixed with lilacs and jacinths. Light straw-colored gloves and white satin shoes. Pearl bracelet on the right wrist.[26]

Considered as a cultural statement, this portrait conveys all the conventional wisdom that linked women with flowers and ornaments; it underlines women's status as beings to be fluffed and primped and put on display like colored gems and porcelains. Yet it was also a reminder to women, both feminists and nonfeminists that they had a preferential right to leisured self-expression in the private sphere.

A culture of fashion would come to pervade much of American life by the end of the century, a culture dependent upon a growing capitalist economy. French styles shaped American tastes throughout this period, as they had done since the 1830s. Everything French, in fact, held a magical appeal for the fashionable middle class. In the 1850s and 1860s the United States was the second-largest importer of French silks. "The whole country," as *Harper's Bazaar* told the story later, "was more or less stimulated to extravagance by our cotton exports."[27] Genio Scott adored French clothes and dwelt on them for his readers, proudly reporting to them on one occasion that he had been appointed the sole American correspondent to the *Commission des Modes* in Paris.[28]

Public places and occasions were created so that "the natural desire to see and be seen" could be fully gratified.[29] "In the cities," wrote an observer, "people are exposed more constantly to view. They see more show. Every change in fashion is displayed before them."[30] Academies of music and design, opera and ballet, restaurants like Delmonico's, great hotels, clubhouses, and jockey clubs, multiplied during the 1850s and 1860s. There were fashionable Episcopalian and Unitarian churches like Trinity, Grace Church, and All Souls in New York, where "to be *married* and *buried* within their walls has ever been considered the height of felicity."[31] Fashionable spas and watering places dotted the purlieus of the cities; parks and amusement places grew up within the urban centers. City planners built wide promenades perfectly suited for fashionable exhibitionism. Broadway itself, though older and not so broad, had become one of the "great shopping streets of the world." "Oh, how exquisitely beautiful are the windows and sidewalks of Broadway!" exclaimed Genio Scott. "Every New Yorker feels proud of this incomparable promenade."[32]

Promenades, spas, theaters, and other new developments hastened the enormous expansion of the dress business and the technological revolution in dress manufacture. With the invention of the sewing machine in the 1840s, followed by its full-fledged adoption in the next decade, the ready-made clothes industry matured into a phenomenon of major proportions. By the early 1850s the cost of manufacturing cloaks and mantillas had plunged by nearly 80 percent; as a result retail prices dropped and demand boomed. The prices of women's hoop skirts and crinolines, as well as men's shirts and collars, dropped to comparably low levels. Wholesale manufacture of machine-made clothing rapidly expanded and reached astonishing dimensions after 1860.[33] Although it threw countless numbers of individual sewing women out of work and although some people doubted its practical use, the sewing machine helped to bring the fashionable world, for the first time, within reach of all classes of people.[34]

Most impressive of all, an imposing number of dry-goods and department stores, and later of mail-order houses and chain stores, appeared on the urban scene. The concentration of merchant capital and of people in the major cities, the beginnings of new selling techniques, the rise in the standard of living—slow and erratic though it was—and the phenomenal growth of transportation systems created the urban fashion industries. Horse-car lines, steam ferries, commuter railroads, and street railways serviced many urban communities in the 1850s, and the post–Civil War period witnessed a great elaboration of transportation networks. The major railroad trunk lines were built in the 1860s; electrically monitored trolleys and New York's elevated railroads were constructed in the 1870s. Short-line railroads allowed women to "take breakfast at home," dine in the shopping districts, "spend several hours and their money shopping and return in the evenings."[35] Even in the midst and aftermath of the 1873 Panic, people shopped and spent.[36]

These new conditions demanded the building of a new kind of store able to turn capital over quickly, to house, handle, and distribute merchandise for much larger and concentrated markets. According to some historians, the revolutionary yet lasting innovations of Aristide Boucicault's Bon Marché, which opened in Paris in 1852, provided the model for American retailers. Supposedly Boucicault pioneered in the small markup to undersell independent retail stores; enforced fixed and marked prices, ending the traditional bargaining transactions; introduced refunds; and let his customers come and go as they pleased without any obligation to purchase the goods on sale.[37] No one appears to know for certain whether or not American store owners consciously emulated these French innovations—one writer makes a very good case for the originality of the American version[38]—but it was clear that American economic and social conditions were ripe for the debut of the great commercial houses.

Marble Palaces

IN 1846 Alexander Turner Stewart, a brilliant Irish entrepreneur who had been in the dry-goods business for twenty years, built on Broadway and Chambers Street one of the most famous stores of the nineteenth century. He called it the Marble Palace. In 1862 he built another New York store on Astor Place and immediately began to enlarge it, gobbling up small retail stores along the way, until by 1869, with the demolition of the last outpost of resistance—Gorpli's art gallery—Stewart's eight-story skyscraper with its cast-iron façade and French-plate-glass windows had the capacity to "hold a dozen farm houses—barns, poultry, houses and all."[39] Architecturally it looked like Stewart's own palatial residence on Thirty-fourth Street and like his Fourth Avenue "Hotel for Women," a philanthropic enterprise Stewart planned which did not see the light of day until two years after his death.[40] The Astor Place store was for retail trade and the Broadway store for wholesale. By the late 1860s Stewart's total sales amounted to more than fifty million dollars a year, the largest volume of any store in the world.[41] Stewart had, in effect, built the most complete and the most successful commercial house anywhere, but his success simply fed his thirst for expansion. Before his death he owned branch stores abroad as well as textile mills in Manchester, Belfast, Nottingham, and the United States.[42] He worked with such mercantile magnates as Henry W. Sage and Horace B. Claflin of Brooklyn to establish a rapid transit system designed to tap a lucrative metropolitan market. For this purpose he backed the West Side Association, an organization of many of the most powerful capitalists in the city, men who owned nearly a third of the taxable property in New York and who planned to build an elevated railroad with their own money.[43] Stewart was eager to exploit the Northern New Jersey and Long Island markets.[44] By 1878 the first train, the Sixth Avenue Elevated, was bringing crowds to Fourteenth Street every day, benefiting Macy's (then at Fourteenth Street and Sixth Avenue) even more than the Astor Place store a few blocks away.[45]

Stewart introduced retail innovations that other stores soon imitated. Although he preferred to employ men, he was probably the first merchant in the United States to bring saleswomen—often English ladies of "refinement and culture"—into full public view, instead of secluding them from the gaze of the people on the street.[46] He adopted three important reforms in retail selling—the one-price system, the small markup, and the cash system.[47] He administered his stores according to the most advanced capitalist methods of the day; his managerial attitude departed from the paternalistic traditions of the past. He vertically integrated nearly all the components of the productive process—manufacturing, foreign purchasing, labor, wholesale and retail sales, technology, management.[48] On the ground floor of the Marble Palace stood seven heavy-duty sewing machines fixed on cars and mounted on seven lines of railway. The operators sat on low-backed chairs and traveled up and down the length of the seam on the track, sewing carpets perfectly and at a speed impossible to match by hand. On one of the higher floors, nearly six hundred women worked at smaller machines, making ready-made clothes for children and women, a market which Stewart was one of the first to predict and to capture.[49] In 1869 he employed more than a thousand men and women as clerks, porters, mechanics, elevator operators, sales people, and messengers, as well as a total of fifteen hundred women making women's and children's apparel.[50] He concentrated into one building a multifarious range of selling departments, each specializing in certain dry goods or ready-made clothing. A tribute to "industry and capital," Stewart's Astor Place store offered

> the treasures of the world . . . the commerce of Europe, Asia, America and far-off Africa. . . . Here are thousands upon thousands of dollars worth of kid gloves, silk gloves, chamois gloves and fur gloves . . . hosiery of every make and description from Belfast and Carrickfurgus; muslins from the thousands of mills of New England; silks from the looms of Italy and from China. . . . Persian and Cashmere shawls running as high as eight thousand and ten thousand dollars a piece; and velvet soft as the cheek of a maiden conjured by a poet's dream.[51]

Stewart packed his store with a staggering quantity of goods. It was his contention that all human problems could find solutions, given an unlimited supply of material commodities.[52]

Such concentration, and its accompanying standardization of labor, yielded two important results. Practically speaking, it reduced shopping to a minimum, eliminating the "weary, day-long tramp in search of varieties and bargains."[53] By standardizing work and thus requiring less qualified personnel who could be hired at lower wages, he helped convert sales into a woman's job. Men continued to hold most of the managerial positions in his business, which was true also of other department stores for the rest of the century.[54]

Stewart constructed a pattern of callous and indifferent relationships between himself and his workers similar to that of the cotton mills of New England.[55] In the early days of retail trade, when both the stores and their work force were small, retail merchants like the New Yorker Arthur Tappan, loyal to a pietistic stewardship tradition, superintended the physical and moral well-being of their workers with an almost paternal affection. Occasionally Tappan even invited his workers to lodge in his own home.[56] Stewart's relationship with his workers, however, indicated both the effects of greater economies of scale and the widening distance between the working class and the capitalist elite. In his business religion and capitalism, management and labor, split asunder. He paid his workers poorly and refused to throw "picnics" for them or to provide them with any form of entertainment. He preferred to know little or nothing about their lives.[57]

Feminist reactions to Stewart's monopoly and to his employment practices were mixed. Such New York City feminists as Elizabeth Cady Stanton, Parker Pillsbury, and Warren Chase condemned monopolies of wealth and the related exploitation of labor. They called Stewart a "social vampire" and his "whole system of business an oppression."[58] But some feminists defended Stewart. Martha Wright, a leading NWSA feminist from upstate New York, privately and starkly criticized her friend Pillsbury for his attack on Stewart in the *Revolution*. Stewart, she said, had no obligation to care for the poor or to support "missions, pastors, charities in a thousand directions, to make the poor richer: so long as some people are born without the faculty of making money . . . somebody must be rich enough to help them and how that is ever going to be helped, when such beget their like continually, I cannot see."[59] Feminist and former abolitionist Oliver Johnson agreed. "Why has not A. T. Stewart as good a right to the dividends accruing from his *large* estate as I have to those accruing from my *small* one?"[60]

Stewart, meantime, did plan his philanthropic "hotel" for single working-class women, but when he died in 1876, with the building still on the drawing board, his will made no clear provision for its use. As the feminist *New Century* reported, "The Stewart Working Woman's Home . . . stands as a living reproach to the dead man whose name it bears."[61] When the hotel finally opened two years later, with its 552 rooms, opulent chandeliers, furniture upholstered in raw silk with gold thread, and fancy restaurant, no working-class woman could afford to live there. Such women made six to ten dollars a week; rooms in the hotel cost five to seven dollars a week. Single middle-class women ultimately reaped the rewards of Stewart's philanthropy—artists, teachers, students, telegraph operators, retail-store buyers, a few store superintendents, and milliners. The hotel also set aside rooms to accommodate middle-class married women visiting the city for lengthy shopping excursions who, according to one account, had hitherto been "obliged to stop at the public hotels, flanked as they are by the bar-rooms." "Here is a hotel," said another, "which fully covers the shopping woman's reputation, as it is a woman's house where character is the qualification for all."[62]

Stewart pursued practices of underselling and outbuying that his nearest competitors admired and that inspired "terror in all small dealers."[63] He cared nothing for the approbation or affection of others. "My method," he said, "is not to obtain popularity, but to compel people to buy of me."[64] On his way to becoming one of the richest and most powerful monopolists in America (so powerful that Grant appointed him Secretary of the Treasury in 1869), he consigned countless small retail businesses to the grave.[65] He encouraged the use of a driven system of wholesale trade called "drumming," a system by which a large, highly competitive commercial house promised its salesmen handsome commissions, apart from their salaries, for whatever commodities they could sell to retailers. Drumming often doubled hours; salesmen were on the road day and night. It drove men to spare nothing—health, time, family, personal taste—to snare customers. According to the *Cincinnati Commercial*, a reaction set in against this practice in 1878, but failed to check it. As one "commercial traveller" warned, "the abolition of commercial travelling would be suicide for New York trade."[66] Apparently Marshall Field of Chicago also saw the matter this way; a year later he adopted this system as a permanent feature of all his wholesale departments.[67]

In the midst of a heavy downpour on a bleak September day in 1869, unable to find a ride across town, Jane Croly, a powerful fashion editor as well as a feminist, took shelter in Stewart's. A manager spotted her and suggested she use the time by taking a tour through the store. Going from floor to floor on a "carpeted, handsomely mirrored and cushioned car," she visited the "carpet manufacturing department" and later the immense sewing rooms where hundreds of women toiled at their machines. The journey profoundly impressed her. Unlike some feminists, she had no pessimistic doubts about the promise that Stewart's represented. She saw in this store a new domestic freedom for middle-class women. "There is no reason," she thought, "why manufacturing houses should not supply every article worn at the cost of material and labor, labor being gauged, not by old methods, but by the application of the sewing machine and other labor-saving machines. . . . The day of family sewing are numbered."[68] Stewart's struck others differently. It "looks like some fairy palace of ancient story," sighed a writer for the *Cincinnati Commercial* in 1878, with its "white parapets extended against the evening sky and its innumerable windows full of lights. . . . With the picturesque marble Gothic Grace Church across the way, the most practical quarter of Broadway, New York, is thus transformed into a beautiful fable, a dream."[69] A transfixed *Hearth and Home* described the Astor Place store in 1869: "Stewart's is a gigantic mass of iron, painted of white color, erected to the worship of dry goods, covering two acres of ground and Theban in its Old World massiveness. This structure from its immensity alone, must paralyze the curiosity seeker . . . for, perhaps, such another edifice does not exist on the American continent." The store loomed as a huge, marvelous symbol of new American wealth and productivity.[70] Under the rubric of a "single price," moreover, Stewart's pretended, like the other stores, to democratize fashion. "There is no distinction," *Hearth and Home* wrote of it, "as to dress or social rank in this caravansary of merchandise. There is only one price asked at Stewart's."[71]

Other stores in New York followed Stewart's in swift succession. In 1853 Samuel Lord and George Taylor opened a new store farther downtown, on the corner of Grand and Christie streets. They hired more than four hundred women to work there and gained their principal claim to fame by sticking fast to their cardinal "rule"—"no deviation from first price." The *Home Journal* described this first store, before Lord and Taylor, like other New York department stores, began the long march uptown as the city grew:

> Among the most important buildings erected may be mentioned Lord and Taylor's establishment which is one of the most significant Corinthian structures in the world being of white marble, five stories high, elaborately ornamental, and substantially built. The main entrance is in the form of a beautiful arch the point of which extends to the third story of the building and is decorated with carving and gilding.[72]

Seventeen years later Lord and Taylor built a four-story store at the corner of Broadway and Twentieth Street, a store distinguished from all others in New York by its intricate ornamental design—mansard roof, Corinthian columns, arches, pedestals, alcoves, balconies, dormer windows, gilded iron railings, and a hundred and twenty-nine-foot tower. Its glamorous show windows faced the streets, and each of them was large enough to hold "the whole stock of an ordinary dry goods store."[73]

In 1857 Arnold, Constable and Company opened a five-story structure on Canal, Howard, and Mercer streets and named it, with no pretense to originality, the Marble House. During the next twenty years New York saw the debuts of Macy's, Gimbel's, Rogers Peet, Bloomingdale's, Stern's, a large Brooks Brothers' establishment uptown, and one of the most fashionable of stores, B. Altman's.[74] The radical feminist paper *Woodhull and Claflin's Weekly* provided a completely uncritical description of Altman's:

> In such an establishment, the idea of an Eastern Bazaar is completely carried out. A lady enters and finds herself in the midst of a lavish display of everything that could be thought of in the way of fancy

goods. No need of running over ten or twelve blocks and into fifteen or twenty shops when making varied purchases. This house is a sample of what can be done by energy, enterprise, and a purpose resolutely carried out in a progressive city.[75]

Macy's was perhaps most purposeful of all in charting new paths for the department store. Founded in 1858 by a Nantucket-born Quaker, Rowland Hussey Macy, by 1872 it had expanded to embrace the ground space of eleven stores. It advertised itself as the "Grand Central Star Establishment." This gigantic conglomeration at Sixth Avenue and Fourteenth Street nearly eclipsed Stewart's in size, and, unlike its contemporaries, by 1877 it truly combined all the elements of a modern department store—great volume of business, wide diversification of departments, and vast choice of merchandise, from textiles, games, dolls, and toys to luggage, china, books, ice skates, velocipedes, bathing suits, and bird cages.[76] Not even Stewart's displayed such a cornucopia of commodities. "One can really buy everything within its limits," feminist Caroline Dall wrote praisingly in 1876, "from a French bonnet to nine cents worth of candy."[77] Macy's created the first window exhibition set in a coherent pattern. It was also the first to employ more women than men, including women as salespeople, floorwalkers or section managers, buyers, and store superintendents. In the late 1870s Macy's became the first to use telephones and to illuminate exhibit windows with electric lights.[78] In still another innovation it stationed a small army of female spies and detectives in all the departments during the busiest hours, and especially at Christmas time, to watch for shoplifters.[79]

These stores made New York City the dry-goods capital of the country and confirmed the statement made by the editor of the *New York Independent* that "intelligent capitalists have so developed the clothing trades that they now occupy the first business rank."[80] Big clothing and dry-goods stores were by no means confined to New York. Benefiting enormously from the prosperity caused by the Civil War, Springfield, Massachusetts, dominated the consumer trade in its region. In the late 1860s and early 1870s its main street was glutted with wholesale and retail establishments.[81] In 1867 D. H. Brigham and Company decided to build a "magnificent clothing house," bringing rejoicing to local patriots. "People who are not alive to the changes and improvements going on in Springfield may read with incredulity the statement that we have, or are about to have, one of the largest and most complete establishments for the manufacture and sale of clothing that can be found in the country; yet the statement is sober fact."[82] The paper did not exaggerate. Brigham's had four floors, the upper floors devoted to the manufacture of ready-made wear, the basement to wholesale, and the first floor to retail sales. It employed more than three hundred workers, sold only the clothing made in the store, adopted the one-price system, and had water closets, ventilation, elevators, and speaking tubes running from floor to floor.[83]

Boston's Jordan Marsh had been in existence since 1851, but not until 1871 did it become the largest, most ruthlessly competitive and powerful retailer in that city. Like the other stores, it showed "rich, varied and captivating goods of all kinds to tempt the eyes as well as the purses of purchasers."[84] Like other stores, moreover, it often resorted to corrupt practices to maintain its competitive position. To undersell others it imported goods cheaply from abroad by evading import duties and submitted false invoices to the government. In Brooklyn Horace B. Claflin and Company pursued a similar policy. In 1875 it bribed a customs official to overlook three million dollars' worth of imported silks, thus avoiding the high tariff, and brought them in illegally and cheaply through a company purporting to sell liquors.[85]

Despite its devastating fire in 1871, Chicago took its place in the 1870s as the imperial entrepôt for retail and wholesale trade in the Midwest. Before the fire, such men as Potter Palmer, Levi Leiter, and Marshall Field had already put their marks on Chicago trade; after the fire, and almost regardless of the 1873 Panic, Chicago became engulfed in a sea of mercantile business. Its major thoroughfare, State Street, became one of the "greatest single concentrated shopping districts in the world."[86] Drawn by the appeals of city politicians and by a big market, a Stewart's representative, G. K. Smith, chose Chicago as the location for a Stewart branch store in 1876. The *Inter-Ocean*, a major Chicago newspaper with strong feminist sympathies exulted over this decision:

> The advent of such an enterprise is of considerable importance to everyone having the welfare of the city of Chicago at heart. It will give a very *perceptible impetus* to business here, catching as it under-

> standably will, a great deal of that Western trade which has hitherto floated Eastward, and the sinking of such a capital means the undoubted revival of financial confidence among moneyed men.[87]

In the throes of its enthusiasm for Stewart's the *Inter-Ocean* assured Field, Leiter, and Company (by 1881 simply Marshall Field's) that this new store in no way threatened its competitive clout. Field, however, had little to fear. After a brief and vicious price war in which Stewart's was vanquished, and especially after the death of Alexander Stewart himself soon after his branch store was built, Field easily regained his hegemonic grip over the dry-goods trade. Marshall Field staffed his departments with more and more women in the 1870s, and followed the Macy's tradition of promoting them to section managers.[88] In 1879 a "grand fall opening" for a new Marshall Field's excited the whole city. On the first day, with the store draped in splendor, more than twenty-five thousand people walked its seven floors of white marble tile, passing through thirty-three departments of what, according to the *Inter-Ocean*, "was confessedly the unequalled display of its kind in the United States."[89]

Department and dry-goods stores symbolized and generated immense productive forces and, in the process, helped to transform the character of social relations and to widen the breach between the capitalist elite and the working class. As a result, in part, of the growth of the stores, merchant wholesalers, retailers, and mill owners began to congregate at textile expositions throughout the country, hoping to make regional alliances or strengthen older ones, and to make political policy among themselves on tariff regulation. The first of these expositions took place in Chicago in 1868, the second and third in Cincinnati and San Francisco in 1869. Organized by local politicians and entrepreneurs, the Cincinnati Exposition of Textiles displayed a great variety of domestically manufactured fabric in large ware rooms open to the public, and was arranged chiefly to spur the development of textile manufacturing in Cincinnati, then third in the nation as a manufacturing center. More important, however, this exposition occasioned a meeting among merchants, wool growers, and mill representatives from both the Midwest and the "new," industrializing South—from South Carolina, Alabama, Georgia, Tennessee, Kentucky, Mississippi, Louisiana, Missouri, Indiana, Ohio, and Illinois. On the second day of the exposition, in fact, the National

Association of Woolen Manufacturers convened to attack unanimously the tariff and taxation policies of the Federal Government.[90]

While department stores helped to usher in an interlocking network of self-conscious regional capitalist elites, they also contributed to the making of a larger and more self-conscious working class. Within the department stores themselves, saleswomen and men began to unite, however primitively, to protest "oppressive conditions." In 1870, for example, a male organization known as the New York City Dry Good Clerks Association aided saleswomen in their fight to win a "dividend policy"—already practiced in two other New York stores. This policy entailed the granting of commissions on "all clear profits" to employees, a policy that reproduced in a more muted way the drumming system of commercial traveling.[91] The department stores were probably instrumental in creating a larger, more vocal working class outside the confines of department-store walls.

The silk industry offers a good illustration of this trend. In the wake of the Civil War, American expenditures on silk fell because of the collapse of the southern economy, a stiff tariff, and the production of cheaper woolen goods in the North.[92] After 1870, however, stores everywhere responded to a growing demand for fancy silk products—indeed, perhaps created it in the first place through enticing advertising and inviting displays. The growth of expenditures on raw silk from nearly three million dollars in 1868 to almost six million dollars in 1876 attested to the increase in demand.[93] This demand partly induced and partly reflected the development of domestic silk manufactures in the United States and brought to life a bigger, more self-conscious working class. By 1873 American factories in Connecticut and New Jersey made all the sewing silk and twist in this country. Twenty-five firms operated in Paterson, New Jersey, alone, producing ribbons, coat linings, dress goods, scarves, and other goods, and employing eight to ten thousand operatives, three-fourths of whom were women and all of whom received extremely low wages. In 1872 the silk weavers of Paterson struck after their employers lowered wages by 20 percent. According to one worker, the "men got only one fifth of what they earned." "These employers," he went on, "*should be seized in the streets and hanged to the lamp posts. . . . Let us commit one act—string up to the lamp-posts these men, and then burn their manufactories down.*"[94]

The department stores and the factories allied with them institutionalized certain already existing cultural and religious formations and reproduced patterns of social relations within the life of the American bourgeoisie. They deepened and organized these formations and patterns in new ways. The department stores, and the fashion industry that underlay them, penetrated into and contaminated the life of established religion, creating a paradoxical marriage between commodity capitalism and religious life that has persisted into our own time.[95] Many Americans became accustomed to thinking of fashion in relationship to religion and religious holidays. In 1856, Genio Scott observed that "real ladies and gentlemen are those who belong both to the Church and to fashion."[96] "No young lady of Boston," wrote an observer in 1870, "can openly make profession of faith in her Saviour unless she has a polonaise white linen suit and a white chip hat trimmed with harebells."[97] And in 1875 another commented with equal wryness:

> To the devoted dames and damsels who have labored faithfully in the vineyard with the choicest hot-house grapes, Lent comes as a benison to their toils. It gives them a chance to rest, reflect and devise their spring wardrobes, and for no reason is it better appreciated than for the last. Indeed, the outsider, judging only from what he sees and hears, might readily believe this to be the sole object of Lent.[98]

The disposition to link religion and fashion often took strange forms. In 1878, for example, the Young Women's Christian Association of Cincinnati made use of a popular fashionable pastime—the display of *tableaux vivants*—to raise money. Designed by local artists, these tableaux arranged beautifully dressed women in settings based on scenes in famous paintings, novels, and operas. Each setting appeared separately, like a fixed but living jewel. The Cincinnati show offered six or seven such tableaux at the city opera house before an audience of a thousand "fashionable" people. The "masterpiece" of the evening was a "gondola scene, 'Les Illusions Perdues,' after Gleyre" in which twelve YWCA women sat motionless, timeless, in stunning finery.[99] The department store magnate of Philadelphia, John Wanamaker, demonstrated "how religion and business might work hand in hand in the service of humanity" and of John Wanamaker. The first paid secretary of the YMCA in the country, Wanamaker was elected president of the central Philadelphia section in 1868, organized two hundred seventy-five open air meetings, and founded the Bethany Sunday School, the "largest Sunday school in the world."[100] It stood as a living advertisement.

Macy's, Field's, Stewart's, Arnold Constable, and Wanamaker's sought to bind this relationship between fashion and religion. As the stores themselves resembled cathedrals, their goods were objects of worship. Tiffany's literally replaced a church, filling the space formerly occupied by Reverend George Cheever's Congregational church at Broadway and Union Square.[101] Most of the stores also set up "holiday departments." Macy installed his famous Santa Claus in the toy department of his Sixth Avenue store in the early 1870s, and all the stores learned to use "openings" in special ways. Often advertised on the front pages of newspapers and organized by store owners as glamorous stimulants to instant consumption, the "openings" were spectacular events. "Such crowds!" exclaimed Madam McCormack, a Chicago milliner, at a Gage Brothers and Company opening in September of 1879. "I don't think there was ever such drawn out in Chicago before. The hard times, I am sure, are all gone."[102] Ordinarily, openings took place in the early fall and early spring of the year; by the 1870s, however, as more stores and new departments within stores proliferated, merchants planned their openings to coincide with Easter week and the Christmas holidays. "Several large openings," the *New York World* pointed out in 1872, "are to take place during Easter week; among others . . . Arnold and Constable, Wilson and Craig and Richard Mearer."[103] When Altman's opened new departments in December 1870, including a "making-up" department, throngs of women packed the store. "Ladies are buying boxes, jewel cases, embroideries and other fancy articles over each other's shoulders."[104]

The department and dry-goods stores not only stitched religion and commodity consumption together into a thickly woven fabric, but also completed the feminization of fashion and institutionalized woman's role as the consumer of commodities even while they created new employments for women in the public realm. Here was one of the greatest challenges to the ascendency of feminism: a new public dimension to the sentimental womanhood feminists labored to reject. Feminization so marked the life of the stores that a twentieth-century historian of Macy's could write in 1943 that a department store was not a department store unless it "catered primarily to women."[105] The act of setting foot in a dry-goods palace for purposes other than saleswork and management, or to accompany women on a shopping tour, became for men a symbolic act of emasculation. "Of all the miserable beings," the *Woman's Journal* noted in 1873,

> a gentleman waiting in a shop, is the most miserable. He looks utterly unhappy and ashamed; he would shrink into a corner if there were one to shrink into. . . . Friends recognize him, pity him and pass on. The wife appears from time to time radiant with success in bonnet, polonaise or suit, and the poor husband, the unhappy scrip-bearer, is willing to pay any price, anything, if he only can get away.[106]

Women, not men, became associated with "fancy goods," with ornamental display, and with "tidies and things. . . . For who ever knew women of thoroughly feminine attributes to willingly dispense with tidies, even though their natural propensities may cause untold annoyance and embarrassment to the men who encounter them?"[107] Whereever openings occurred, whatever expositions of textile fabrics were given, "ladies," not "gentlemen," were "especially invited to be present."[108] It was one of Alexander Stewart's dreams to see "two acres of ladies all shopping at one time."[109] So too Marshall Field and Potter Palmer conceived of their stores as shrines dedicated to the interests of womanhood.[110]

American department stores fixed the very meaning of femininity as consumption, powerfully rigidifying the sexual division of labor, and they did so in ways that powerfully influenced the women and girls they served. For one thing, the stores used all the seductive devices at their disposal to attract and secure the patronage of women. Stewart employed the handsomest young men he could find as salesmen to lure the eyes and purses of his female customers.[111] Arnold Constable constructed an alluring "shadowy interior," visible from the street, to draw even the indifferent shopper into the labyrinth of consumption; and Lord and Taylor used a similar tactic by showing their silks in the remotest part of the store, in a "dark, elegant room" illuminated day and night by glittering gaslight.[112] Merchants tried to make the stores into places of "popular resort" where women might talk, eat, and safely be social together during their shopping hours.[113] They became homes away from home. In the 1860s Stewart was probably the first to put toilets and writing tables in his store.[114] In 1878 Macy's installed the first "ladies' tearoom," an innovation copied by Wanamaker's and Field's in the 1880s.[115]

The stores took women out of the home and into public spaces, where they could "see and be seen," shop and compare, delight in the spectacle of colorful dry goods and show themselves off in a new social circumstance. "In no other country do women of all classes make such a business of dress," Jane Croly wrote in 1869. "The great thoroughfares . . . from twelve in the morning to 5 o'clock in the afternoon, are thronged with beautifully dressed women, nearly all of them young, and most of them unmarried, who seem to have no object in life but to put on elaborate attire and go out and display it."[116] The stores were one of the few public places that gave women the freedom to be alone without escorts, a freedom compounded by the absence of industrial time, which placed such nagging restrictions on male work. Shopping, in fact, may have fostered a new sense of independence in upper- and middle-class women. In spite of efforts to convert them into social "clubs," the department stores reconstructed shopping into a radically depersonalized phenomenon with few reminders of the family intimacy of the country or neighborhood store. The dry-goods palace was a far cry from the friendly atmosphere of Arthur Tappan's small paternalistic retail store and an even farther cry from the Yankee peddlers who unburdened their splendid wares in the privacy of farmhouses, who told stories to their customers and indulged in "sharp talk."[117] The unfolding public space of the department store really tended to discourage social encounters.

Under these conditions, shopping produced its own deviant middle-class and upper-class pathology—shoplifting. One of the most famous arrests of a "respectable" woman for pilfering occurred in 1871, when a Macy's detective spotted Elizabeth Stuart Phelps, the wealthy philanthropist, novelist, and feminist, stealing an inexpensive package of candy from a store counter. Phelps was taken to the police station, booked, and released on her own recognizance; the charge of theft was later withdrawn for want of sufficient evidence.[118] Whether a genuine case of shoplifting or not, the Phelps incident was symptomatic of a shoplifting "mania" that had chronically plagued the stores since the 1850s. In *Husband and Wife; or the Science of Human*

Development Through Inherited Tendencies (1863), Hester Pendleton wrote:

> The tendency to petty larceny among women of . . . respectability is often remarked by merchants. It is considered much less safe to trust them with a variety of small merchandise than it is to trust men. . . . A merchant told me that some of his best lady customers had from time to time been seen to take small articles secretly, yet he dared not think of speaking of it, as it would ruin his business.[119]

In the 1870s the *New York World* reported the theft by women of everything from "entire pieces of cloth" to silks, laces, and camel's-hair shawls, especially the most expensive and ornamental articles on display.[120]

Undoubtedly a variety of factors promoted female pilfering, among them the impersonality of stores and the intense competitive character of fashion, which may have impelled women of middling means to steal. Three related factors certainly helped create the need to steal: the deep seductive appeals to female narcissism, the dependent status of women in the homes of their fathers and husbands, and the selfless, self-sacrificing mode of living that their own mothers had prepared them for. It appears that a woman shoplifter ordinarily did not steal what she could justify purchasing as a housewife, daughter, or mother, such as household items and clothing for her husband or child. But to spend her husband's or father's money on herself could have meant sacrificing the interests of the family for the interests of the self.[121] The shoplifter commonly stole more frivolous luxuries, things to adorn herself. Stealing may have been a substitute for impulsive, reckless, selfish spending that would have tapped reserves of resentment against her father, her husband, or above all her mother. Many women could not cope with such feelings. The stores offered a subversive and ever present narcissistic ambiance, mirroring the need of this deviant woman for unqualified love. Thus they may have produced the psychologically safe but compulsive resolution to steal.[122]

The shoplifting woman became a part of a new microcosm of social types brought to public attention by the marble palaces; she joined the female consumers, the doting salesmen and saleswomen, the fashionable promenaders, the skulking, embarrassed husbands. It was a world of excitement and charm as well as structured roles and pathologies.

Set against it was the feminist world with its emphasis on healthy female and male bodies, its doctors, preachers, and teachers, its rational intellectuals, and its earnest activists intent upon serving the public welfare.

Fashion, Stereotypes, and Etiquette

BY THE TIME the first issue of the feminist magazine *Revolution* was off the press in 1868, Americans had seen the dramatic extension of older trends and the creation of new ones in fashion. Department stores were beginning to cement the alliance among femininity, sentimental religion, and fashion. So, too, the ready-made business had grown as never before in every major city. Men, once mainly dependent on self-employed tailors for their clothes, increasingly relied, as women now did, on factory-made goods.[123] Fashionable colors were christened with exotic names—cassis, sphinx, souris, jujube, chausseur, and corbeau. Corset making took a leap into absurdity; in 1873 one factory made twenty varieties of corset, including one for babies.[124] Fashion magazines proliferated; *Demorest's Monthly* (1865) and *Harper's Bazaar* (1867) took their places with *Peterson's*, *Godey's*, and the *Home Journal* as the reigning organs of fashion news. *Harper's Bazaar* called itself the "first Weekly Journal of Fashion ever published in the United States" and introduced its first issue with a round complaint against uniformity in male dress.[125] Both *Demorest's* and *Harper's* kept up a "costly correspondence with the best sources of information in Europe."[126] Finally, the invention of paper patterns by Ebenezer Butterick in 1871—the inevitable outgrowth of a long tradition of home manufacture started by *Godey's* and *Demorest's Monthly*—permitted women anywhere to reproduce the prevailing fashions cheaply for themselves.[127]

As the fashionable world flourished, similarities of dress between the sexes had all but disappeared. It is true that, after the war, female fashion seemed to offer a multiplicity of plain and practical as well as ornamental styles unheard of during the 1850s. "If freedom in dress," wrote Croly in 1869, "is to constitute part of the millennium for women, we must be gradually approximating toward it for never was greater latitude allowed than now."[128] Croly no doubt knew her times. The attempt, for instance, to inflict on women the "Dolly Varden" costume, with its bulging bustle and gaudy pannier, met a fatal resistance. Furthermore, no market could longer sustain the fading life of the great steel and whalebone crinoline or the tilting hoop of the 1850s, which by the 1870s had almost passed into oblivion.[129] Still, fashion "over-ruled everything."[130] Tightly fitting garments with fantail trains, accentuating the physical line of the female body, replaced the geometrical extensions of the earlier period; ornamentation never lost its appeal. Feather fans, parasols of all descriptions, and most of all, hats trimmed with flowers, beads, berries, birds, feathers, and animal fur came into vogue.[131] The English reformer Emily Faithful was "dismayed" by the new upsurgence in fashionable display that she noticed during her trip to the United States in 1873. She agreed with George Eliot that women—especially American women—submitted to standards established for them by men, and were willing "foils" for their husband's wealth.[132]

In 1875 the *New York World* marked the passing of the "King of the New York Bohemians," Henry Clap, and with him the end to an old way of male living and male dressing. By that time men had lost, it seemed, any pretense to the fashion consciousness of the 1850s. The variety in male apparel that technological advances had helped create for masses of men ceased to characterize male dress, confirming the fact that technology often does not shape or bears little relation to changing patterns of social behavior. "For some reason," Higginson noted sadly, "the whole male sex among human beings has now suddenly dropped into plain and almost colorless costume. The gorgeous tints of the past linger still only in the necktie, and are vanishing there. . . . If men have lost display, they have reason to regret the loss. Let women hold it, while they may."[133] Black was the dominant color in male dress. Men wore stiff, uncomfortable collars, woolen coats and trousers, and flannel underwear. "Baggy unmentionables" superseded Prescott's "white inexpressibles."[134] After 1865 designers created the creased trouser leg to conceal completely the line of the male leg: "tubularity" identified the typically dressed man.[135] Male fashion highlighted function and utility and suggested power, at the same time erasing the male body.

"Man," wrote the twentieth-century psychologist of clothes, John

Flugel, in a classic statement on this development, "abandoned his claim to be considered beautiful. He henceforth aimed at being only useful. So far as clothes remained important to him, his utmost endeavor could lie only in the direction of being 'correctly' attired, not being elegantly or elaborately attired."[136] Flugel did not add that man had been culturally conditioned to abandon beautiful clothes over a long period of time. To beautify the self for its own sake implied a radically different cultural context, a different concept of time, and a different ordering of human priorities. It meant, for one thing, a preoccupation with the erotic (rather than the exclusively genital), which by its very nature was profoundly present oriented if not free of time altogether. In this sense women still reflected an older world view, or at least, in a larger culture hostile to the erotic, the attenuated remains of another way of life.

For many Americans the relative timelessness and sensuousness of sentiment and fashion lay at the very heart of fashion's defense. Such arbiters of taste as Nathaniel Willis were drawn to sentiment and fashion because they made living a rather tolerable and enjoyable thing. Fashionable people wished to please others by their appearance both in public and in private. The fashionable man "bound" himself to appear "agreeable to those with whom he is thrown in association" and the fashionable woman "bound" herself to "present to her friends and society the most pleasant exterior she can . . . to appear gracious and lovely to the last so far as nature allows."[137] For Willis in the 1840s and 1850s, and for such fashionable feminists as Mary L. Booth, Jane Croly, and Mary Mapes Dodge in the 1860s and 1870s, the salon became the vehicle for the very best fashionable activity—good dinners, good wine, and, above all, "the noble art of conversation" presided over by the most urbane and informed women.[138] Such behavior helped to stop, or at least temper, the "exhausting and accelerated" movement of contemporary life. Fashion caught time in a series of "unrepeatable moments . . . insulating women (and the men in the shelter of their company) from the material world, from time and space."[139] This was the point of view that made Martinique, Bermuda, and Havana so attractive to that American *flaneur* Nathaniel Willis; these were places where men and women gathered and talked together in the streets, elegantly and indolently, with an "air of Creole grace and *laissez-aller*."[140] At the same time, however—and here lay its principal contradiction—fashion itself undercut time, subverted social stability and fixed traditions, and was rooted in an economy that flattened differences and often transformed the fashionable civilities Willis and others so admired into social savagery.

Most men looked alike and most women, too, had succumbed to homogeneity of dress. Cocksure of its powers, *Godey's* in 1870 took responsibility for this remarkable diffusion:

> As the Lady's Book penetrates to the fastness of the Rocky Mountains and the Pacific sands, it is not strange that a homogeneousness greater even than that produced by speaking the same language should be the result. Two ladies from opposite poles meeting each other on the plaza or boulevard of the city recognize with lightning-like quickness the true fashionable height of the hat, the length of the plume, and the dress with or without the trail, as the case may be.[141]

Such a sameness of middle-class female style offered convincing evidence that beauty, color, and curving lines were, indeed, intrinsic to the very soul of women. Homogeneity in dress, furthermore, proved for some that cultural priorities had been satisfactorily ordered through gender.

These sartorial differences were so great that they drew wide-ranging reflections by contemporaries as well as by later historians of costume. It seemed to those who thought about the matter that Americans—women in particular—could not resist fashion's magnetic attraction. One man insisted that the craze existed deep within women themselves, that they could not help themselves. "It seems to me," he concluded, "that a weak and foolish desire to 'adorn' themselves underlies it all."[142] Another wrote that "there is something quite ludicrous in the precipitation with which American women follow all the fashions emanating from Paris. . . . The wonder of it is, how the women of all grades can so speedily conform to fashion."[143] Some called this infection the inevitable "bondage of the furbelows" while others bemoaned the "fashion makers" as the "powerful stimulators of the clothes mania."[144]

A few observers welcomed this subservience to patterns of dress as a means through which Americans might be effectively constrained and even controlled in the public realm. E. L. Godkin of the

Nation, for example, elaborated on the social and political advantages of similarities of costume. "National costumes were very pretty things," he wrote in 1867,

> but they were the outward signs of things that were not so pretty. It is a lamentable fact that the more diversity there was in attire of the different nations of the world, the worse they behaved. . . . As long as similarity of costume helps to draw men together, as long as they feel less compunction about blowing out the brains of a person with a turban . . . than a person with a stovepipe, we cannot as moralists regret to find the population of West Arkansas wearing as their daily working clothes "full evening dress" of more civilized regions. . . . It may safely be asserted that the general diffusion of the same costumes the world over as part of the graceful process of assimilation going on among the nations of the earth is something that friends of humanity and civilization rejoice over.[145]

Godkin's words expressed a general nineteenth-century attitude toward dress in a democratic context. It rested on the belief that dress reinforced, shaped, and even changed social behavior, and even that similarities in dress expanded the range of opportunity for all people: equal dress made everyone equal.[146] Godkin, however, desired homogeneity for different reasons. He was afraid of the social and political dangers inherent in a free, diversified society, and he believed that only strict limits put on individual expression could insure the survival of a democratic society acceptable to all.

An etiquette of fashion accompanied the growth of the industry, an etiquette that complemented even while it contradicted the cosmopolitan etiquette propounded by Nathaniel Willis, Genio Scott, and other "theoreticians" of fashion. All men and women were expected to study dress as an art form. Fashionable people knew which clothes to wear for the right occasion, the right complexion, and the right age. Women dressed appropriately to their "seasons." In youth, or the "spring of life," they wore a wide range of colors and styles, while in middle age, or woman's "prime," they abjured both ostentation and "dowdiness." In the fading autumnal years, black, white, lavender, and gray became suitable to women. "These things," Scott declared, "are every woman's duty to observe as long as she lives."

Men, on the other hand, were "bound to make conscience" of clean boots, fingernails, and linen, and to locate their "dress centre" or "nucleus," from which the rest of their dress whorled and unfolded in tasteful harmonies.[147]

The etiquette of fashion included more than good taste or rules of dress. Fashion made sense, for some, only as it supported the sentimental domestic configuration, or, to use Sarah Hale's words, only as it prevented women "from doing what men do."[148] In a long tribute to the glories of clothing, Hale wrote in 1868 that "a woman who is careless about her personal apparel will not be apt to make her home pleasant. She must dress for her husband as she would have done for her lover, and be as agreeable as possible in her own house."[149]

This implacable etiquette marked the magazine and newspaper writing of both antebellum and postbellum periods. Historians have often observed it. After the war, however, a relatively new form of etiquette emerged that showed the effects of important changes in the social relations of men and women of the middle and upper classes. According to Jane Croly, the practice of receiving and sending cards to the houses of friends and acquaintances in order to gain entrance gradually became a burdensome necessity of social life. "The freedom with which large parties of unknown persons formerly went from house to house is not tolerated," she wrote in 1878. "Gentlemen rarely associate together in greater numbers than two or three, and very many will only call at the houses of their personal friends, or upon ladies from whom they have received cards."[150] On the face of it, the system of cards, like the Comstock laws, seemed to betray a new and tighter rigidity in the relations between the sexes; in reality it constituted a move to enforce order at the very moment when these relations had become confused and volatile in unprecedented ways. The industries of fashion themselves, and most of all the dry-goods palaces, had contributed to this volatility, not only by bringing women into the public space to shop in great numbers but also by employing them in positions from saleswomen to department store superintendents. In the 1870s women and men had begun to mingle together in new ways—in places of work and social pleasure, in schools and universities, and, of course, in the forum of reform and politics. The feminist movement took hold, as never before, in this decade. Even while Mrs. Grundy challenged this unusual integration with a new weapon of constraint—the card—feminists responded with a greater challenge.

Early Dress Reformers

NINETEENTH-CENTURY dress reformers loved to tell a story so old and worn that variations on it existed as far back as Juvenal and Martial. It told of a young man who accidentally meets a beautiful woman on an afternoon promenade. She has black, lustrous hair, white teeth, red cheeks and lips, a full bosom, and enviably perfect posture, and wears an ornamental costume of satin and silk. The young man is overcome by her beauty and proposes marriage on the spot, a proposal she eagerly accepts. The man's haste, however, seals his fate. On his honeymoon night he discovers he has married a grotesque and misshapen witch with false teeth, false hair, no color in her cheeks or lips, a humpback—a deeply curved spine which had been obscured by the weight and fullness of her dress—and an India-rubber contrivance for a bosom. She sleeps in dirty clothes and smokes a pipe at night.[151]

Such a tale amused the initiated and warned naïve men and women who had not yet confronted the full power of fashion. It was also one of the several feminist morality tales that showed how much the dress reformers themselves feared and parried with the imperious mistress fashion, how much they were aware—as the fashionable were aware—of the pervasive influence of dress. For the early reformers especially, fashion was a poison, a "pestilential excrescence" of a new society, "a vile breath of pollution" that spread slowly until it struck and disabled the very heart of a woman's identity.[152] The president of the National Dress Reform Association, Lydia Strobridge, declared in 1863, "We must grapple with the monster Fashion, with false habits of dress."[153] According to some accounts, subduing the "monster Fashion" often changed women overnight into vigorous, renewed, and rational human beings. "I find life desirable once more," one woman wrote after she rid herself of fashionable dress. "Weakness and debility" fled, wrote another, "once I learned that I must seek out and obey the laws of life." Lydia Sayer Hasbrouck, the editor of *Sibyl*, solicited these reports and received many claiming remarkable changes. Repudiating fashion always resulted in a mixture of psychological and physical exhilaration.[154]

The first phase of the dress reform movement crested in the 1850s, the decade of new stores and a new wave in fashion. In the 1840s and early 1850s, Amelia Bloomer's upstate New York temperance journal, *Lily*, and the New York City *Water-Cure Journal* represented the principal organs of dress reform; both introduced the famous Bloomer dress and the less famous Weber dress. A grotesque blend of Turkish and Quaker—a simple woolen skirt and merino pantaloons—that offended the eyes of even Gerrit Smith (a fanatic on dress reform), the Bloomer dress was originally a gymnastic uniform used at the water cures in the 1840s, first worn publicly by Elizabeth Smith Miller and others around 1850.[155] The Weber dress had a fleeting popularity in this country, although it apparently found a lifelong advocate in Dr. Mary Walker, the famous Civil War doctor, who had a penchant for wearing male clothes. Usually accompanied by "clipped hair done-up in the male style," it "resembled a man's suit in every particular."[156] In 1856, with the demise of Bloomer's *Lily*, Lydia Sayer Hasbrouck —who wore the Bloomer costume herself—took the reins of dress reform journalism with *Sibyl*, a feminist paper that survived seven years, specializing almost wholly in dress reform. By this time reformers had begun to meet at dress reform conventions in western Pennsylvania and especially in upstate New York. In 1856 they organized the first National Dress Reform Association at Glen Haven Water Cure in Glen Haven, New York. By the late 1850s seven hundred "practical dress reformers" were known to exist in the United States, and in 1863, when the association met for the last time, in Rochester, eight hundred men and women attended.[157]

The early dress reform movement was not centered in the largest cities. The bulk of its membership was in the large towns and cities of New York (upstate), Wisconsin, Ohio, Illinois, Massachusetts, Pennsylvania, and Iowa, and it had a smaller number in Michigan, Nebraska, New Hampshire, and Minnesota.[158] Most of these states were being transformed by a transportation revolution, the growth of manufacturing, a population boom, a dramatic expansion of markets, and the disappearance of household production.[159] It was a region struck by a revivalist-perfectionism that reflected the excitement of these changes and that tried to cope with them in moral terms. Many of the early reformers, including Hasbrouck, Gerrit Smith, Harriet Austin, James C. Jackson, Mary Walker, and Mary Tillotson, were noted for their radical sectarianism, believing as they did that dress reform represented the most important of all feminist reforms.

Lydia Sayer Hasbrouck, wife of a wealthy Whig journalist in Middletown, New York, was a most radical individualist. She believed that married and single women should have "freedom in every relation with man," that both should be "individualized" and "free to choose their own occupations in life, mingling their works and counsels in every department of life." Like Lucy Stone she refused to pay her taxes, and as a result some of her own possessions were confiscated and put up for public sale.[160] For her, however, the key to freedom for women was dress reform and the reform dress was the only dress. She wore the Bloomer costume throughout her adult life, nearly relishing the ridicule it invited. "I have walked firmly through the ranks of the rabble," she announced. "I have met the fury of the mob. . . . What are the sneers of neighbors . . . to the purer metal of your soul."[161] In 1857 Lucy Stone gave up the reform dress as an intolerable nuisance, and many who had worn some form of it did the same, out of frustration or necessity. "If I put on this dress," Paulina Wright Davis wrote to Stanton in 1851, "it would cripple my movements in relation to our work at this time and crucify me ere my hour had come."[162] As Mattie Jones aptly put it, reformers were "watched" wherever they went, and could not walk a decent distance for fear of being pelted with "snowballs in the winter and apple cores in the summer."[163] Hasbrouck could not endure the moral "cowardice" of those who feared to wear reform dress and called Stone and others "traitors," women unworthy of the cause of woman's rights.[164] In 1861 she put capital into her own business, Sibyl Ridge, a hygienic retreat in Middletown, New York, complete with baths, electro-magnetic equipment, and hot-air furnaces; and in 1863, her journal defunct, she departed the ranks of the main feminist movement forever.[165]

The early dress reform movement was directly tied to the utopianism and secular, hygienic perfectionism that characterized the reform experience of the antebellum period. *Sibyl*, for example, published articles and advertised the books written by Fourieristic communitarians, free-love sexologists, advocates of Rochdale cooperative stores, and "positivistic sociologists." "It may be said," one writer offered in an article on the state socialist Louis Blanc in 1860, "that when Positivistic Sociology takes its place amongst the established sciences, and the Religion of Science shall have been practically inaugurated, then mankind would have begun to enter a new stage in progress and development."[166] Most of the early reformers were also strict Grahamites; as Sylvester Graham urged, they ate coarse and simple foods and eschewed all alcohol, tea, coffee, tobacco, and meat. After the meeting of the Cayuga Dress Reform Convention in 1857, the reformers attended a dress reform ball where a "Graham supper . . . was gotten up in a good style, purely Graham, with honey, apples and cold water."[167] Hasbrouck, Tillotson, Austin, and others had severed their attachment with established religion and turned to "nature" and science as the basis for the social organization of the future. As an official sign of this secular bias, the National Dress Reform Association of 1860 met in the assembly rooms of the Progressive Friends of Waterloo, one of the most rationalist of upstate New York reform associations. Also known as the Friends of Human Progress, it supported "no Church but the Church of Humanity."[168]

Early dress reform ideas reflected this rural, sectarian, and secular-perfectionist heritage. Dress reformers often argued from an extreme "natural rights" position for an American costume suited for republican citizenship, a new dress that expressed the "sameness of humanity" everywhere and that erased "distinctions of physical force, birth and rank," all things "fashionable, aristocratic, and European."[169] Men and women, both Elizabeth Cady Stanton and Gerrit Smith believed, had a "common nature;" therefore their dress should be as common and similar as possible, consistent with good sense and decency. Stanton wrote to Smith in 1856,

> Believing as you do in the identity of the sexes, that all the difference we see in tastes, in character, is entirely the result of education—that "man is woman and woman is man"—why keep up these distinctions of dress. Surely, whatever dress is convenient for one sex must be for

> the other also. Whatever is necessary for the perfect and full development of man's physical being, must be equally so for woman.[170]

James C. Jackson of Dansville took this position as far as it could logically go. Dress styles, he thought, should be the same for both sexes. The only distinction might be established by "badges . . . to make plain the difference of sex . . . the men wearing something on their clothing that would indicate their sex, and the women wearing something which would indicate their sex."[171]

The natural rights argument, however, marked the edge of a much deeper position, a physiological theory widely shared by generations of physiologists, phrenologists, utopians, reformers of all kinds, and even such "fashionables" as Genio Scott and Nathaniel Willis. A measure of the enormous contemporary interest in nature and the natural, the infatuation with physiology reached a high point in the 1850s because many Americans feared losing touch with the natural—the land and the primal earth—in a society that every day seemed more "artificial" and "civilized." "We breathe a freer, if not purer, atmosphere," one woman wrote from the Rochester suburbs, "here among the mountains, than do the dwellers in the cities, have more independence, are less subject to the despotism of fashion."[172] It was an air reformers wished to continue to breathe.

The natural had also become an alternative source of moral and religious energy that had nothing to do with traditional religion and offered a fresh moral context to a generation of religiously disaffected men and women. Dress reform rested on this physiological base, on an abiding belief in nature's laws. "By robing ourselves, exercising and living physiologically," Mary Tillotson said in 1858, "we know that we can develop harmoniously provided that civil and social equality come within our scope."[173] One obeyed nature's laws by avoiding the "aesthetic" laws of fashion. As Hasbrouck put it, "If only a woman would study the laws governing her physical being, instead of the fashion laws to adorn her outward seeming, we would hear less of the catalogue of fearful consequences resulting from civilization."[174] What precisely did fashion's laws do to a woman's body? For one thing, bodices, corsets, stays, and tight lacing prevented the free circulation of the blood, causing pain, congestion, headaches, and general torpor. For another, both the underwear and the overdress threw the body into a chaos of imbalance. Cold drafts from beneath and hot air trapped in the mass of petticoats from within set up a conflicting cross-current of air, endangering the physiological harmony of the body. This disequilibrium, combined with the heavy weight of crinolines, petticoats, and flounces hanging at the waist, drove the circulating blood to the internal organs, there to engorge, enflame, and overstimulate. Disease was the fatal denouement. The "lower organs" became overexcited and an abdominal region once unconscious was now conscious, creating an unhealthy condition in an area that (as one reformer baldly put it) already had a "hankering after something to stimulate and satisfy."[175] Given these conditions, dress reformers were neither shocked nor surprised by such common female ailments as prolapsed uteri, puerperal convulsions, hysteria, constipation, leukorrhea, and "uterine neuralgia."[176]

What did a healthy woman do to escape the nemesis of the fashionable? The healthy wore less clothing and distributed it more evenly over the entire body. As one utilitarian dryly remarked, dress was designed to cover the body.[177] Full coverage promoted a consistent body temperature to the tip of the extremities. Women should leave the body organs ungirded by dispensing with all artificial devices and removing the weight of their clothes from their hips, supporting it by suspenders or, more commonly, from the shoulders. The standard here was to be natural—"the only unchangeable standard of beauty we possess. . . . A dress can only be beautiful in proportion as it approximates the form it covers."[178] The standard was conventional male garb. "Man's dress," one critic said, "is allowed to fit his body; woman's body is compelled to fit her dress."[179]

Influenced by the model of Quaker costume, which "left all the extravagances off," by a general Protestant dislike for all gewgaws and artificiality, and by a utilitarian naturalism, early dress reformers also demanded an overdress free of all superfluous ornament. As Bloomer wrote in *Lily*, "The Lord God designed males and females to dress alike. . . . Clothing was not given to mankind for the purposes of ornamentation" or to "gratify personal vanity."[180] If adults had to adorn their clothing, then ornament should serve some "social use" or express "interior individuality," not "public opinion" or "civilized standards."[181] Early dress reformers scorned the cult of the ornamental for another reason: it threw woman's dependent status in relief with undiminished clarity. "An air of fashion," Mary Wollstonecraft wrote, striking the key for later feminists, "is but a badge of slavery."[182] "Self-reliance," "independence of thought and action," and "freedom of body," a married woman told *Sibyl* readers, cannot be "attained in fashionable dress."[183] Fashion reinforced woman's domestic dependence not only by restricting her movements and replicating the domestic limits in which she lived; according to reformers, it also appealed to her sexual "vanity," making her constantly alive to it and constantly aware of the role she must play in marriage. "Every part of a woman's dress," Stanton said in 1851, "has been faithfully conned from some French courtesan."[184] The courtesan was a symbol of dependency par excellence, and of every wretched feature of that dependency—a fashionable prostitute tied to the purses of men who, by her power to influence and inspire fashionable modes, transformed other women into prostitutes by imitation if not in behavior.

Mary Tillotson of Vineland raised this issue directly at dress reform conventions. Indeed, it remained central to her position into the 1870s, when she became concerned with social science and organized the American Free Dress League.[185] "The manacles of despotic fashion," she wrote to *Sibyl*, bound "woman to dependence and degradation;" they bound her to the "ministry of man's lowest wants," to the exigencies of "physical attraction," and they bound her through the agency of man himself, who used fashion to stimulate and control her sexual life. "At this period of revolution in the condition of women, it is important to teach woman that it is through the excesses of her passionel department that she has been ruled and degraded,"[186] she told a dress reform convention in 1858.

Twenty years later Lester Ward would develop this point at greater length in his *Dynamic Sociology*. According to Ward, man got the "pleasure craved" by imaginatively arousing in women "passion which nature does not spontaneously supply." Man accomplishes this feat by assiduously segregating woman from the outside world, from the world of male activity and attainment, or, to use Ward's words, by enveloping her "in a veil of secrecy . . . seconded by the material facilities which clothing and houses afford" and assisted by "all the prurient charms of society." Secrecy, fashion, and prurience: these elements fed the male and female imagination and thus created passion in both sexes. "By appeals to the imagination [man] actually creates the passion which Nature declares useless and withholds, but which he declares useful for satisfactions of his own."[187] The social outcome for women was to consign them entirely to the sexual function, to render them ill equipped to adapt to public life or to perform "useful" functions, and to "reduce the conjugal relation to mere animal gratification."[188]

Most early dress reformers believed that they could plead for "individuality," independence, and variety while at the same time driving the superfluous color, ornament, and theater out of dress.[189] They thought the self needed nothing but the self—the natural self—to depend on, that it could stand alone in austere purity like a diamond. It is not surprising, then, that reformers reinforced the deepening sentiment against expression in male dress. Nonfeminists, too, disapproved of male "butterflies." "The Dandy," wrote the editor of *Harper's*, "is the sum total of coats, hats, boats, vests, etc. . . . One is puzzled to tell whether he is a female gentleman or a male lady. He combines the weaknesses of both sexes, but knows nothing of the good qualities of either. *He does nothing*—either for himself or others. . . . In fine, his soul lies in his clothes."[190] Another critic agreed that dandies were "splendidly useless human beings, glittering simpletons, who follow the animal rather than the intellectual propensities." A correspondent for *Godey's* wrote of these "male creatures," "What did they conceive, we could not help thinking, was their *raison d'être?*"[191]

Dress reform parables often mocked "butterflies" mercilessly. In a dialogue between a "bee and a butterfly," which appeared in *Lily* in 1855, the butterfly addresses the bee somberly:

> There is one thing that has puzzled me. . . . It is true—you are such a favorite with mankind, they build you houses and even palaces, they feed you in the winter, if your provisions chance to go short, do everything for you that kind friends ought to do. But they will let me starve, and it not infrequently happens that the children chase me, till I am nearly out of breath. I know that if they could get hold of me, they would kill me for my gaudy clothes.

And the bee responds with resolute wisdom:

> The reason is simply this—they look upon you as an idle good-for-nothing fellow, vain and conceited—puffed up with that big head so that you can barely stand, and in my opinion they are not far from the truth. . . . If you wish to gain their good opinion . . . pull off that holiday suit—put on good stout clothes and go to work; in other words, become a useful man.[192]

The death of the butterfly became, in fact, the symbol of the metamorphosis of mankind to a higher, more pure, and more rational state of civilization: "After the political system has done its work, then comes another, religious and social, that shall . . . tear the gewgaws from the wicked great like the gaudy wings of the butterfly, and leave the crawling worm to be the disgust and loathing of all to see."[193]

Early reformers disputed with *Godey's* over the value of the separation of the sexual functions, but they did not dispute the significance of the deeper cultural values within the fashionable world view. Idleness was bad, but it was bad for both sexes. In some ways the fashionable man, even more than the fashionable woman, dangerously threatened the future of women, for by his dress he challenged the utilitarian priorities of American society. Such narcissistic backsliding might be excusable in those weak women who had been handicapped by their fashionable upbringing, but it could not be tolerated in men. After all, if men preferred display to action, private to public life, what could justify the feminist renunciation of display? Must feminists reject traditional feminine behavior only to have to face it again, like a haunting mockery, in the dress of men?

Butterflies Defended

IN NOVEMBER OF 1868, Harriet Austin, one of the passionate sectarians of the antebellum period, rebuked the *Revolution* for "having so little to say" about the "tyrant Fashion—the greatest oppression from which women suffer." "To dress beautifully," she wrote for the *Laws of Life*, "to make the exterior showy, attractive, charming—is the great thought instilled into girls, from their first consciousness; and it does more to ruin all true womanly character, to make them vain, useless and empty-headed, than any other simple idea ever taught them."[194] Five years later she censured the Dress Committee of the New England's Woman's Club for failing to attack every abuse in fashion from overdress to underwear, make-up to jewelry.[195] To some extent Austin's criticisms were justified, for although certain antebellum features remained basic to it, the feminist dress reform critique had undergone important changes and entered a new phase. For one thing, the center of reform had shifted from smaller to larger cities, from Rochester and Cayuga to Boston and New York. For another, it had become accommodationist, reflecting the rapid growth and development of the fashion industries and the existence of a class structure less visible in the rural districts.

Two important feminist women—Jane Croly and Mary L. Booth—became editors of two important fashion magazines—*Demorest's Monthly* and *Harper's Bazaar*—which drew this comment in the first volume of the *History of Woman's Suffrage*:

> In the United States the list of woman's fashion papers is numerous and important. For fourteen years *Harper's Bazaar* has been ably edited by Mary L. Booth; other papers of similar character are both owned and operated by women—for example, Madame *Demorest's Monthly*, a paper that originated in the vast pattern business which has extended its ramifications into every part of the country and given employment to thousands of women.[196]

Croly and Booth both tried to establish standards of class behavior for feminists and fashionables alike. "Nothing serves to indicate class and the habits of class," Croly wrote in 1878, "more than the details of walking and travelling dress."[197] Giles Stebbins of Detroit, Michigan, a feminist and later the chairman of the Business Committee of the state Woman's Suffrage Association, pointed out with pleasure in 1869 that women and girls in Massachusetts earned twenty-five million dollars yearly working in the textile mills and factories. A manufacturer himself and a member of the National Association of Woolen Manufacturers, Stebbins championed the independence and diversification of American industry. Like many who combined a zeal for capitalist enterprise with feminism, Stebbins saw significance in the proliferation of industrial employments for women. Women, he said, "can now command respect, meanwhile remaining womanly in character and conduct. No small item this, in these days when women are looking for their support and independence."[198]

To be sure, many feminist journals portrayed the miserable working and living conditions of women in the mills, millinery stores, and dry-goods palaces.[199] At the same time, however, many feminist women could not escape the lure of an exciting and bedazzling fashionable world; nor did they wish to. "We are still women, with womanly tastes," said the wealthy Brooklynite Laura Curtis Bullard. "The desire to please," echoed the equally wealthy Louise Chandler Moulton, "is a natural characteristic of womanhood. Each woman should have her own scent—to this woman, rose belongs, to this other, violet."[200] Caroline Dall loved Macy's and the elegance of Stewart's. Susan B. Anthony, who always had a sneaking fondness for dress, patronized a "tip-top" dressmaker in Rochester.[201] Mary Putnam Jacobi, Kate Gannett Wells, Julia Ward Howe, Laura Curtis Bullard, Lillie Devereaux Blake—all these women followed the prevailing modes, all probably detested the "unutterable garments" of Dr. Mary Walker, and all doubtless enjoyed shopping in the new stores.[202] "Even the hand of the strong-minded," remarked a correspondent for the *New Century* in 1876, "instinctively grasps her pocket-book before the bewitching array."[203]

Famous for her coiffure, Stanton wore black velvet and expensive mantillas and "did not pretend to have a soul above point-lace."[204] After some years of city life and a wider familiarity with the habits and conflicts of fashionable women, she lost some of her earlier hostility to the "butterflies of society" and proposed an alliance between feminists and fashionables. "Suppose we form a partnership," she said. "You imitate what is worthy in us, and we what is worthy in you, and by such a combination make an order of women that shall be the pride and glory of the nation."[205] Feminist women could invariably be found at their conventions throughout the 1870s dressed in "rich and elegant silks, trimmed with tasteful laces and cut in the latest fashion."[206] The New York Sorosis often met at Delmonico's, and in 1880 a hundred and fifty "elegantly dressed" women gathered there to dine on a fare of oysters, fillet of veal, boiled whitefish, French peas, croquettes, and duck with mushroom sauce—a far cry from the Cayuga dress reform ball where Spartan men and women feasted on honey, apples, and cold water.[207]

In the lives of many well-off feminist women, fillet of veal and cold water, elaborate attire and a refusal to wear a corset, feminism and fashion, could be combined without much conflict. For others, however, the attractions of fashion, the seductive invitation to a life of pleasure and bodily delight that only a few women had ever known before, tipped the balance and drew them away from the feminist movement. The glamorous Anna Dickinson was one woman who took this path. Her mother was a pious Quaker daughter of a Delaware slaveholder and her father a Quaker merchant and ardent abolitionist who, in the early 1840s, lost much of his property in a financial panic. In the midst of his failure, Dickinson wanted another child; his wife refused but, "overcome probably by threats of separation," consented and gave birth to a girl she did not want. "From the moment of her pregnancy," Caroline Dall wrote in her journal in 1862, after hearing the story from Anna Dickinson's own lips, "Mrs. Dickinson hoped the child might be a boy, that its soul would never be violated as hers had been. . . . Of course, Anna inherited the smothered wrath her mother bore at her heart, for nine long months."[208] Soon after her birth, Anna's father died, leaving her with four siblings and a mother who learned to love her, to educate her in "classical learning," and to rely on her—once Anna had become famous—for support. Yet her mother's original "wrath" and her father's early death scarred her psychological life, making her subject to "great emotional depressions" and to a yearning for love which could not be fulfilled.[209] Still under twenty, Anna Dickinson entered the reform ranks and lectured for four years on abolitionism and, after the war, on feminism. By the late 1860s she rivaled only one or two men on the lecture circuits in earnings and influence, grossing on the average twenty thousand dollars a year.[210] Independent, rich, arrogant, brilliant, an indifferent Quaker turned freethinking Unitarian, she became the subject of female adulation and feminist pride. Stanton, Stowe, Laura Curtis Bullard, Lucia Runkle Calhoun, and Jane Croly competed for her company, and Susan B. Anthony claimed her as a daughter, "my darling Anna . . . my chick a dee dee."[211] In the early 1860s, however, Dickinson began to buy clothes and jewelry in great quantities and developed a "passion for extravagant living that was beyond her power or will to curb."[212] She lavished gifts on friends and relatives, saved nothing, and became so well known as a fashion plate that

Charles Tiffany offered her a guided tour through his store: the diamonds she purchased there glistened on her fingers as she spoke to her admiring audiences.[213] "Her jewelry would make many a rich lady jealous,"[214] wrote James Redpath. In a sense Dickinson dressed according to feminist principles: she renounced corsets, crinolines, and elaborate headgear; but she also threw principles to the wind and indulged her passion for silk, satin, diamonds, and gorgeous colors. Her favorite shade was flamingo red.[215] "I heard Anna Dickinson last Tuesday," Louise Nell Grays wrote to Amy Post in 1865, "and I thought she was dressed in rather strange taste for a Public Lecture. She had on a red silk gown with a gored cut and trail."[216] Two years later Caroline Dall exclaimed to herself, after hearing Dickinson speak in Boston: "dressed, alas, like the Scarlet women of Rome!"[217]

Anna Dickinson thirsted after the stage and longed to join a growing number of American women—Laura Keene, Ann Stephen, Marian Foster, Charlotte Cushman—who not only had the power to play upon and move the feelings of a multitude of people, not only lived daring, autonomous, unconventional lives, but dressed more fashionably "than any other women in the world" and, indeed, set prevailing modes in fashion.[218] Dickinson's reformist friends—including Whitelaw Reid, the editor of the *New York Herald* and the only man for whom Dickinson felt any serious affection—pleaded with her to desist in her plans. "Oh, Anna," wrote one friend to her in 1872, "don't give up your usefulness and influence to gratify a passion of your own."[219] Such pleas fell on unreceptive ears. In 1876 Dickinson made her debut as Anne Boleyn in a play by that name, and in the early 1880s she took up male roles, including Hamlet.[220] In our own time Anna Dickinson may be considered, by some, to have been the most emancipated of women; resembling the "new women" of the early twentieth century, she ached for new experiences, captured some of them, and tried to live absolutely on her own terms. Yet by throwing off religion, domesticity (the maternal tradition), and reform (the paternal tradition), she left herself unprotected from the full weight of the fashionable world. Only moderately successful as an actress, never repeating her older conquests as a reformer, and always vulnerable to deeper and deeper depressions, she ventured into the theater and into fashion and found only misfortune, estrangement from friends, and bitter isolation.

In the dress reform movement as a whole, the compromise between fashion and feminism played itself out in less dramatic, less sad, and ultimately more productive ways than it did in the life of Anna Dickinson. To be sure, many antebellum sectarians—especially Mary Tillotson, Harriet Austin, and James C. Jackson—still gave shape to the reform critique. Others merely nagged the movement. Appearing often at feminist events in male evening dress and top silk hat, Dr. Mary Walker was "the Queen of the Utterly Ridiculous," the marplot of every women's convention.[221] While Dickinson thrilled her reform audiences in flamingo red, Walker enraged them in stovepipe black. At the NWSA Washington Convention in 1873, Walker nearly made Elizabeth Stanton lose control of her temper. "I endured untold crucifixion at Washington," she wrote Martha Wright. "I suppose as I sat there I looked patient and submissive, but I would have boxed that Mary Walker with a vengeance."[222] It was a tribute to the tolerance and openness of the feminist platform that for twenty years women like Stanton endured Walker in silence without repudiating her in public. By the late 1870s, however, feminist tolerance for Walker had grown very thin. At the 1879 NWSA Convention she was hustled off the platform "as an unwarranted intruder," and a year later she was elbowed to the back-row seats of the convention hall.[223]

Walker's eviction from the feminist platform marked both the rise in the number of feminist women and the decline of sectarianism in the feminist critique of fashion. All attempts to revive the moribund National Dress Reform Association, with its dual focus on overdress and underwear, failed, although Tillotson's short-lived American Dress Reform League appeared in the early 1870s and local societies (mostly anticorset) sprang up time and again in the 1870s and 1880s. Dress reform, however, did not expire with the National Association or with sectarianism; according to one report, more than two thousand dress reformers were active in the United States in 1873.[224] Moreover, reformers worked within the larger, more organized postbellum feminist societies, including women's clubs, the AAW, state social science affiliates, and social science congresses.[225] In 1873 and 1874 both the New England Woman's Club and the New York City Sorosis set up dress reform committees, appointed women doctors to staff and direct them, and together organized a coordinated series of public lectures to be given in "free chapels" from Boston to

Brooklyn. Nantucket Sorosis, Brooklyn Woman's Club, Woman's Club of Orange, Business Women's Union of Brooklyn, and Isis participated in this series and invited the women doctors to speak. All the doctors lectured in their designated places at precisely the same time, launching a "simultaneous attack" on hygienic abuses in dress.[226]

Dress reformers now "proposed no marked alterations in our present appearance" or "any radical change . . . in the externals of dress."[227] Some, of course, continued to rake selfish fashionable women over the coals and appealed for similarity in dress for both sexes. "The true idea of fashion," Stanton wrote for the *Revolution* in 1869, "is for the sexes to dress as nearly alike as possible."[228] Reformers also disparaged excesses in female overdress. The popular magazine writer Lucia Runkle Calhoun loved beautiful clothes, but she would not yield to vagaries of fashion. She wore a "peach bloom" dress with "brown over-dress" for years, after it lost its vogue. "It is very becoming," she wrote Anna Dickinson, "which few things are, and so I keep it not only by philosophy but with a calm content."[229] To all feminists excessive jewelry, ribbons, streamers, and cumbersome hat paraphernalia were signs of bad breeding, bad taste, female uselessness, and, according to those who read Darwin and Spencer, evolutionary backwardness.[230] *Woodhull and Claflin's Weekly* had this to say about a dandy: "He is a lad with his throat enveloped by the most exquisite neckties. His feet, neat and tidy, are encased in the most elegant of gloves."[231] Despite such scorn, dress reformers' proposals for women had a good deal in common with the style of dressing they found so reprehensible in the dandy—cleanliness, gracefulness, efficiency of line, natural simplicity, and unobtrusiveness.

Mary Safford Blake wanted dress reform booths set up at all industrial expositions, where "doctors of dress, specialists in the profession" would advise women how to combine "self and clothes into an harmonious whole" and "see to it that we are so well dressed that no one can tell what we wear."[232]

Nevertheless, in spite of continued attacks on excess and overdressing, reformers made distinctions between private and public dressing nearly unheard of among reformers in the earlier period. At home and at play women should "fling utility to the winds and feed their sense with sweet satisfactions of beauty and grace." "Though forced to play the role of the bee," wrote Abba Gould Woolson, by 1874 the new leader of dress reform, "we are never to forget that of the butterfly. . . . Let us work when we work and play when we play, arranging our lives according to cooperative principles." Above all, avoid dressing like men, who have chosen the same dreary clothes for all occasions. "In casting aside their inconvenient fineries," Woolson continued, "when gentlemen were more ornamental than useful, and by adapting a suit fitted for business and work, they have forsworn all richness and variety of color and ornament. If woman's dress-reform were to mean this, we might well dread the sober look which the world would wear. . . . While we retreat from one extreme, let them retreat from the other."[233]

Public dressing, however, was another matter. Women must wear the "raiment" that work in public "demands."[234] Reformers concentrated on woman's dress in the public realm and in the areas more and more populated by women—colleges and universities, factories, business, and, to a limited degree, the professions. Of women in factories and business, Woolson wrote in 1874: "Now that women have gone from the home to the workshop and counter, the demands of business will force them to attire themselves as not to impair the market value of their labor."[235] By 1878 Lydia G. Bedell, a doctor from Chicago, could write to the *Inter-Ocean* that "there is a crying demand for simplicity and comfort and convenience in business suits for the thousands of teachers and doctors and artists and saleswomen and bookkeepers and workers in all sorts of industries."[236] Croly observed of professional women that they "are gradually forming a class of their own and are compelling their dress to obey the modest and sensible requirements of an anomalous and difficult position."[237] In the public space, therefore, male dress set the right example: it was perfect in a rational way because it called absolutely no attention to itself.

Reformers focused most of their fire on female underwear, attacking the citadel of fashion from inside out. According to Woolson, reformers had been searching for years for an underwear that would permit "feminine" dressing and remain hygienic. In June of 1874 they displayed some of the fruits of their search at the Freeman Chapel in Boston, and invited Mrs. O. P. Flinyt of Stewart's, "the caterer to the ultra-fashionable," to criticize their handiwork and help them design some sensible underclothes. Mrs. Flinyt quibbled, scoffed, and subsequently came up with the "chemiloon," the "chem" or the "chemlin" (as the Chicago journalist Jane Swisshelm called it), a garment that almost completely covered the body, combining flannel or cotton pantaloons with a long-sleeved flannel or cotton chemise. In effect, this garment was really a hidden Bloomer designed to subvert the overdress by forcing it to conform to the lines of the body. Mary Safford Blake discussed the advantages of the chemiloon before a mixed audience at the Second Woman's Congress in Chicago; it was a discourse notable for its frankness. "I will say legs," she declared, "whenever I mean legs." Loud applause greeted both the chemiloon and Blake's candor. Her speech, coupled with the exhortations of others, triggered genuine demand. Mrs. Flinyt was innundated with orders, and, as if to highlight the friendly relations between fashion and feminism, both Arnold Constable and the Dansville Health Home became distribution centers.[238]

The emphasis on underwear broke with the past, but the commitment to hygiene tied Woolson and others to the legacy of the 1850s. "What is needed now," Woolson announced, "is not to assail Fashion, but to teach hygiene."[239] The hygienic argument was basic to the critique, the strongest argument available. At every public lecture the dress reform doctors had the same message: wear nothing that overstimulates the body or threatens its equilibrium. Overexposure, crinolines, corsets, the weight of petticoats—all these things rendered certain parts of the body more sensitive to stimulation than other parts. Hygienic dress, however, set "every part in its true relation to every other part," allowing "all the functions to go on without consciousness." Hygienic dress liberated the "graceful, natural woman," the walker, the activist, the selfless servant to the cause of human betterment, the "true woman," from the ennui, false modesty, sexualized currents, and subterfuge of the fashionable world.[240]

Urbanism and the rejection of sectarianism distinguished the postbellum from the antebellum feminist critique of fashion. The willingness to work within larger organizational structures and relative conformity to the fashionable world and, by implication, to the definition of femininity embedded within it, also set this movement apart. Constant, however, was an abiding rationalist perspective, the desire to remove extremes, to direct human energies into channels "both innocent and useful," and to strip the body of "masks" and "deceptions" that concealed the pure and natural self. Like most middle-class Americans (only more so), feminists believed that freedom, individuality, and access to the public realm required an end to all masquerade and the rationalization of the sensuous life. Masks, costumes, all forms of personal theater, brought an uncertainty, an ambiguity, a mystery—even a "frenzy of excitement"—that reformers hoped to remove.[241] What they wanted, then, was what E. L. Godkin wanted, a "similarity of costume" that would draw people together, so that "friends of humanity and civilization" could rejoice.

The dress reform critique was neither bold nor brilliant. The feminist attraction to fashion, combined with the new freedom and economy of shopping and the new employments for women generated by the dry-goods and department stores, inevitably blunted its effect. Nevertheless, reformers did level a consistent, if shallow, attack on female commodity consumption, and they did herald the streamlining of women's dress that would mark the fashions of the twentieth century. Feminist emphasis on elegance, good taste, function, and natural simplicity, and rejection of extravagance, ostentation, and display would later become established standards of middle-class dress. Moreover, even though the political and economic content of feminism seemed buried within the ideology of symmetrical hygiene, the dress reform critique still carried the weight of that content. Alone, the critique would have challenged little: seen in relation to feminist ideology as a whole, it challenged a great deal.

Part 3

THE RELATION OF FEMINISM TO SOCIAL PRACTICE

CHAPTER 10

FROM PERSONAL DISUNION TO SOCIAL COMMUNITY: THE FEMINIST CAREER OF CAROLINE DALL

"I HALF WISHED myself *unsexed*," seventeen-year-old Caroline Healey wrote to a friend, "that I might sway the sympathies and passions of the mighty multitude at will. . . . I know that there are hearts whose trembling chords would vibrate with the thrilling power of my touch."[1] "I don't wish to be a Sarah Hale or a Lydia Sigourney, I want to live as *men* lived who aided their race."[2] As the eldest child in her family, desperately trying to escape her mother's invalid fate and to meet her father's expectations, she managed a nursery school and taught two Sunday-school classes. She wrote constantly —novels, letters, essays—and at fifteen had a minor literary reputation. In 1842, when her father went bankrupt and she was twenty, she left home with his approval to take a job as vice-principal of a young ladies' school in the District of Columbia. It was, her father told her, a responsibility "calculated to harden, and not to hurt."[3]

Throughout her life she worried obsessively, often to the point of hysteria, that her life was wasting away; she agonized over the future.

As an adolescent she had feared reproaches even when she was ill. It was wiser to be healthy than sick in a family where only her mother was permitted frailty. Her father shrewdly, and almost wickedly, cut through her excuses. "As to your health," he once told her, "that is all in your imagination. You will be well enough—if you had enough to do."[4] She came to hate people who dawdled and squandered time. "I hate a man," she wrote in 1866, "who does not know his own mind ten minutes at a time."[5]

Dall's anxieties mark a woman who, almost against her will, had moved into a world known only to men. Dall was an individualist. She felt the pressure of that individualism in painful ways. As a result, she labored to make for herself and for others a more communal world that would, while still supporting her right to personal autonomy, lift the weight of her anxiety.

In the 1850s Dall became an abolitionist, a preacher, and a feminist of major importance. After the Civil War she pioneered in the organization of the American Social Science Association. In many ways her career, her journey from individualism to community, was representative of the careers of other feminist women. Yet her experience of being separate from others and, above all, of being estranged from the world of women, set her apart from many other feminist women who had known, and never lost, a strong sense of female community. However, it is because of the intensity with which she experienced her strangeness, her feelings of being broken apart or disunited, that we can chart so clearly the decline of individualism in the feminist critique and the rise of a world view based on the principles of social science.

A Born Transcendentalist

CAROLINE HEALEY thought of herself as having been born dead, abandoned by a mother who was too sick to nurse her and by a father who had wanted a boy. Her account of her birth in 1822 is pathetic and symbolic. "My mother, Caroline Foster," she wrote, "was three days and nights in labor and her life was despaired of. At last instruments were used—they compressed my head and clipped a bit off each ear. I was born without a cry—and black and bleeding was thrown upon a pile of soiled clothes in the corner; no one having the time to think of me or dreaming that I could live."[6]

Her father, Mark Healey, was a self-made Boston merchant capitalist with more than fifty merchant vessels and great wealth amassed by exploiting markets in India and the American South. He ruled his wife's family and his own with almost unchallenged authority. Both Healey and his wife, rich, cultivated, religiously liberal, came from a long line of English clergymen, including Dean Whittingham, translator of the Geneva Bible and husband of John Calvin's sister, but both husband and wife practiced a rather indifferent Unitarianism.[7] In the introduction to her journal, which she kept all her life, Dall described her mother as a 'lively, beautiful woman—a most notable housewife—of tender feeling and more than ordinary intellect" and her father as a "strong-willed man" with an "unsocial nature."

Like many feminist women of her generation and class who willingly or unwillingly broke from the maternal bond, Caroline Healey Dall spent her life struggling to reestablish the maternal connection in some compensatory way.[8] In large measure she owed the making of her personal independence, her strength, her individualism, to her father. Walking at six months and spelling at twenty months, she read the world's masterpieces under his guidance and by her fifteenth year could speak and translate five languages—French, Spanish, Italian, Sicilian, and Portuguese. At sixteen she received for her birthday "two presents," one a "violent headache" and the other "a dozen volumes of French literature from my dear father, and the first, I trust, from my heavenly Father."[9]

Healey preferred and relied on Caroline's domestic services, not those of his wife, whom he shielded from public responsibility and conflict in the sentimental bourgeois manner. Always frail, ill, and also ineffectual as a household manager (hardly "notable" as a housewife), Caroline Foster Healey wasted half her life recovering from pregnancies; she was confined to bed for eighteen months after Caroline's birth.[10]

"I was mistress," Caroline wrote of her fifteenth year, "over two wetnurses, a chambermaid, and cook, and seven servants. I managed the affairs of business and education of my sisters and did many things which perplexed my mother's friends, because there were days when she seemed perfectly well and although she was not allowed out without me, she never went when there was the slightest doubt as to her fitness."[11]

Dall was given full power (at least in her own mind) to care for her youngest brother, Charles, the last of ten children and the second boy, whose birth had further crippled their mother. Like many nineteenth-century women, both feminists and nonfeminists, Dall worshiped her brother, seeing in him the boy she could never become; she loved Charles as she herself wished to be loved, and resented him for receiving the unambiguous attentions and favors of her father. That her father had apparently honored her with the task of his care doubled the weight of her conflict. Tragically and unexpectedly, the child died from a strange fever and, to assuage her guilt, she herself "carried the coffin to the funeral gathering."

"Into my arms he was thrown," she wrote her future husband in 1843, in a letter that must have been meant to elicit his pity and love. "I implanted the first kiss upon his brow. I devoted myself to loftier studies that I might be prepared to fit him for his college course and when he grew to a beauty so strange and startling, artists would seek him on the street. . . . He was my own—as much as if he had drawn life from me." Caroline married a man who bore her dead brother's name, possibly in part because of a need to make reparation, or to find another Charles with whom she might more successfully test the power of her love.[12]

In the role of companion and confidante to her father, Caroline Healey educated her sisters, tended her brother, and mothered her mother. No Oedipal resolution reduced her to relative dependence within the family. She acquired a preternatural sense of her own personal power, of her ability to control and shape, for good or evil, the character of other people's lives. As a young child she acted out violent fantasies built on hatred for her younger siblings and on her need to retain a favored place in her father's eyes. At the age of four, after she saw on the street a parade of prisoners on their way to be hanged, she played a game called "hang" with her younger sister Ellen. "Poor little Ellen," she wrote in her journal years later, "was senseless when my mother was able to cut the sash."[13] She never resolved the conflicts that aroused such anger, and never forgot the incident.

Dall's extraordinary sense of power in the home cast her in upon her own resources and into a private world of conflicts and resentments. Judging from several entries in her voluminous journals, she was overstimulated narcissistically by the emotional life of her parents, and by the intensely sexualized domestic currents in which she lived. She remained conscious of strong sensuous feelings—of what she called, in 1858, the "hot current in the blood . . . of all women."[14] She both acknowledged and sought to suppress such feelings. As a child she felt "intense love of beauty in Nature and Art," and as an adult she responded with sensitivity to the beauty around her. In 1865 she recorded a "sight to see and remember," a visual experience of three women: a black woman with "a crimson turban and long, drooping gold rings in her ears . . . a splendid Celtic peasant woman with black hair and eyes, warm color and long earrings . . . with two magnificent light-haired blue-eyed children. It was a picture for Murillo . . ." The third woman was "dressed in violet moire with a hat of same, with a white and violet plume, long earrings made of gold coin and some sort of green and gold scarf worn girdle-like."[15] She appreciated and took "pleasure" in the sight of male "physical beauty" also, and after the collapse of her own marriage she was constantly drawn into painful infatuations with young, handsome, and inaccessible men.[16]

Her dream life gave an even deeper insight into her sexual nature. "I had a queer dream," she wrote in 1879, "a dream generally *hinges* on something." Caroline dreamed of a man she knew whose son had just become engaged to marry. In the dream she greeted him in a drawing room, which "suddenly changed into a bed chamber and he was lying naked on the bed with a sheet so arrayed as to hide everything except the side line from hip to heel. It was a very beautiful line which no line of his body could be—and I said, 'How like one of the statues he is so fond of.' " She congratulated the man on his son's engagement, whereupon he "threw his hand over his eyes, seemed confounded and said 'Am I gone off my head!' " At this point the man's wife appeared and "it seemed to be immediately necessary that she should go up stairs—I put my arm round her to go up with her and immediately we moved onward floating as it were, with such exquisite rhythmical pleasure as I cannot describe. I said, 'This is like ascending to heaven with my own soul!' " When Dall awoke, "I was still conscious of a physical delight such as no enthusiastic dancer ever experienced."[17]

The influence of Caroline's parents bequeathed to her all the yearnings, feelings of power and powerlessness, miseries, and joys associated with the romantic state. As a young woman she identified with Margaret Fuller and thought Charlotte Brontë "the only woman who ever lived to whom it would be safe for me to lay my heart open."[18] By the age of forty she had read Brontë's *Villette* ten times.[19]

If Caroline's alliance with her father gave her much of her strength, her distance from her mother made her always skew and twist the effects of her love. She fought for fame largely to curry her father's favor and win his love. "My mother scarcely felt I was her child," she wrote Charles Dall in 1843, "but the love that grew up between my father and myself was different from any I have known."[20] At the end of her life, long after he had died, she was writing poems to him on his birthday. For her mother Dall felt a melancholy mixture of yearning, anger, and fear, and what she wrote of Anna Dickinson in her journal applied with the same force to herself. In her own mind, Dall too "inherited" a "smothered wrath"—against the mother who withdrew from her when she was born and against the father who had wanted a boy.[21]

At eighteen she wrote her mother: "Is there anything which I can do to please you? I will make any sacrifice and will give up—dearly as I love them—my books and pens. I will promise never to write or read another line, if you will give me the love a mother owes a child. Oh mother! am I your own or not?"[22] More often, however, she feared her mother's emotional power, a power compounded by the cult of domesticity, the sexual division of labor that made mothers the source of all nurture. Crushed by original maternal neglect and later by maternal lack of interest, she unconsciously knew the danger to her ego in projecting herself upon that treacherous screen. She feared her mother because she wanted to hold on to the power, energy, and direction she got from her alliance with her father. Her mother's love and all it meant—"confinement," frailty, and dependence—threatened to destroy her own dearly earned sense of personal autonomy. She did not behave in a conventional feminine way as an adolescent; she disliked "coming out" parties, opened her own doors and put on her own cloaks, and refused to dress up in "fancy muslin dresses." As an adult, she came to define "femininity" as "cowardice," and "womanhood" as strength tempered by tenderness.[23] "To have the strength of a man," she wrote in her commonplace book in 1877, "without losing the delicacy and sensitiveness of a woman, would be to conquer the world."[24]

Yet Dall understood reasonably well what her mother's passivity and her father's power had done to her. By singling her out for success (and none of the other children ever achieved any success, to say nothing of fame) while at the same time distancing himself from her in order to make her strong, Healey not only cut off one source of unambiguous love, he also made her a powerful stranger in her own home, stirring up bitterness within the family and a guilt in Caroline that always haunted her. Her sisters, her brother, and her mother resented her authority, withheld their love, did nothing to support her ambitions, and often mocked her "masculine" pretensions to fame. Caroline resented them for their "jealousies," their pettiness, and mostly she resented her mother for the power she, Caroline, had over her, power that made it seem to her that she herself had caused her mother's pain and forced her into confinement. In her own mind Caroline deserved to lose her mother's love. Above all, she hated while she loved her father, "who did his best," she wrote after years of reflection, "to embitter my life by making me love him to distraction and teaching me to distrust everyone else. My father's proud and devoted love actually did the work of hate. . . . I inherited peculiarities —disagreeable to others—that had no real root in my character."[25] Conflicts raised her out of the oblivion of history, but weighted her down with rage and self-contempt.

As a man Dall felt strong; as a woman dangerous. She never lost this sense of power, a sense that later gave her the feeling of being immune from pain or, more exactly, of being immune from the ordinary consequences of pain. She often wondered at her capacity, in fact, to "feel many things" one moment, and "nothing" the next, to experience "torture," self-doubt, and depression one night and then, the next morning, to become absorbed in the "perfectly pleasurable contemplation of the capers of crabs and colors of mosses, as if nothing else existed in creation." Dall attributed this "special endowment" to her self-sufficiency, "her independence" from the "need for sympathy," and compared it, ingenuously, to the behavior of "children and maniacs."[26] In effect she accepted, even exulted in, her narcissistic tendency to move, over the course of a single day, from feelings of great strength to a miserable invisibility.

In a democratic, individualistic society, psychic choice by parents rather than an inflexible system of male primogeniture potentially determines which children, male or female, shall inherit property or wear the family mantle. As in the case of many first-born girls who became leading feminist women, Dall's development was forcefully plotted out by her father; her mother established only a feeble (but for Dall, fiercely threatening) counterpoise to his influence.

Psychologically prepared for a boy, Mark Healey did not change his mind set once he discovered the sex of his first child. Indeed, he wanted more from his daughter than a precocious demonstration of managerial skills; he wanted her to succeed brilliantly and independently as a writer, and he cultivated in her a compulsive need to produce and act. As Caroline put it, simply, in 1897: "He loved himself in me."[27] He supervised Caroline's education in almost every detail, and also supervised her social life. He bolstered her family pride by giving her family possessions in her youth and pressed her to pay careful attention to her genealogy.[28] Although slightly uneasy over its implications (except when she felt like a martyr, which was often enough), Dall boasted of her descent from John Calvin on her mother's side and from the famous British cleric John Rogers on her father's. Later she sought to instill the same pride in her own children. When her daughter, Sadie, reached her eighteenth birthday, Caroline Healey celebrated the events with these lines:

> Kate Calvin's blood gives a strong will,
> Her husband's faith and sweetness.
> Saint Rogers lends a holier charm.
> The Martyr—austere meekness.
> Doubly descended from these four
> Should not my daughter honor
> By woman's firm and tender walk
> The stress they lay upon her?

No better we, for *rank* and *power*
Borne to the years behind us;
But better surely—we should be
When nobler lives incline us!
The blood which honorably coursed
Through double generations
Should have some power to shield *one* soul
From foolish iterations.[29]

She wrote these words in part for herself: she, far more than her daughter, suffered from the burden of "foolish iterations" and from the struggle to "honor . . . the stress" of what, in her own mind, was an onerous family inheritance.

In every facet of Caroline Dall's behavior we find the driving edge of what the historian Page Smith has called the "Protestant passion." It was a passion, strengthened by paternal power, that channeled her narcissistic energies. "This is what I want," she wrote at seventeen, "I want one young man and woman to be producing results if not on the world, on each other. . . . Action! Action must be our gathering cry."[30] In the same year she wrote of her desire, her grandiose dream which she half-fulfilled, to speak before thousands and to capture their minds—a desire born in part of her own thwarted need to express love and to be loved, but also of her need to make a mark on the world.

In the fall of 1844, after many delays, letters, and expressions of doubt, Caroline Healey married Charles Dall, a Harvard-trained Unitarian minister-at-large. Tender, feminine, "childlike," often sickly, and absolutely selfless, Charles Dall differed in every way from Mark and Caroline Foster Healey, and to that extent represented the ideal, but potentially treacherous, marital choice. Unlike Caroline's father, he was without vanity and without the need to judge, superintend, or compete; in the early years of his marriage he could only love, respect, and observe with pride the intellectual gifts and the splendid strength of his wife.[31]

"She finds or rather *makes* time," Charles wrote to his parents soon after the birth of his second child, "to do *twice* the amount of reading I do even now. She *writes* more pages than I do, week by week. . . . As you might suppose, her rule is 'never give up.' "[32] The next year he was writing, "For myself I feel continually that I am among the most favored of men. Wedded to one of the noblest souls of Earth, the *very* cheerer and helper of my better life, that I had always prayed for."[33] In the beginning of their marriage his temperament liberated Caroline's combustible productive powers. Wherever they moved in his ministry-at-large, Caroline worked not as a helpmate but as his equal: she contributed page after page of her own to the contemporary press, bore two children, and ran the household; she taught his Sunday schools, supervised his ministry, and directed the church finances. In 1855 Charles Dall left America without his family to run a missionary school in Madras, India, and never returned home permanently. Caroline did not go with him. She did not want to leave her family or Boston and had increasingly found her husband's pliant, deferential ways vexing. She persisted in building an independent life. She coedited a feminist newspaper, *Una*, for two years, traveled, published countless articles and books, and wrote, in a fifty-year period, nearly 85,000 letters. Until she received a sizable inheritance from her father, she supported herself and her children by writing, taking in boarders, and a small stipend from Charles.

Dall's heritage and the even greater influence of her father's ambitions could not have given birth to her autonomous spirit had they not been filtered through the individualist culture of antebellum Boston. In Dall's life, as in the lives of Elizabeth Peabody, Louisa May Alcott, Margaret Fuller, Lydia M. Child, Georgiana Bruce, and many others, individualism took the form of transcendentalism, a fusion of Protestant antinomianism and democratic liberal ideas. Not only did transcendentalism place an extreme emphasis on self-development; it also removed the self from its social context and beyond the limits of traditional authority. In this guise, the self found meaning and virtue only through direct contact with God, or, in the transcendentalist vocabulary, unmediated contact with the natural world. "I am a transcendentalist," she wrote Charles in 1843, and indeed, given certain aspects of her psychological development, the label suited. In the late 1820s, she attended and recorded Margaret Fuller's "conversations" and read—but did not always agree with—Emerson's essays. In the early 1840s she was deeply influenced by the transcendental Unitarianism of Theodore Parker.[34]

Solitary, self-absorbed, "habitually sad" and prone to fantasy, "avoiding all" her "own age," Dall found in transcendentalism her own personal romantic language. With coaxing from the Boston education reformer Elizabeth Peabody, she kept a detailed journal devoted to self-culture, through which she escaped into introspection, a journal, she wrote, "not of emotion" but of "thought stripped of feeling."[35] As a young adolescent she received much spiritual satisfaction from intimate communing with the natural world; she preferred it, she said, to the "dull and dead . . . spirit of humanity," and she wrote her friends of her intense love of caterpillars, spiders, rats, and mice—transcendental and "masculine" beings, perhaps, but also beings whom Dall saw as like herself, unattractive, uninviting, fending for themselves in dark and ambiguous places. "I strove," she wrote to Mrs. Sarah Choate, "to find out their natures . . . to master the enigmas of science."[36] "I saw life and light and love written upon every leaf and every pebble—upon the head of a dragonfly—and the wing of the butterfly as plainly as upon the great scroll of heaven."[37] To Theodore Parker, always a willing ear for such matters, she wrote that she had discovered the "Infinite Cause, Being, Conscience, self-compelled" and the "proofs of Love . . . in the soul of Nature."[38]

Like all transcendentalists, she acted out an inherent dynamic within individualism: her repudiation of human community compelled her to project her need for love and meaning upon a natural screen. Like an insatiable child—and, indeed, the child stands at the center of the transcendentalist world because he/she can possess the world completely without compunction and without guilt—she longed to "find out" the nature of things, to appropriate, and finally to become at one with the very heart of maternal nature. Such a compulsion gave an imaginative richness to Dall's generation's perception of the world, and in the following century it has continued to explain, in part, both the meaning of modern conservationist cults and the exploitation of the natural environment. It was such a compulsion that made Dall's own son, William, into one of the great naturalists of the nineteenth century.

"The Broken Vase of My Young Life"

IN THE SPRING of 1866, feeling unusually drawn and depressed, Dall paid a long deferred and desired visit to the Boston home of her close friend, Henry James, Sr., who had never failed to praise her or to aid her in times of personal trouble. She came ostensibly to talk of her falling out with Elizabeth Peabody, who had once chosen Dall as her protégé but had now turned against her for personal reasons. She spilled out her deepest feelings, revealing her need for warmer companionship, for understanding, for a "richer, fuller life;" she talked of all the travail she had known living in this world. When the time came for her to go, James said to her, "My dear child! *you* a born ruler to feel so!" "And then he sitting [James was hurt in a fire, resulting in the amputation of one leg] and I standing holding his hand with a light farewell clasp, he drew my head down with a gaze as compulsory as a blow and kissed me on the mouth, prolonged, steady and compelling as I might have received an electric shock—all the time making keen intellectual inquisition into its purpose. . . . I felt as if the kiss sealed a compact. . . . No man had ever treated me so before."[39]

This episode says much about James, who clearly had more sympathy for feminist claims than Henry James, Jr. He believed that "every" sexual "barrier" should go, freeing women's "activity in any line they themselves pleased."[40] The episode also says a great deal about Caroline Dall. It points—perhaps in a theatrical way—to the conflict between her individualism and her need to identify with a community beyond herself; for in spite of the fact that she insisted upon her right to independence and found exhilaration to an almost stony self-reliance, her desire to exercise power over herself and over others gave her lifelong misery. Almost every page of her journal, in fact, documents the pain of her "aching, weary heart." Nothing, she wrote in a typical passage, feeling like the despised spiders, rats, and mice she collected as a child and like the prostitutes she defended as an adult, can fill "the broken vase of my young life. . . . If there were only somebody on earth to care for me . . . to take off a little of the weight of fatigue and depression."[41] Dall also expressed these feelings in poetry:

> Let me enter into rest,
> I have worked my whole life through.
> Can no one come to love me
> For myself, as lovers do?
> Forgiving all shortcoming
> For the purpose in my soul.[42]

James touched her because he understood both her longing for love and her need for independence—"*you* a born ruler"—and tried to resolve her conflict, however briefly, with a kiss.

Dall's "need to be rocked in somebody's arms," and simultaneously to assert her dominion over others, penetrated, even poisoned, almost every relationship.[43] Her nagging feeling of never being understood or acknowledged often inverted her desire to serve into a quest for possession. "If I demanded of others," she snapped in 1866, using a verb that embodied her fears, "that they should annihilate themselves before I could receive them, as they have demanded it of me, I wonder what they would think."[44] "What is there in my love, my self-sacrifice, my care for others," she asked herself more self-critically in 1863, in a journal remarkable for its unsentimental character, "that makes them seem to resent it so?"[45] In the first year of her marriage she praised her husband for his "lady-like" tenderness, his selflessness, his otherworldliness—he was a man "willing to work hard for very little, disdaining money, willing to be hungry that others may be fed"—and longed to "give up control over her future" to him, hoping he would "soften her hardness" as women were supposed to soften the hardness of men. Yet she also attacked the very heart of her attraction to him: "Stand on your own two feet," she had written him before they married. "I must respect you, darling, as much as I love you;" and, as the date of their wedding drew near, she cautioned him to develop "perspective," "symmetry," and a strong, practical grasp of reality, to budget his time scrupulously so that he would not be emotionally robbed by his parishioners, and to stop wondering in "amazement" at the world like a child. "Wonder," she wrote him, "is childlike and men who have had some experience of life are not so easily astonished."[46] Her ambivalence led to the ultimate destruction of her marriage. It also infected her relationship with her son, whom she fought to keep from her husband and early persuaded to leave her home "to earn his own bread and his own clothes" while still demanding his unquestioning love, sympathy, respect, and obedience. According to his sister, William treated his mother like a shrewish "wife," calling her "morbid," "complaining," and "upbraiding." He bridled at what he described as her cold egotism, and she bemoaned his "ungratefulness," like Lear, and not without some justification.[47]

Dall's ambivalence also affected her relationships with the young

Unitarian men she occasionally took into her home as boarders and whose lives she attempted to mold and manage. Behind her efforts lay the original drama between herself and her dead brother, Charles, the boy her father had given her to make into "a beauty so strange and startling, artists would seek him in the street." She developed romantic attachments for two young men in particular, a Unitarian minister, Edward Towne, and a Boston physician, David Lincoln. The story of her affair with Towne from its beginning to its disastrous end unfolds painfully in her journal from 1861 to 1864. In almost every respect Dall superintended Towne's life while he lived with her: she filed and read his letters, taught him French and "Natural Philosophy," wrote up for him "lists of calls . . . most needing to be made," and "no matter how sick or weary . . . always sat up till he came home."[48] On June 2, 1863, unable to endure such supervision any longer, Towne left Dall's house and never spoke to her again. She did not know whether to adore men or destroy them, so she did both.

Like men and like some other feminist women, Dall suffered the consequences of being thrown from the secure sanctuary of the home, of being forced to reject the mother's influence, the relative passivity of a woman's life, and the reassuring bonds of female friendship, for a world of individual success, competition, and aggression. Like them she had to bear the weight of a democratic, individualist system of time that rejected the past, that put the burden of existence on the present generation, that inflated the power and promise of every child. Men, however, were always reared with the pleasant expectation that they would inherit a home, a wife, a mother of their own as reward for exposing themselves to the exigencies of a free-market capitalism. As a woman Dall could expect no such recompense; the world expected her to give that comfort to men. Uprooted or removed from the traditional world of women, she experienced the exhilaration and the agony of individualism as no nineteenth-century man could have.

To remain sane, to survive the passage from the psychologically safe world of domesticity into a chilling world of competition, self-reliance, and independence, she needed in some way to confront and resolve the dilemma. Caroline Dall needed, in other words, to shape her independent destiny in harmony with her life as a woman. Before her lay several options. First, she could have discarded individualism and returned to the sentimental world of her mother, or else have taken her father's place and established a home with a man or woman more "feminine" than herself. In a sense Dall had already tried this option by marrying Charles Dall, but she failed to find it fulfilling. In her early years she often waxed poetic about the "sweet-home circle" and the mother's place within it. Yet she had contempt for other women (and men) who reproduced her mother's patterns and who lived selflessly on the fringes of the real world. "I cannot fear your father," she wrote Charles Dall about his family in 1843, "for it is only women who misunderstand me and who never forgive in their own sex a self-reliance and independence which they always seek in the other."[49] She looked to men for close friendship.

Dall could have given up individualism, the sexual division of labor, and competitive capitalism and turned to some form of utopian socialism. She was no revolutionary but a merchant's daughter (many feminist men and women had merchant fathers), with a warm respect for property, a disdain for exangelical enthusiasms, a horror of mobs. She had an abiding need to retain her social status. The third choice for a woman like Dall was to integrate her desire for independence and dominance into a compensatory, reformist-feminist system of service, rational harmony, and communal order. She could try to feminize her father's influence without destroying it and thus save her mother (and her own autonomy and identity as a woman) from extinction. She did choose this option, gaining a variable mastery over her romantic impulses. The resolution provided the firm basis for her feminism and made it possible for her to keep a measure of psychological stability.[50] She became a Unitarian minister, an abolitionist-feminist, and an active member of the post–Civil War American Social Science Association. She committed herself to the reconstruction of American society into a more feminized, corporate, and rational system of relations.

Feminist and Social Scientist

BY THE LATE 1860s and early 1870s many educated bourgeois women had begun to enter previously all male professions. Many studied medicine, a few law, and a small, though not inconsequential, number preached in the liberal ministries. Enough women preached, in fact, for Julia Ward Howe to call a special "Woman's Preachers' Convention" in 1873, the first such convention of its kind in the Western world.[51] Universalists, Unitarians, and, to a limited extent, nonevangelical Congregationalists unofficially supported the integration of women into the established ministries and numbered in their congregations a high proportion of the male feminists. These denominations supported women not only because, as Ann Douglas has said in *The Feminization of American Culture*, the feminization of religion had invaded liberal thought by this time, but also because liberalism had been injected, in some respects, by a powerful reformist, evangelical, and democratic spirit of the antebellum period and had generated within itself a highly rationalist, almost fully secularized sexual cosmopolitanism.[52]

Universalists were the first denomination to ordain women formally as equals with men; twelve of its thirteen theological schools opened their doors to women, and in 1870 twelve or so women had graduated from such institutions. "Universalists are rather proud," wrote one Universalist spokesman in 1870, "that they are in the van of the woman movement, particularly in its relation to woman's work in the church."[53] In 1875 Pennsylvania Universalists forced the president of the Universalist state convention to resign "solely" because of his "non-belief in the ministry of women."[54] Unitarians demonstrated even greater enthusiasm for women. By 1870 women could preach in almost any Unitarian pulpit and attend most Unitarian schools; local Unitarian societies throughout the East and elsewhere gave women equal voice with men in parish affairs; and in that year the American Unitarian Association elected Lucretia Crocker to its Board of Executive Officers.[55] Crocker herself also served as one of the first woman members of the Boston School Committee in the company of other Unitarian women (among them Abby May, Elizabeth Peabody, and Kate Gannett Wells), and she shared with most feminists a desire to introduce science into the American school curricula.[56] In 1871, in spite of objections from some of the leading Unitarian men, the American Unitarian Association elected two more women to its board.[57]

Dall never went through the formal process of ordination, but she thought of herself as a member of this group of female ministers. "I consider that I *am* and *have* been a minister of the Gospel," she wrote in 1869 to an inquiring Unitarian minister, "and my principal business has been to reconcile congregations to the appearance of women in the pulpit by giving them a service likely to make them forget whether the preacher was a man or a woman."[58] From the 1840s onward she began to prepare herself for what later became an active Unitarian ministry. She had grown up in the bosom of Unitarianism, admired the leading, progressive Unitarian lights of Boston—Charles Lowell, William Ellery Channing, and Erastus Bartol—and in her maturity relied on the liberal ministry for advice and support. In the 1840s and 1850s she organized Sunday schools wherever her husband's ministry-at-large took him. Later, she began to substitute as a preacher for such feminist men as James Freeman Clarke, Robert Collyer, John Sargent, and Charles Ames; and, by the late 1860s she had become a well-known figure within the liberal Unitarian community. She preached wherever she could and was finally elected in 1868 as one of the two women delegates to the National Conference of Churches.[59]

Her ministry blended rationalist free inquiry with the Unitarian ethos of public service and self-sacrifice. On the one hand, Dall had absorbed the advanced, freethinking tendencies within Unitarianism that produced not only the brilliant rationalist scholarship of Samuel Johnson, Octavius Frothingham, and Moncure Conway, but also positivist-oriented Free Religion. Stimulated by the "muscular" scholarship of Theodore Parker and by the work of other scholars, she studied the literature of comparative religion, ancient languages, and the latest biblical criticism; she published two pieces of her own biblical scholarship, a sentimental, occasionally erotic assessment of Jesus called *Nazareth* and a learned if turgid examination of the work of the German Egyptologist Christian Bunsen.[60] In the latter Dall attacked the "stupid dogmatisms of the past" and labored to distinguish irrational "mysteries, discrepancies and dreams" from the universally compelling moral content in the Scriptures.[61] She became something of an expert on Confucianism and Buddhism and lectured on these subjects before the Second Radical Club, a progressive, sexually integrated organization given over to the theoretical study of religious questions.[62] She entered the ranks of Free Religion with many other feminists after the Civil War. "I consider myself," she told her friend Bartol, "as good a free religionist as Frothingham."[63] In the 1870s she contributed a long series of articles to the *New Age*, perhaps the most advanced Free Religionist journal ever published in the nineteenth century, and was interested in starting a Radical Religion Association in Boston that would be open to all Unitarians, male or female, radical or not, "who have anything good to say for the benefit of the community."[64]

Dall's ministry tended to be intellectual and rationalist and she respected such great rationalist women as Harriet Martineau and George Eliot. Still, she disliked the religious skepticism of her friends and "averted" her "eyes from the painful atheism" of the women she admired.[65] Even though she did not believe in "the prexistence of the soul" and rejected "any immortality into which I may not carry any conscious results of individual life," she never stopped believing that the crucified Jesus had a transcendent place in the religious history of the world.[66] She chose the practical, spiritually untroubling ministry of Christian service over dangerous intellectual speculation. Influenced especially by the social ideas of Joseph Tuckerman, a Unitarian minister whose work resembled the social gospel of the late nineteenth century, Dall recommended to other women ministers that they take a "parish" rather than a "preaching" ministry. In modern times, she wrote, the male minister had ceased to play his former role. Aloof, removed from his congregation except on Sunday, and increasingly given over to bureaucratic duties, he no longer spent time with his people intimately tending to their emotional as well as religious needs. In an uncertain, competitive age in which industrial forces had transformed social relations, men and women needed this very tending more and more; they "needed someone to hear and consider *who is not in a hurry*." She envisioned a new kind of ministry, a nonsectarian, nonproselytizing, therapeutic "committee of comfort" dedicated to the care of the "restless and unhappy" so that the "wheels of life would move more smoothly."[67] As she imagined for herself a powerful therapeutic role in society, so she imagined the same role for other women: given the chance, the power and authority, women could touch, salve, and rehabilitate the psychic lives of men and women in social and economic distress—the poor, the forsaken, the sick, the emotionally torn, confused, and alienated.

Dall was an abolitionist by the 1840s. She subscribed to and occasionally wrote for the *Liberator* and the *Liberty Bell*; she reorganized a fading antislavery society in Portsmouth, New Hampshire, with the aid and advice of Theodore Parker. While living with her husband in Toronto she distributed relief to fugitive slaves in that city. She made plans (never to be materialized) to lecture in the field as a Garrisonian abolitionist.[68] These acts—and especially the last, which she embraced in spite of her family's frenetic warnings against it—constituted a clear-cut rebellion against her merchant father, who had financial ties in the South and abhorred her abolitionist career. "I am sorry you intend lecturing on slavery or anything like it," her mother wrote, "as it would annoy your father more than anything that could happen."[69] An uncomfortable but needed estrangement between Dall and her parents lasted until after the Civil War. "An old merchant and his wife," Dall wrote Charles Sumner in 1856, "have not spoken to me for years, ever since I took the *Liberator*."[70] Regardless of the pain, Dall enjoyed her abolitionism and considered Emancipation Day the happiest of her life. And she felt this way not because the Garrisonians bolstered her independence and heightened her self-reliance, but because they gave her a way to act independently without rejecting paternal authority or social order. But she was no radical, and most Garrisonians by the late 1850s were not.[71] Garrisonian abolitionism appealed to Dall and to other women like her because it charted the path to a more scientifically ordered, harmoniously organized capitalist system based, in part, on the full integration of women into the social system.

Dall's feminist career, which began in the early 1850s and ended in the late 1870s, was checkered with notable successes and failures. Although she participated in all the major feminist organizational activity, from the national conventions of the 1850s to the state and local conventions, woman's congresses, and moral education societies of the 1860s and 1870s, she had less influence in the leadership than most feminists of her stature. Other reformers criticized her for personalizing her feminism at conventions and for failing to contribute to the movement as an equal. Anthony, for example, warned her throughout the 1850s to stop introducing facts about her own life into public feminist debate—a warning given in bad faith since such debate often intermingled private and public materials.[72] Moreover,

Dall's irrepressible desire to dominate alienated many feminists; because of it the Boston Woman's Club and the American Woman's Suffrage Association blackballed her out of the leadership in 1868. Deeply wounded, or "annihilated," as she might have put it, Dall reacted to this treatment with an almost antifeminist broadside in the *Woman's Advocate*. She argued that certain "ambitious" women had overorganized the movement and endangered the sanctity of the home, corrupting feminism with dangerous impurities.[73] Her charges only deepened the rift between herself and the movement she loved. Her importance as a feminist rests, rather, on her written output, an output so important that Carrie Chapman Catt wondered in the twentieth century how such a "great pioneer" had disappeared from the feminist ranks.[74] Her contributions to *Una*, which she edited with Paulina Wright Davis (and again with much conflict) in the 1850s; her articles and books on famous women, among them two women doctors, Marie Zakrzewska and Anadabar Joshee, the first foreign woman to graduate from the Woman's Medical College in Philadelphia; and two major feminist works, *Woman's Right to Labor* (1860) and *College, Market and Court* (1867), which sold widely in the United States and abroad, placed her in the front ranks of the feminist movement. In her review of *College, Market and Court*, Elizabeth Cady Stanton called it "a work of great value," and the *Woman's Advocate* described it as a document "for which all women should be grateful."[75] It was the clearest, most organized feminist statement of book length written by an American up to this time.

The historian Barbara Welter has made the extraordinary assertion that Dall "dismissed the so-called woman's rights movement" and "judged herself according to the conventions of True Womanhood." On the contrary, Caroline Dall threw her intellectual weight behind every position held by any major feminist in the nineteenth century.[76] "If you know anything of me," she wrote Catherine Sedgewick in 1857, "you must know that I am what is popularly called a Woman's Rights Woman. . . . I am willing to stand where Lucretia Mott has stood so long."[77]

She concentrated almost exclusively on education and labor, especially on discrediting theories against higher education for women. "No one knows yet," she declared in *College, Market and Court*, "what women are; or what they can do," and since what we did know was twisted by hearsay and myth, we should open all the doors of education to women and let nature test itself. Besides, women needed knowledge to achieve self-mastery, practice self-help, and develop "perfect symmetry."[78] Suspicious of sectarian, sexually segregated private schools, she maintained that "women will never be thoroughly taught until they are taught at the same time and in the same classes with men," and preferably in places with progressive curricula. She refused to speak at Antioch because, coeducational appearances notwithstanding, Antioch's president, Horace Mann, divided the sexes by curricula, discouraged women from openly competing with men, and frowned upon women speaking in public.[79]

Equal access to work followed logically from equal access to education: education unemployed in productive labor made no sense to Dall or to any other feminist. Only "ladies" raised objections to woman's work. "Now I believe," Dall declared, "that work is good for ladies: so let us look at the truth."[80] The truth pointed three ways: ladies had to work for reasons of health, necessity, and humanity. First, productive labor made women healthy by freeing them from idleness and from the unnatural repression of their "faculties," both of which resulted, in Dall's very hygienically biased mind, in "disease, depression and moral idiocy."[81] Second, many women worked because they had no other source of income or because they feared the degradation of prostitution. Here Dall questioned one of the principles of political economy which ascribed the low wages of women to a glutted work market: employers paid women less not because they foolishly competed with men in a crowded market, but because most men had an insatiable "lust for gain" and would brook no competition from women.[82] Third, woman's status as a human being depended on her ability to produce goods and services, to "own herself," and to buy, share, and sell property. Dall believed that without property or wealth women would not only be less than human, they would also be unable to change the condition of their own lives or to affect in any way the lives of other women. "Why be educated," she asked Catherine Sedgewick, "if not to labor—why labor if not to acquire capital of some sort—including money—the great motive power."[83] With this in mind she lent her support to woman's industrial schools;

advised the "very best women" to enter the fields of medicine, law, and pharmacy; and begged women to become (like her father) hard-working, high-level capitalists, "the heads of firms, the movers of great undertakings, the contractors of supplies, persons conversant with large interests."[84] "Now I should rejoice," she wrote, "to see a large Lowell mill wholly owned and operated by women."[85]

Such ideas appear again in Dall's comments on marriage, for marriage too had value only as it expanded the possibilities for "self-development." For nineteenth-century feminists the principle of growth, of ever developing youth, was at the heart of life because it measured the worth of all institutions. "We outgrow all things," Dall asserted in *College, Market and Court*, "most of all we outgrow, have outgrown our laws," and to her close friend, the utopian reformer William F. Channing, Dall wrote that "I do not feel that marriage is here an end. I think like other relations it is a means to growth."[86] In exchange for the traditional sexual division of labor, therefore, she offered women the model of "two workers" carrying on equally the private and public functions of life: "occupied with real service to men and each other, how happily will they meet at night to discuss the hours they lived apart."[87] She came close to suggesting that "exceptional" men and women might need to find fresh "counterparts" throughout their married lives, to feed their natural passion for growth. "Great souls," she said in 1859, "creatures of genius, are rarely happy in marriage—they are young, growing to the last, and the counterpart soon becomes an analogue."[88] Extremely wary of speaking publicly on divorce and personally suspicious of most people's motives for divorcing, she nevertheless counted William F. Channing and Mary F. Davis, a divorcé and a divorcée, as two of her good friends. Often aware of her own failings, she did not readily condemn the faults of others, especially others who had no immediate emotional connection to her. During a conversation with some friends she once remarked that "we must love whom we could and could not stop to count the years more or less."[89]

Unlike Elizabeth Cady Stanton, Caroline Dall was not known by her contemporaries as an open advocate of marital and sexual reform. She ignored, she said, that messy and sensational field for the championship of woman's right to unfettered professional advancement. In 1876 she refused to speak on sexual subjects before the

Washington Moral Education Society on the grounds "that silence and a natural life with little morbid dwelling on sexual themes is as good a remedy for present trouble as any I see suggested;"[90] and fourteen years later she wrote in her journal that "fifty years has taught me that no audience can be trusted with sacred private themes. The attempt to make physiological knowledge popular has already done great harm."[91] Yet she was temperamentally drawn, both willingly and unwillingly, into the vortex of sexual discourse, the unrelenting sexual debate and reform that stood at the heart of nineteenth-century feminism. From the 1840s onward she never failed to keep herself up to date on the latest sexual theories or to attend public (and private) scientific lectures on "sacred private themes"—the first being John Quincy Adams's lecture on physiology, population, and procreation called the "Foundation of Society," which she heard in 1840 at the age of eighteen.[92] In the next two decades Dall assimilated everything she could on the principles of rational physiology and hygiene and lectured on these subjects, as well as on prostitution and the medical education of women. In 1859 Elizabeth Peabody rebuked her for "one-sided" absorption in sexual matters, and in particular for publicly mentioning "eunuchs" and explaining their "peculiarities."[93] During this time she also developed her own theories of "impregnation" and "venereal disease," borrowing from current theories. Of impregnation she wrote in 1859: "The power of impregnation is precisely the power the women consent to give, no more. . . . Did resistance continue to the end, conception would be impossible. . . . I would myself defy any power to give me a child not born of my own love."[94] Of venereal disease, ten years later: "The deteriorated male secretion—the result of excess—is itself the poison. . . . The uterine surface absorbs the poison. Happy healthy coition in which pure souls and bodies meet is nutritive, if childless—I have often seen it prove so—but it is only to absorb corruption that woman yields to a debauchee."[95]

In 1868 Dall implicated herself in a plan to stop a local outbreak of "chronic" masturbation. In January of that year John Worcester, principal of a private school in Newtonville, Massachusetts, asked Dall's help in reforming young Willie Elmer, a boy who not only constantly "abused" himself but also corrupted other "little boys and girls" with "indecent, boastful and false conversation. . . . I do not wish that he may die soon," declared Worcester, fully aware of the "degenerative" consequences of masturbation, "but if Providence sees best to order it so, I shall rejoice for him."[96] Dall's letters about this episode do not exist, but it is clear from Worcester's letters to her that she agreed with him and played a large role in determining the future of this unfortunate boy. "It is my duty to send him away at once," Worcester wrote Dall in the last letter, "waiting only for your directions as to his journey."[97] One wonders what Willie's parents thought of this matter, why they apparently so willingly deferred to the advice and guidance of a total stranger.

In the 1870s Dall's sexual education deepened and broadened. She attended the Agassiz' Cambridge lectures on the "determination of sex" in fish populations, discovering that no generalizations could be made based on animal data about the "physical superiority" of either sex. She wrote in her journal:

> Among the bees the males are unfecundated females but there is a small crustacean—a billipod—of which Liebold had a pond full. . . . Among 1500 he did not find *one* male. The females in this case are *unfecundated* males and of course lower in the scale of development. . . . Agassiz says that in a century hence he thinks the limits of sex will be sufficiently understood for us to produce a husbandry whichever sex is needed. The obvious case as to the human race he said nothing about.[98]

In the summer of 1871 Dall spent a month at a popular water cure in Elmira, New York, for the treatment of exhaustion and depression. Its proprietor was Rachel Gleason, a great favorite of feminists. Gleason lectured to her patients on female physiology and showed, according to Dall, "how women deceived themselves about their diseases and how often the remedy lay in their own hands rather than the physicians'." On one occasion Gleason unveiled a "curious collection of pessaries" which she had removed from women's bodies. "Well," exclaimed Dall, "how many married men could ever imagine that women could endure such instruments of torture without fatal injury passes my comprehension!"[99] At Saratoga Springs for a meeting of the American Association for the Advancement of Science, she made a special point of attending sessions on biology and sexual differentiation by George Minot, Lester Ward, and Burt Green Wilder. Of Minot's lecture she wrote: "Nothing could be better than Minot's account of the reproductive organs of the grasshopper."[100]

Dall dealt with sexual issues directly and in openly feminist ways. She pursued one central feminist goal: to demystify woman's sexuality, to free it from older "fictions" that had been used to justify woman's dependence and inferiority.[101] Contemptuous of female sophisticates who consciously flourished their sexual charms ("Nothing," she wrote, "is so hard as the heart of a worldly woman."), of women who "died for love," and of men like Michelet who reduced the substance of women to a tubercular sigh, she regarded tenderly women who were unwilling victims of sexual convention and singled out the prostitute as the quintessential victim.[102] As a lecturer in the 1850s and later as an active member of the American Social Science Association, she often said before sexually mixed groups that absurd ideas about woman's "angelhood," coupled with prejudices against women working, had thrown many women into prostitution. At the risk, she believed, of being "called a prostitute herself," she argued that virtuous women should empathize with prostitutes—women often "nobler and more disinterested than many who remain pure"—for the reason that they too had the potential for such degradation. She warned her audiences that "the 'perishing classes' are made of men and women like yourselves."[103]

Dall wanted to prove in her public reflections that most women, even the most degraded women, had an innate capacity for personal transcendence, and therefore for personal autonomy. She and other feminists focused obsessively on woman's desexualized qualities: her purity, her virtue, her seemingly natural ability to transcend the seamy side of life. This partly explains why Dall's own life assumed, at least on the surface, a plainly ascetic complexion and why she felt such conflict about making public appearances devoid of reformist significance. She declined, for example, to give "dramatic readings," not because they were inherently vulgar but because they implied an older, static, barbaric role that had forced "all women to dance, sing and play for men" and not to participate as independent, rational people in the progressive experience of the world. "Even the magnificent Fanny Kemble," Dall wrote to a man who had invited her to read in public, "may read Shakespeare for years without carrying the world a single step."[104]

In 1865 Dall placed the capstone on her reform career by cofounding the American Social Science Association, one of the most important reform organizations of the post–Civil War period. She had tried independently to build the basis for such an organization by establishing an "Institute for Social Science" within the Boston Public Library and for years had admired the British Association for the Promotion of Social Science for making use of the services of the "very best" women—including Mary Carpenter, Isa Craig, Louisa Twining, and Florence Nightingale—who worked in the four central fields of social science: popular education, punishment and reform, public health, and social economy. "Will you not open a way for our women," she asked George Tichnor and Edward Hale, both trustees of the library, "to become associated with the noblest men of these times?"[105] In 1859 Dall sent a memorial to Tichnor and Hale signed by several Boston notables from upper-class Unitarian backgrounds—Ezra Stiles Gannett, Edward H. Clarke, Samuel Gridley Howe, Cyrus Bartol, Samuel Sewall, and John A. Andrew, among others—as well as by members of the "middle ranks of society." She outlined in it a plan for the construction of "an apartment or alcove of the City Library devoted to the deposit of all Reports, whether of Governments, Parliamentary or Philanthropic Committees, or Societies for the Prevention of Pauperism, Idleness or Crime, especially the reports published in England, Scotland, Ireland, France and Germany."[106] Such an institute was needed, she wrote, since social reform could no longer be solidly grounded on mere philosophers "who only know vice in the closet, and who pass it every day not only in the streets but in the best society without suspicion." Rather, it must be based on "statistics . . . results . . . probabilities," or on the "hard facts of life, the hard figures as in the columns of such writers as Duchalet and Legouvé."[107] The trustees rejected Dall's proposal for want of library space. Nevertheless, her plan paved the way for the American Social Science Association.

Elected one of the original twelve members of the ASSA and its principal librarian, Dall worked as hard as or harder than any other member, although she stretched a point when she claimed that the

ASSA "owed its existence largely to myself."[108] She drew up the original constitution and helped write the bylaws. She was given the authority to appoint her own assistants, a power she did not squander, and served on several committees throughout the thirty-five years of her tenure as both director and vice-president. She took a healthy part in debate at the social science meetings, and enjoyed nothing more than matching wits with male members on important issues, often to the point of irritating her opponents. At one meeting she rebuked the venerable Amassa Walker for advising women to learn how to make a "pudding" well before they learned Latin and mathematics. "Woman's brain should be educated as much as possible," she retorted, "and Latin and Mathematics would do her as much good, and as little harm." She also blamed "female debility" on the "infidelities" and "excesses of men."[109] On another occasion she cornered and defeated the secretary of the ASSA, Franklin Sanborn, on a question of procedure, a defeat she gloated over in her journal: "I deserved a service of plate, for not one man would speak his mind, and but for me, the meeting would have come to death."[110] She proselytized for the ASSA, traveled widely with other members to examine conditions in prisons, hospitals, and factories and delivered papers on a range of subjects from nearsightedness in children and the adulteration of milk to lodging houses for women. In perhaps her most audacious effort, she conceived a plan, similar to one proposed by John Ruskin in England, for the establishment of cooperative lodging houses and restaurants to be owned and operated by working women. She defended women's rights to equal wages with men, their right to become "capitalists," and their right to "combine as men do against unjust employers."[111]

Priding herself on her work in social science—"a sphere," she wrote in 1873, "for which I have carefully fitted myself"—Dall wanted her children to remember her for that if for nothing else. She was especially proud of bringing women into the ASSA—among them Abby May, Zena Fay Peirce, and Mary S. Parkman—and of compelling the ASSA to take feminist positions. She told Sanborn that she would not "yield the woman question" and that "it is necessary to *fix* whatever good we have gained."[112] In a sense her whole reform career came to fruition in her work as a member of this organization.

Dall hoped not only to free American society from the burden of an uncontrolled and subversive individualism, from irrational prejudices and myths, and from social and economic disharmonies, but also to master the guilt she had known so long as her father's daughter, to recover or reconstruct for herself and for other women what she had lost in childhood. "Is there anything which I can do to please you?" she had asked her mother, as other feminist women may have asked their mothers, feminist women who, like Dall, were torn between the power of mother love and the quest for independence. Dall never relinquished her books and pens, but she did seek to build a new society in which she would become anew her mother's child.

The mid-nineteenth-century feminist ideology of social relations took form within a clashing convergence of two opposing or contradictory modes of social behavior, one highly individualistic, competitive, and materialistic, and the other communitarian, cooperative, and, to some degree, ascetic. On the one hand, it sought to explain and justify the social and economic changes that had already taken place in the lives of many women during the late antebellum period and that made it possible for them to behave in quite different ways from their mothers and grandmothers. Thus feminism demanded that women be allowed to become what some of them were already becoming: self-reliant individualists in a capitalist society. On the other hand, feminism drew much of its power from a larger reformist-reconstructionist ideology—the ideology of social science, seeking not only to reunite the sexes on a new and more lasting basis, but also to humanize and order capitalist society and to rid it of the bleakest and most disruptive features of competitive individualism. Feminism, like early social science, wanted to free women for individual productive careers while at the same time making a new communal life founded not on older religious, patriarchal standards but on rational, scientific, and egalitarian principles.

The relationships between feminism and science, feminism and educational and health reform, feminism and rationalism were not simply indigenous to feminism. They were determined by another, broader relationship between feminist ideology and the ideology of the progressive, educated bourgeoisie. As one of the principal founders of the American Social Science Association, Caroline Dall clearly illustrated this broader relationship. She was a significant member of a generation of feminist men and women who linked the interests of both sexes with the interest of class and who worked toward the development of a new ideology. The most important instruments of the formation of this new ideology were the organizations of the post–Civil War period that were inspired and shaped by social science.

CHAPTER 11

FEMINISM AND SOCIAL SCIENCE

SOCIAL SCIENCE perspectives colored every facet of the feminist movement and tended to unify the vast array of feminist organizations that appeared after the Civil War. The principal ideological signatures of the feminist movement were these: rejection of individualism, of "excess" and "selfishness," in favor of symmetrical order and group action; adherence to "nonpartisan," "scientific" social organization as the means to cope with the changes going on in an otherwise "sound" political and economic system; and a preoccupation with the body—with physical and social organisms—as the road to social order and advancement. Such ideological affinities clearly marked organized feminism—that is, the feminism of the National and American Woman's Suffrage Associations—and linked it with a more encompassing project to reconstruct American society. But organized feminism did not stand alone in its kinship with social science. Other feminist groups and societies drew deeply from this new ideological reserve: feminist spiritualists, moral educationists, the Woman's Christian Temperance Union, and female and male sociological clubs and societies. Explicit social science organizations, of course, reflected these themes, including the Association for the Advancement of Women, the American Social Science Association, and a myriad of feminist social science associations, affiliates of the American Social Science Association, which appeared in several cities in the 1870s and early 1880s. Most of these groups were made up of both men and women. It is one of the most remarkable aspects, in fact, of the reform experience of the post–Civil War period that many men were as devoted as women were to the reconstruction of society along more harmonious, scientific, cooperative, and feminist lines.

Feminist Spiritualist Reformers

THE PROGRESSIVE feminist spiritualists of the 1850s through the 1870s were perhaps the most curious group of reformers to adopt informal social science ideas.[1] Many of them—Lita Barney Sayles, Mary F. Davis, and Lucinda Chandler, for example—practiced spiritualism and social science simultaneously.[2] Progressive spiritualism closely resembled the Free Religion practiced by many members of the American Social Science Association (in spite of the fact that "academic" Free Religionists often felt contempt for the typically lower-middle-class or not extensively educated spiritualist), the Friends of Human Progress, and the Liberal League.[3] The editor of the *New Northwest*, Abigail Scott Duniway, in line with other feminist spiritualists, called herself a "scientific spiritualist" and zealously admitted into her columns articles on science, comparative philology, comparative religion, and the other progressive-rationalist subjects of the day.[4] Like Free Religionists, the "scientific," reformist spiritualists—those who had broken away from the established churches—tried to integrate "Hegelian" idealism with "Spencerian" positivism, psychology with physiology, facts with ideals. Seldon J. Finney, a noted spiritualist, called spiritualism "the Catholicism of Rationalism, with a fact, an idea, a reason, and a symbol for every possible mood of man." It represented the "middle ground of philosophy where sense and soul touch and unite."[5]

Spiritualism had its "spiritual" side, although it had no real religious or theological content. Its secular and postauthoritarian cult of guardian angelhood, mediumships, and clairvoyance, and its extraordinarily maternal moral system reassured reformist men and women disillusioned with established religion but unable to throw off its claim completely. The spirits often divulged what people already thought, gave direction to their desires, or justified unconventional or even radical activity.[6] The freethinking feminist Lucy Colman described the behavior of spiritualists in Michigan during the 1850s, a state where "spiritualism rioted like some outbreak of disease," as she described it. "The subject more generally dwelt upon was the disharmonies of domestic life and more than one couple released themselves from each other by and through the advice of the Spirits."[7] Many spiritualists swallowed the unsettling pill of free-love anarchism and a minority still suffered from its influence after the Civil War. In the 1870s this volatile and determined minority contributed to the collapse of an already unstable spiritualist organization.[8]

Many other spiritualists, reputed free-lovers as well as more conventional men and women, rejected this single-minded obsession with sexual individualism. If the pages of the *Banner of Light* are any indication, a diminishing number of them could be found poring over Emerson's "Self-Reliance" or waxing lyrical over the merits of unlimited freedom.[9] As loyal to laissez faire republicanism as their more upper-class counterparts, spiritualists attacked the "northern monied interests" and "conservative religious organizations" and, like social scientists, fell to talking and writing of "moral and strictly social topics"—the new fabric of "political" discourse.[10] Luther Colby, editor of the *Banner of Light*, wrote, "We now discuss temperance, social evils of various sorts, the *welfare of woman*, universal suffrage, progress and various other concerns in which man is so profoundly interested. Our politics . . . are not altogether questions of economy and wealth, but also of the true modes of applying economy and disbursing wealth."[11] Spiritualists too looked to social scientific means to strengthen the symmetrical harmony within society (what Andrew Jackson Davis called the "interdependency of things in the world"), which was seemingly endangered by the tumult of modern life.[12] Victoria Woodhull, president of the American Association of Spiritualists as well as a free-love feminist, wrote in her *Weekly*, "What we must come to in the end is the scientific organization of society, under the leadership of and devoted allegiance to the best thought in the world consecrated to the highest uses. Politics, in the ordinary, or vulgar sense of the term, in preponderance, will give way to social science."[13] The creation of national and state spiritualist organizations in the early 1870s may have shown the depth of this new interest.[14]

The spiritualist reform agenda bore the mark of the times. Progressive spiritualists embraced the multifarious work of prevention. "The work of the reformer," wrote one spiritualist to another, "is not with the outcast, the Magdalen, but with the causes that make outcasts—better save future generations than twenty fallen women."[15] They worked for prison reform. "The penal system is one of vengeance. . . . Our prisons should be converted into reformatory schools," declared Mary F. Davis, a spiritualist who was also a member of the New York City Sorosis.[16] Spiritualists founded industrial schools and business cooperatives and took up the banner of harmonious health. "The Trouble of the Times," said the *Banner of Light*, "is that we are all of us sick—sick of physical and mental disease which we have both inherited and begotten."[17]

Above all, progressive spiritualists threw their energies into their own independent educational institutions—a sure sign that they marched with their times and had renounced, at least to some degree, the heritage of an individualist age. "Daily we are taunted," Warren Chase wrote for the *Banner of Light* in 1867, "with the lack of practical utility—no institutions, no places of instruction"—no colleges, no universities, no children's schools.[18] That year Andrew Jackson Davis tried to turn the tide by expanding a new system of progressive children's lyceums he had already established in New York City. He organized the Lyceum Missionary Fund, modeled after Peter's Pence of the Catholic Church, and prodded spiritualists to build their own libraries; he wrote and published the standard textbook, the *Children's Lyceum Manual*.[19] By 1868 the lyceums had mushroomed everywhere throughout the East and Midwest—Chicago, Philadelphia, Buffalo, St. Louis, Springfield, Ill., Worcester, Mass., Bridgeport, Conn., Vineland, N.J., and other cities.[20] Most of them flourished at first but soon began to fail. Too expensive to equip and plagued by inexperienced teaching and confusion over proper curriculum, the children's lyceum disappeared by the early 1870s.[21]

Strikingly similar to the kindergarten system proposed by the American Social Science Association, the lyceum set forth a hierarchically structured program designed to "harmonize" and expand the mind of the child, to "address" the child's "reason through the social affections," to sever education from "any creed or dogma," and to give a "natural," "symmetrical," and hygienic training suited for "real life."[22] As an institutional structure it emulated the military and Catholic examples. The typical unit was divided into twelve tightly organized groups for children four to fifteen years of age. Each group had a different name and colored badge to identify its "spiritual" position in the hierarchy of groups. Ten-year-old children in the Shore Group wore green badges and studied the physiological facts of the human body; thirteen-year-olds in the Star Group displayed azure badges and learned, among other things, that "physical subordination was rewarded with perpetual health and cheerfulness."[23] These badges and names tied each child directly to the lyceum and the spiritualist cause and indirectly to their families and society at large. Spiritualists keenly understood the binding psychological "power of badges and emblems." "None can dispute" such power, wrote Mary F. Davis, "not only over children, but full-grown men and women. . . . They indicate the devotion of each member of a common brotherhood to some great affection, cause or principle." To "intensify" and "broaden . . . this feeling," spiritualists used the national flag in their parades and public gatherings.[24]

According to Elizabeth Cady Stanton, progressive spiritualists were "the only religious sect in the world, unless we except the Quakers, that has recognized the equality of women."[25] Several female mediums —among them Victoria Woodhull, Emma Hardinge, Maria M. King, and Elizabeth Lapierre Palmer—were prominent in the feminist movement.[26] Spiritualists believed that the emancipation of women would harmonize the social sphere, and for this reason they supported those who wished to "incorporate the female element in the governing and advising boards of benevolent institutions."[27] They divided the world equally between the Father-God and Mother-Nature and thought that the interfusion of these two principles would remove the damage done by individualism: man would lose his aggressive compulsions and woman would become more rational and masculine. As one spiritualist said, "Our Fathers are not dead;" they had rather hitched their power to "Truth, love and charity." In the world of "Tyranny," "good and highly developed spirits withdraw."[28] The Rochester spiritualist Isaac Post even speculated that this integration might eventually eliminate sexuality. "Once the life principle progresses far enough," he wrote, "a new class of beings male and female" would result, "and without the classification of sex."[29]

The American Social Science Association

IN 1865 a small but influential group of Boston reformers, among them Caroline Dall, called into existence the American Social Science Association, an organization informed by the spirit of Radical Reconstruction in the North and inspired by both the British National Association for the Promotion of Social Science and the initiatives of Samuel Gridley Howe. Howe had labored to rationalize and coordinate the work of various relief agencies, and to develop ameliorative techniques for dealing with social and economic problems.[30]

Throughout the 1860s and 1870s the American Social Science Association was a focus for institutional reform. It spawned such independent organizations as the National Prison Association, the National Conference of Charities, the National Conference of Boards of Health, the American Public Health Association, and the National Association for the Protection of the Insane and the Prevention of Insanity. It guided the creation of state, local, and regional social science organizations, as the social science leader Franklin Sanborn wrote "in order to get social science a foothold in as many sections of the country as possible."[31] Many of these groups soon eclipsed the functions of the association, constantly drawing upon its membership. Nevertheless, the ASSA remained a principal entrepôt of ideological debate and institutional coordination well into the late nineteenth century. It brought men and women together in common cause.

Whiggish Radical Republicans, men and women, many with Garrisonian abolitionist histories, joined the association—among them Garrison himself, Thomas Higginson, Abby and Samuel May, and Henry Villard.[32] Three Republican governors of Massachusetts—Andrew, Boutwell, and Hawley—presided over it in the early years.[33] Most of these men and women had contributed or were contributing to social reconstruction in their various states—Massachusetts, New York, Illinois, Wisconsin, and Michigan—bringing reformist zeal into the Association.[34] Such reconstruction efforts peaked in the 1860s and early 1870s; later they suffered some retrenchment but no fundamental break in direction.[35] Four association members served on the Massachusetts Labor Commission and on the state's boards of health and charities.[36] The association attracted institutional reformers and the most important liberal economists of the period.[37]

For the most part these men and women were Republicans. They practiced law and medicine, preached in the liberal ministries, taught in the most prestigious institutions. Most of the members came from liberal Protestant backgrounds, and most were linked, like the Radical Republicans in Congress, with industrial or merchant-capitalist wealth. The relationship between merchant capital and reform had a long history in nineteenth-century America.[38] Caroline Dall and some of the leading women—Mary Parkman and Mrs. John Lodge, for example—came from mercantile wealth, and all of them supported the development of northern textile industries after the Civil War. Villard, the association's secretary in the 1860s and Garrison's son-in-law, had accumulated much wealth as a financier; Gamaliel Bradford, treasurer of the association, descendant of William Bradford of Plymouth Colony and father of an equally famous son and namesake, was also a banker; Elizur Wright, life insurance executive and treasurer of the Indian Orchard Cotton Mills in Massachusetts, backed expansion of the railroad system in the 1850s and, later, prided himself on discovering the method of sending greater quantities of cotton goods to England via the steamship.[39] Samuel Bowles of the powerful *Springfield Republican*, which had served as an unofficial organ of the United States Government during the Civil War, now made his paper the unofficial organ of the American Social Science Association.[40] Bowles was also one of the associate directors of the Longmeadow Railroad Company in the 1850s. Amassa Walker helped found Allen, Harris, and Potter, at one time the largest shoe factory in the country, and retired wealthy at the age of forty-one to become a lecturer in political economy at such colleges as Oberlin and Amherst.[41]

The American Social Science Association marked the movement of reform away from the romantic, individualist, and laissez faire dispositions of the antebellum period and toward a new institutional, ameliorative reformism. It conceived of government increasingly in positive, not negative, terms. It represented a concentrated effort by the reform-minded sector of the American middle class to come to grips with the pauperism and crime, the mental and physical disorder, the disorientation of family life, the instability and expansion of city and village populations, and the class conflict that resulted from the industrialization and urbanization of American society and the "unrestricted" and "unadministered growth of the last two and a half centuries."[42]

"From 1865 onward in the United States," Sanborn, the principal secretary wrote in 1885,

> it has been more and more the affair of government, both local and national, to busy itself with school funds and school boards—with boards of health and boards of charities, commissioners of lunacy, of valuation and taxation, tariff commissions, Indian commissions . . . the investigation of railroads and telegraph traffic, and its partial regulation; the establishment of parks, museums, libraries, art galleries, etc., wholly or in part at public cost . . . Our Association, at its very outset, seems to have contemplated just this expansion of the powers and interests of a popular government, and not to have dreaded disaster therefrom.[43]

Sanborn carefully pointed out that the extension of government would not lead to the "slavery and socialism" so much feared by Herbert Spencer.

The five departments (originally four) of the association clearly underlined the ameliorative format, touching on every aspect of institutional reform:

1. Education: bureaucratic organization of the public school system; higher education; coeducation; public school curricula; the construction and management of male and female reformatories; industrial schools; adult and evening education; and the elementary and secondary education of the "wage earning classes."
2. Health: sanitation; population increase; management of hospitals, public baths, gardens, parks, and asylums; the adulteration of foods; ventilation; tenement housing; architecture; the diffusion of hygienic knowledge; and all questions relating to the duration of life.
3. Social economy: pauperism; hours of labor; relation of employer to employed; woman's work and working conditions; prostitution; "the relation and the responsibilities of the gifted and educated

classes toward the weak, witless, and ignorant;" and civil service.

4. Finance: national debt; the habits of trade; the subjects of tariff and taxation; the quality of manufactures; the control of markets; the value of gold and all questions connected with currency; the monopolies in sale of food or the production of articles for common use.
5. Jurisprudence: principles of law and legislation; international law and municipal civil law; criminal law; and the amendment of the "law of the land."

Association members worked hard in all these departments, although they expended much of their energies on education, health, and social economy, meeting "oftener than once a month" to discuss these matters.[44] In 1866, for example, they examined such topics as art education and design and their relationship to American manufacturing; prison holidays and their influence on convict discipline; the employment of women and children in industrial establishments; the eight-hour day and "cooperation" between labor and capital; and the early training of children as preventive to pauperism.[45] Every state, local, and regional association replicated the ASSA departments in its organization, and all hoped in some way to influence the "law of the land."

Male Feminism and Social Science

IN SPITE of quibbling by some men who wanted to keep the American Social Science Association detached from apparently partisan causes, the association itself showed a strong feminist bias almost from its inception, confirming the alliance between men and women that already existed in the feminist movement. Because of its relationship to marriage, social reconstruction, and social order, the women's movement attracted a great number of educated, professional, and progressive men—lawyers, doctors, journalists, and businessmen—as well as *petit-bourgeois* property owners, shopkeepers, and artisans, who, in the words of the New Jersey senator and politician James Scovel, wanted to "work out woman's salvation and ours and at once."[46] "This is no woman's movement," one man wrote to Thomas Higginson. "It is a reformation intended to advance and improve the happiness of *men* and *women* equally."[47]

Throughout the period 1850 to 1880 men gave much of their time to woman's cause. They lobbied in state legislatures, fought on the floor of Congress, argued for coeducation and for the admission of women to the Protestant ministries. They organized the Young Men's Suffrage League in New York City and elsewhere. In 1869, for example, the young men of Auburn, New York, started their own "lecture course for the benefit of Woman Suffrage," inviting Wendell Phillips and the actress and feminist Olive Logan, among others, to speak before them. Even Horatio Alger, that supposed paradigm of American masculine self-creation, got up a series of lectures in his home parish to which he drew such feminist luminaries as Lucy Stone and Wendell Phillips.[48] Moreover, Thomas Higginson, Lester Ward, Moncure Conway, Parker Pillsbury, not to mention John Stuart Mill, Victor Hugo, Salvatore Morelli, Charles Renouvier, and Frederick Engels, wrote some of the most important feminist literature of the nineteenth century.

Men attended and managed conventions and often held the highest executive offices in the major societies—with the single, notable exception of the National Woman's Suffrage Association. The NWSA objected to the election of men to positions of authority for two reasons, both of them historically grounded. First, many "respectable" feminist men who had supported the feminist cause before the Civil War abandoned it temporarily after the war for the cause of Negro suffrage. Elizabeth Cady Stanton called this a betrayal and established an all-female executive for the NWSA. Second, according to Matilda Joslyn Gage, men who chaired women's rights societies habitually lacked the devotion to get things done. "Women," she wrote, "can work more successfully for their own freedom than anyone else can work for them. It was frequently the case that WS parties officered by men failed to accomplish much work."[49] Nevertheless, NWSA women never expressed extreme hostility to male participation and often enthusiastically encouraged it. Stanton adored Parker Pillsbury, one of the editors of the *Revolution*, and she and Victoria Woodhull insisted in the early 1870s that men be present at the conventions. "To tell the truth," Stanton said, "I should prefer a place in the affections of my countrymen; to live only in the hearts of women would be like talking to one's self all the time."[50] Gage's own New York State Suffrage Association continued to have men vice-presidents, although they did not serve on the executive committee.[51]

Men acted as presidents throughout the 1870s for the Wisconsin, Michigan, California, Illinois, Oregon, New Jersey, Ohio, Pennsylvania, and New England associations as well as for local organizations within these states.[52] They also frequently presided over the American Woman's Suffrage Association, an organization that divided functions equally between men and women. Men opened the drive in Maine to bring women into school committees. Men alone participated in the formation of the Vermont Woman's Suffrage Association; and, according to Henry Blackwell and the *Woman's Journal*, men exceeded women in numbers at the New York City American Woman's Suffrage Convention.[53] In many states feminist husbands joined feminist wives in the feminist cause. Such couples included Giles and Catherine Stebbins in Michigan; Abraham and Mary Putnam Jacobi, Charles and Elizabeth Smith Miller; Theodore and Elizabeth Tilton in New York; Francis and Virginia Minor in Missouri; Lucy Stone and Henry Blackwell and William and Harriet Robinson in Massachusetts and New Jersey; James and Myra Bradwell in Illinois; Margaret and Robert Campbell in Colorado; and Sarah and William Knox in California.[54] The New York lawyer William R. Martin said in a speech at New York University Law School in 1871, only working-class men resisted or opposed the feminist movement, while "thinking men . . . the more intelligent classes" supported it.[55]

The most important American male feminist of his generation was Thomas Wentworth Higginson. Higginson's feminism, like that of so many male feminists, grew from a fusion of two primary and related influences: the transformation of American Protestantism in a humanitarian climate and the daily, pervasive presence of women in his life from his early years to the moment of his death. Like his Unitarian mentors, William Ellery Channing, Theodore Parker, Ralph Waldo Emerson, and James Freeman Clarke, Higginson had no sympathy with the patriarchal rigidities of the Old Testament, with the inhospitable grimness of John Calvin, or with traditional institutional religion. Like them he preferred the sunny religion of the heart with its more gentle accents on Christian love, remission, and forgiveness.

Higginson escaped, in his own words, "almost entirely those rigors of the Old New England theology which has darkened the lives of many."[56] Reared in a bland upper-class Unitarianism, he knew only the New Testament, not the rigors of the Old, and easily absorbed, in his own fashion, the transcendental idealism of Emerson and Thoreau.[57]

Most feminist men gradually, or in some instances abruptly, even violently, tore themselves away from the older Calvinist world view. They were rebelling against traditional authority, shaking off the patriarchal heritage passed on to them by their parents. For John Hooker, Gerrit Smith, and Parker Pillsbury the rebellion may have caused chronic illness; in most cases it inspired a persistent fear of personal excess and violence, acute self-consciousness, and the need to maintain self-mastery.[58]

Partly because of his desire to play a major public role and partly because of his mother's influence, Higginson became a Unitarian minister in 1847 and received an appointment as pastor in Newburyport, Rhode Island. For him religion was not dogmatic theology but simply the love of God and the love of man: "It is plainer and plainer that the age is about to try the utterly novel experiment of living without faith in ritual and mythology."[59] Restless in the pulpit, he changed his pastorate to the Free Church at Worcester, Massachusetts, and in 1857 took the inevitable step of resigning his ministry altogether, severing all ties with institutional religion.[60] Higginson continued to retain strong faith in the nonauthoritarian, "feminine," and Christian virtues of love, charity, and benevolence. He lectured on spiritualism and remained attached to it for the rest of his life. In a lecture in New York City he said, "Fifty thousand mediums have been tested by the hardest skeptics. Two out of three of these skeptics have ended by becoming Spiritualists themselves."[61] After the Civil War he preached the Religion of Humanity and wrote editorials for the feminist-inclined *Index*, one of the principal organs of Free Religion. Like many other reformers, he saw no contradiction between his affinity for spiritualism and his allegiance to Free Religion. Free Religionists considered the *Banner of Light*, *New Age*, and *Index* all as organs of Free Religion and easily combined—so they thought—sentiment with positivism.

The secularization and feminization of religious ethics created the intellectual frame for Higginson's own feminism. The powerful impact of women on his life provided the most intimately formative source. Higginson was the last of ten children and the sickly pet of the family.[62] After a long series of economic failures and personal reverses, his father died unexpectedly, leaving his family in debt, when Thomas was nine years old. Steven Higginson had once been a good-natured, affluent merchant gentleman, the product of a respectable and prestigious family of shipowners, but the impact of the War of 1812 wiped away his fortune; he became steward of Harvard College only to be dismissed eight years later for negligent and incompetent administration of college funds. He was ineffectual as a conventional paternal authority in the home, eclipsed from the beginning by his wife's strong, consistent, and intelligent domestic management. His death left his son in the shadow of his father's personal humiliation but in the able hands of four devoted women—two sisters, his aunt, and his mother.[63]

A woman with respectable colonial roots and a rich respect for the past, Louisa Higginson gave her son the sense of personal power within his own class that his father had lost through his failure. "He is the star that gilds the evening of my days," she wrote of Thomas, "and he must shine bright and clear—or my path will be darkened."[64] Higginson warmly acknowledged his mother's influence, called her the "Lady Bountiful of the household," and praised in her those qualities—refinement, intelligence, and "sweetness"—which he called the cardinal virtues of womanhood.[65] He considered his mother "one of the most fascinating persons I ever saw. . . . I wish the world could have a chance to know her loveliness before she passes away from it. She is the most wonderful being I ever knew."[66] Under her aegis he first learned literature, art, and music, and in later years he traced to her his own "love of liberty, of religious freedom and of the equality of the sexes. . . . Seeing the uniform respect with which my mother and aunt and elder sister were treated by the most cultivated men around us, I cannot remember to have grown up with the slightest feeling that there was a distinction of sex in intellect."[67] He was convinced that his mother's power, and the power of women like her, enhanced the nature of his manhood.[68] His mother did for him what Caroline Dall's father did for her: she strengthened his ego and gave contour to his life.

Many feminist men replicated Higginson's experience although their feminist positions did not always follow Higginson's. Many of their fathers were victims of economic failures, weak, nonexistent, unapproachable, cruel, or rigidly authoritarian. William Lloyd Garrison's father abandoned him. Parker Pillsbury, Moncure Conway, Andrew Jackson Davis, Gerrit Smith, and Warren Chase suffered from demanding, authoritarian fathers; and the fathers of Henry Blackwell and James Freeman Clarke failed dismally in business. All these men praised their mothers, or other women in their lives, for their care, compassion, energy, and benign influence. Henry Blackwell summarized their debt:

> All that I am I owe to women. For when my father died in Cincinnati, fifty years ago . . . leaving his wife and children almost destitute, it was my mother and sisters who organized a school, kept the family together, sent me to college, and gave the best years of their life to the education of the younger children. . . . The woman's suffrage movement is not new; it is not exclusively a woman's movement, it is a movement of women and men for the common interest of all.[69]

The democratic, economically unstable climate favorable to the expression of stable feminine virtues was also a source for Higginson's attitudes toward both men and women. Lacking a strong male figure to emulate in his youth and nearly smothered in the love of several doting, solicitous women, he continually sought ways to demonstrate his manhood, although he never consciously understood or recognized the source of his need to do so. Relishing moments of courageous daring, he trained and developed his body to the limits of its physical capabilities. He loved the life of the solitary rebel, the transcendentalist outlaw, the brave maroon who had the conviction and strength to stand outside the purview of civilized authority.[70] Of all the radical abolitionists he was the most capable of violence and the one most likely to defend and justify fearless acts of insurrection. When the war came, he fulfilled a dream and enlisted in the Union Army, where he rose to become commander of one of the first black regiments of the Civil War. He gloried in his leadership position and, more especially, in the excitement and dangers of his wartime life.[71]

Higginson's intense sensual nature and his acute and persistent awareness of being profoundly alive amplified his pleasure in daring acts of personal courage. In *Army Life in a Black Regiment* he described the erotic exhilaration of a nighttime exploratory mission during which he swam alone through a southern "marsh" past the enemy lines to observe the condition and deployment of Rebel outposts. "In retrospect," he wrote, in one of the few truly beautiful passages he ever composed,

> I seem to see myself upon a horse's back amid a sea of roses. . . . I do not remember ever to have experienced a greater sense of exhilaration than when I slipped noiselessly into the placid water, and struck out into the smooth, eddying currents for the opposite shore. The night was so still and lovely, my black statues looked so dreamlike at their posts behind the low earthwork . . . the horizon glimmered so low around me . . . that I seemed floating in some concave globe, some magic crystal, of which I was the enchanted center. With each ripple of my steady progress all things hovered and changed: the stars danced and nodded above; where the stars ended the great Southern fireflies began; and closer than the fireflies, there clung around me a halo of phosphorescent sparkles from the soft salt water.

Higginson loved these moments of imminent danger, of romantic transfiguration that brought to the surface of his consciousness the sensuousness of his own being.[72]

What today we might call Higginson's libidinal or erotic powers he called—in a language both Antoinette Brown Blackwell and Caroline Dall probably understood—an "untamable gypsy element in me."[73] "Health," he wrote, almost hedonistically, "is to feel the body a luxury . . . to feel one's life in every limb." No one more than Higginson knew the pleasures of such feeling or was more partial to the rewards of the senses. He "thrilled" to the sounds of the human voice, and responded with an exquisite sensitivity to every existing form of color:

> The thistles are out, beloved of butterflies; deeper and deeper tints, more passionate intensities of color prepare the way for the year's decline. A wealth of gorgeous goldenrod . . . the superb beauty of the Cardinal flower, the brilliant blue-purple of the Vervain, the pearl-white of the Life-Everlasting. . . . The Cardinal Flower is best seen by itself. . . . There is a spot whence I have in ten minutes brought away as many as I could hold in both arms; and I could not believe that there was another mass of color in the world.[74]

Nature's profusion gave him daily pleasure, and his communing with it, often alone, reawakened the "heart" of his "manhood" and stimulated the "untamed wildness" of his senses.[75]

In his biography of Higginson, Tilden Edelstein writes that Higginson's love of nature reflected his disenchantment with men, with society and action, and tended, in its final effect, to emasculate him.[76] But Edelstein exaggerates the feminine side of his nature and may be mistaken when he sets Higginson's "passive" love for beauty against his need for action and "outlawry." Higginson certainly felt the isolation that solitary excursions into nature occasionally brought him—as he felt it in his "outlawry"—but more often nature aroused and refreshed his senses. Both nature and his essentially solitary and transcendentalist acts of courage and defiance worked together to activate his personal feelings of power and to stimulate his sexuality. "I seem to see myself adrift upon a horse's back amid a sea of roses" beautifully expresses this relationship between nature and action. The tension in Higginson's life can more accurately be rendered by juxtaposing nature and action against the demands and attractions of a civilized, rational, and egalitarian system of relations.

Like other feminists, therefore, Higginson rejected male dominance, feared the sexual and romantic potential inherent in individual self-expression, and looked for new ways to discipline, restrain, and rationalize social behavior in a democratic society. Like them he thought that good health partly served this function, that, far from expanding sexual power, the health of the body channeled and subdued it, permitting the development of the equally important moral and intellectual faculties. "A vigorous life of the senses," he wrote, distilling the wisdom of the hygienic tradition, "not only does not tend to sensuality in the objectionable sense, but it helps to avert it. Health finds joy in mere existence."[77] Higginson linked bodily health with the emancipation of refined and educated women in the public and private spheres and believed that both would offer the most effective antidote to the abuse of power and to possible outbreaks of sexual license.

The second significant conflicting current in his own past was the maternal heritage. Higginson identified in his mother—and in all women—the "finer forces" of civilization. Because he believed so completely in the efficacy of these virtues, he considered the home, "the domestic tie," the most important of all "civilizing agencies."[78] Without women, men degenerated; untempered by tenderness, compassion, modesty, and refinement, man's natural "rowdyism" and "sensuality" descended, as in Whitman's case, into a vulgar "phallicism" or, as in Oscar Wilde's, "into the forcible unveiling of some insulted innocence."[79] He "dreaded" the consequences of sexual segregation and thought it an "evil for all concerned . . . on the mere grounds of sexual purity and refinement."[80] Excited by and at the same time fearful of his own solitary impulses, he continually returned to the reassuring, domesticating world that ironically stimulated those impulses. What he lost in "intensity" and "exhilaration" as an "outlaw," he gained in comfort and sanity as a "civilized" human being. Throughout his life he enjoyed quoting one of Emerson's less timeless aphorisms: "A gentleman makes no noise; a lady is serene."[81]

Although Higginson occasionally felt suffocated by the constraints of domestic space (constraints compounded by his first wife's invalidism), and in some deep sense must have resented women, he never consciously experienced sexual conflict or, for that matter, conflict of any kind. Unlike Caroline Dall, who suffered from lifelong inner torment and struggled to retain her identity as a woman by rebelling against her father, Higginson felt neither torment nor the need to rebel against the maternal source of his strength—at least not consciously. Perhaps a strong resistance to rebellion, stronger than any rebellion he may have felt toward his father, prevented him from turning against his mother. In any case, Louisa Higginson made her son feel like a man: Dall's father tended to undermine her identity as a woman even while he gave her support.

Higginson's road to feminism was thus the easier one by far. He sought only to create other Louisas—strong, resourceful, creative, and independent women to be sure, but also women who reinforced and stabilized the lives of men. More than likely Higginson had homosexual tendencies (as Edelstein implies and as Higginson's thoughts on Whitman and Wilde suggest), but no sign exists that he ever experienced their presence.[82] He was married twice, first to the witty, perceptive, but frail Mary Channing, and after her death to Mary Thacher, a doting, loving woman, the mother of his only child and cast from the mold of "cultivated womanhood." Higginson had found a way to balance his deepest needs with his social role. If he struggled for anything, it was to establish himself at the apex of the evolutionary ladder, where desire and passion did not intervene. Inconceivable and even incomprehensible, sexual deviance did not appear in his life, or in the lives of the reformers he knew.

The conflict between a strong, intense "gypsy nature" and an equally strong desire for self-mastery and social harmony established the limits of Higginson's feminism, which was the most sustained theme in his reform career. Like Lester Ward and Moncure Conway, he considered the woman question "the one great discovery of modern times," and in a letter to Caroline Dall he declared that the "discovery" of women made it possible for historians to rewrite "all history."[83] During the 1850s, when Higginson's abolitionism had reached its most violent phase, he attended woman's rights conventions and contributed intelligently to the growing feminist literature. In *Woman and Her Wishes* (1853), *Women and the Alphabet* (1859), and *Outdoor Papers* (1862) he defended woman's equal right to, respectively, suffrage, education, and physical well-being. Although he rarely took a public stand on woman's right to self-ownership, he nevertheless privately and warmly supported it. "The Progressive Friends are noble," he wrote Dall in 1859. "They adopted calmly and unanimously a protest . . . against the right of the husband over the wife's right to impose on her the sacred privileges of maternity against her will. I framed it."[84] Higginson's feminism was so well known that Anthony could write Stone: "It seems to me his opinion would be worth almost more than any other man's. He seems *specially* interested in all that pertains to the W. R. [Woman's Rights] movement."[85]

After the Civil War Higginson became an important leader of the American Woman's Suffrage Association and twice served as its president. Editor of the *Woman's Journal* from 1869 to 1883, he wrote numerous articles on women for other journals, closely followed all the scientific literature on sex, and by 1875 had written more on the relations between the sexes than on any other question.[86]

Perhaps most important of all, Higginson personally encouraged women to become feminists and supported countless others in their efforts to make independent careers. "Be true to women," he wrote Isabella Beecher Hooker in 1859, "to the cause of woman. Be true to it, not merely as represented by men, but by *women*. . . . It costs a man nothing to defend woman—a few sneers, a few jokes, that is all; but for women to defend themselves, has in times past cost almost everything."[87] Women who benefited from his praise sent him letters of profuse gratitude. "You have bereft me of all words. Praise so precious, in a largesse so lavish, has simply taken my breath away. . . . Your generous letter has put new spirit into tired wings." "In an inconsistent womanish way I *feel like* crying when I contemplate your generosity, your belief in me." "I do not know which to thank you most for—the careful, critical reading which you have been good enough to devote to my play, or the enthusiastic tone of your letter which itself must stimulate me to better work." "Your friends ask but one thing of you and that is to live forever." "I have had few pleasures so deep as your opinion, and if I tried to thank you, my tears would block my tongue."[88]

Although Higginson often objected to the idea that the sexes could be divided into separate "moral categories" and often attacked the positivists for doing so, in his feminism the emphasis was on "the fact of sex" and on idealizing woman's civilizing function, her capacity to "repress violence" and "restrain egotism."[89] This view circumscribed his position on marriage as it did not that of Conway, Pillsbury, Tilton, John Weiss, and others. Although Higginson respected woman's right to a single life and often insisted that marriage and children did not constitute the sole purpose of woman's existence, he held that "no theory of womanly life is good for anything which undertakes to leave out the cradle. . . . And so clearly is this understood among us, that, when we ask for suffrage for women, it is always claimed that she needs it for the sake of her children." When Sarah Orne Jewett's *A Country Doctor* appeared in 1884, with its portrait of a woman doctor who chooses her profession over marriage, Higginson disparaged "celibacy" as a sorry and "abnormal" alternative.[90] Hostile to agitation about divorce, he condemned Woodhull and the "anti-legal" marriage faction in the woman's movement and could not understand Stanton's intense hatred for traditional marriage and her need to raise the matter at conventions.[91] Higginson was torn between his wish to "free" women and his own need to rely on them as a source of tenderness and civilized social cohesion. The tension reflected his own inner conflict, and perhaps the conflict in the lives of many of the reformers he knew.

"I like to have full swing for my impulses," he wrote in his diary, yet his impulses warred at every point with his need to master his impulses and to assume the Emersonian posture of a gentleman who made "no noise." Higginson's contemporaries thought of him as a "radical" and he coached them to do so. He played a role in almost every major reform and pleaded, he thought, the cause of the oppressed. Fearful, however, of repeating his father's failure, and of not meeting his mother's expectations, he never acted the fool or made mistakes. He could not "scream." In a feminized society that constantly invoked the power of his mother, he could not act as daringly as he could alone or at war. No wonder, then, that self-mastery weakened his creative powers and turned an otherwise fertile, intensely alive human being into a literary pantywaist. "There is an air of drawing-room elegance about Mr. Higginson," observed a reviewer of *Malbone*, a novel Higginson completed in 1869, "a suggestion of rustling brocades and rich laces and ribbons . . . something which suggests the dancing master and dressmaker." Another reviewer, less malicious and more accurate wrote that "Higginson never quite loses himself in anything, never touches the high-water mark of his power." Such criticisms struck their target. Higginson lamented the absence in his work of the "freshness," "abundance," and "zeal" that characterized, he believed, the "egoistic . . . exuberance" of the Beecher family.[92] Higginson's romantic impulses fell into disuse, immured within an alliance between sentimentalism and rationalism. They contributed little to his writings and nothing to his vision of a better world.

Higginson paid a heavy price for self-mastery, but he rarely expressed any regrets about it. His whole nature "glowed" with a warm complacency. Without the least hint of irony, he thought of his life as one long series of "cheerful yesterdays." "For myself," he wrote, in a voice that echoed in the lives of hundreds of educated middle-class men and women of his generation, "the universe is all clear and sweet; nor do I know why it should not be, to all healthy natures."[93] In a colloquy with Calvin Stowe in 1880, at a meeting of the Boston Radical Club, he said: "The more I know of fellow creatures, the more I am impressed with the general preponderance of good in them." "Here in Boston?" Stowe asked sarcastically. "Yes," Higginson responded with a smile, "but I am thinking of Five Points, New York."[94]

The cultural balance that Higginson helped to make and guarded so well became the suffocating heritage of a later generation of American intellectuals who longed to let go, to experience and to act.[95] This balance characterized, to some extent, the feminist ethos as a whole. If woman's public function was to "repress violence," "restrain egotism," and establish a rational, civilized social order, then the feminist movement had to remain free of dissidence, free of private idiosyncrasies, class biases, and racial hostilities. The "omnibus" of suffrage had plenty of "room inside" for everyone—atheists, Jews, Protestants, and Catholics—but only to the extent that "everyone" set aside individual differences and worked for the higher cause of woman's emancipation. Under no circumstances should feminists ally themselves with political extremists or speak out on such "side issues" as divorce and labor unions. Higginson wanted the road to woman's "freedom" paved in "sweetness and light" and guided by the spirit of a mother's love.[96] His feminism was therefore both an important asset and an important liability to this, or any, feminist movement.

Thomas Higginson joined the American Social Science Association in 1865. On May 14, 1873, he delivered the most important paper on the coeducation of the sexes in universities ever given by a member of that organization. He began by listing the "fallacies" that had poisoned arguments against it, isolating, in particular, the "fallacy" of woman's physical and intellectual inferiority. "The question of intellectual education," he said, "is not one thing for Man and another for Woman. . . . All the testimony of the experts is in favor of joint education." Because we cannot prove that men and women differ radically in mind and body, we cannot deny either sex the right to an equal education. "When we separate treatises on the laws of digestion for the sexes, it will be time enough to have separate treatises on the education of Women."[97] Higginson was not the only member of the ASSA to argue in this fashion and he was by no means the only feminist in the organization. Many of its original male members—among them George William Curtis, Edward Jarvis, Amassa and Francis Walker, Henry Carey, Francis Bird, Andrew D. White, Henry Bowditch, John Bascom, William Robinson, William T. Harris, and Samuel

Eliot—espoused the cause of woman's rights. Many of them, moreover, came from the same pool of men who worked for sexual equality in the feminist societies.

Of the one hundred ninety-three men listed as members of the association in 1874, forty, at least, openly locked hands with the feminist movement.[98] John Bascom, who became president of the University of Wisconsin in 1874, had demanded in the *Woman's Journal* the year before, the "free interplay of woman's influence" in every area of life. He aligned himself with the "Woman's Preacher's Movement" and was strongly in favor of coeducation.[99] In the 1860s Henry Bowditch defended the right of women to become doctors, and in the next decades he argued for their right to attend Harvard Medical School and to join the American Medical Association.[100] In a lecture before the Boston Woman's Club, William Harris argued that girls and boys should "pursue the same studies. . . . I would have the two sexes together in the same schools," he said. "With the increasing conquest of nature by man comes a change in the sexual relation, and women, as well as men, are set free from the necessity of providing for mere subsistence, and allowed to cultivate their minds."[101] "A large believer in the capacity of women," Edward Jarvis embraced all of John Stuart Mill's theses in *The Subjection of Women*; and both Samuel Bowles and William Robinson worked for more equitable marriage laws in Massachusetts and, like almost every important member of the American Social Science Association, expressed enthusiasm for equal sexual representation on the various boards of health and charities.[102]

Bowles and Robinson belonged to the New England Woman's Suffrage Association in the early 1870s, and Robinson even wrote the suffrage memorial to the Massachusetts legislature in 1871.[103] A woman's rights pioneer since the 1830s, Amassa Walker wrote to Lucy Stone in 1867 with pleasure that "no cause has made such prodigious strides as yours within the last fifteen years. . . . To me it is truly surpassing anything I should have believed possible."[104] In 1873 Walker's son, Francis, supported woman's right to labor at equal wages with man. He not only attacked the "wages fund theory" as inapplicable to modern productive conditions; he also believed that the traditional theory of "domesticity"—"the sentiments of an older age" —was contradicted by new economic conditions ("the decay of household manufactures") and interfered with "woman's preparations for active pursuits" as well as "with her prosecution of them. . . . Just so long as girls grow up in the belief that their mission is to adorn a home, their education will take shape accordingly." Walker felt certain that women could work outside the home and that the economic system could easily absorb their work. To him the problem was the "want of respect for woman in her work and . . . social constraints upon her freedom of movements as a laborer." Here, he concluded, "we find the explanation for her lower wages."[105]

Other association men agreed. As strongly as Caroline Dall, Franklin Sanborn "insisted, from the beginning, that women should have a place in the management" of this reform organization.[106] In the 1860s he was the treasurer of the American Woman's Suffrage Association while at the same time carefully reporting to the social science meetings the progress of "woman's franchise" and praising the National Labor Union for admitting women. In July 1874 he reported that the "work of social science" had so far culminated "in the movement to give full occupation to the aspirations and capacities of women."[107] Andrew Dickson White and Daniel Coit Gilman championed coeducation; and A. J. Peabody and others made a strong case for women being equal members of the state school committees.[108] Social science men also believed in the absolute upgrading of woman's education to eliminate all female illiteracy in the United States.[109] Indeed, in 1867 the corresponding secretary—later president—of the association, Samuel Eliot, told the ASSA that it was "the business of Social Science to meet all the old arguments . . . against equal rights for women . . . to meet them as they block the way for women on every side . . . Whatever chance man has, woman ought to have in society." The world waited for the "blended qualities" of both sexes.[110] Such an appeal moved Caroline Dall, who thought she had played a role in making up Eliot's mind. "I doubt whether he would have ever read a woman's rights book if I had not asked him, and my *Rights of Labor* was the first he had read," she wrote in her journal.[111]

The American Social Science Association demonstrated its feminist character by having women in positions of power from its inception. As the authors of the *History of Woman's Suffrage* pointed out, the association and its local affiliate, the Boston Social Science Association, "were the first large organizations in the country to admit women on an absolute equality with men."[112] Mostly because of Caroline Dall's unflagging diplomacy, two of the original twelve members of the Executive Board were women—Mary Eliot Parkman and Dall herself. As librarian of the association, Dall assembled special reading lists for the ASSA. Her 1870 list included five feminist or feminist-inclined books—John Stuart Mill, *The Subjection of Woman*; Virginia Penny, *Men and Women, Work and Wages*; C. H. Ward, *Women in Prisons*; Thomas Lecky, *History of European Morals*; and Josephine Butler, *Women, Work, Women's Culture*—and for provocative contrast, two weak antifeminist polemics, Horace Bushnell's *Woman's Suffrage* and Carlos White's *Ecce Femina*.[113] Fourteen other women were regular members of the ASSA, including a graduate of Dio Lewis's hygienic school in Boston, Abby May (recording secretary), and a lifetime advocate of medical education for women, Caroline Severance. Freethinking, abolitionist, and upper-class women—among them Maria Chapman, Mary J. Quincey, Mary E. Stearns, and Lucy Goddard—also did active work.[114]

The number of women in the association fluctuated during the years 1865 to 1880, but it never dipped below the figures set in the first year. In 1874, for example, seven men and six women were elected to its Executive Board—May, Parkman, Dall, Alice Hooper, Mrs. John Lodge, and Mrs. Henry Whitman—and twenty-five women out of a total of two hundred eighteen were listed as regular members.[115] William Harris was instrumental in getting Dr. Emily Talbot, one of the founders of Boston University, to commit her time and labor to the Education Committee; Talbot pioneered the establishment of the kindergarten system throughout the country and also helped build a university that gave the first doctorate to a woman in the United States and administered, according to her, examinations in medicine "more severe than are given in any medical school in the country." In 1877 she proudly wrote Dall that eighteen women had passed such examinations.[116]

In 1874 a major affiliate, the National Prison Association, voted under pressure from a feminist contingent to divide officerships equally between men and women. "I am perfectly agreed to equal numbers of representation," said the president, Enoch Wines.[117] As early as 1870 twenty women attended the National Congress on Prisons as official delegates.[118] In 1878 the Social Science Association elected a kindergarten reformer, Elizabeth Peabody, and a wealthy philanthropist, Elizabeth Thompson, to prestigious honorary memberships. Thompson was an interesting but not startling choice, given the intellectual disposition of this organization: an agnostic reformer and a member of the Loring Moody Institute of Heredity, she donated all her money to the cause of scientific research.[119] In 1880 Dr. Mary Putnam Jacobi of New York and Emily Pope of Boston became members of the association's Health Committee—a major coup for feminists since earlier, in 1865, Drs. Marie Zakrzewska and Lucy Sewall had been refused positions on grounds of political "precaution." Many other women served on the association's committees on social economy, education, and health, and three women and eleven men held vice-presidential office.[120]

Social Science as Woman's Work

"THE WORK of social science," Franklin Sanborn reported in 1874, "is literally woman's work, and it is getting done by them more and more; but there is room for all sexes and ages in the field of social science."[121] That so many women had entered it made considerable sense to him. After all, he called social science the "feminine" side of political economy.[122] Reformist women took him very seriously on this point. "Whatever may be said of politics," wrote one in 1876, "as a study appertaining to women, it is certain that Social Science is *her field*."[123] And another, in 1879: "I believe that of all the new avenues of usefulness opening to women, social science is to be their peculiar field."[124]

Feminist moral educationists of the 1870s and 1880s turned to "social organization" or the "Science of Society" as the key to the future. "The science of society," *Alpha* declared in 1878, must be founded "first, on principles, or organic compacts; and second, on statutes of administration or modes of analyzing." It must exchange "acquisitiveness," "worldly success" and class antagonism for "individual fitness" and the needs of "community," and give "systematic direction" to individual and communal life.[125] In the late 1870s Frances Willard, the brilliant feminist president of the nonsectarian WCTU who was also active in the Association for the Advancement of Women, constructed thirty-nine WCTU departments that followed the pattern laid out by the American Social Science Association, including Health and Hygiene, Labor, Social Welfare and Social Purity. In her speeches Willard imagined the future triumph of "organized, harmonious and vigorous womanhood" over "isolated, segregated, passive womanhood."[126]

Throughout the 1870s several locally based and often short-lived clubs and societies appeared in such places as New York and Philadelphia, all reflecting social science themes. They included the Sociologic Society of America, the New York City Sociology Club, the Women's Progressive Association of New York, the Century Club of Philadelphia, and the feminist Humanitarian League of New York City. The Sociology Club and the Women's Progressive Association attracted feminists interested in social science and were so similar in outlook that they merged almost at once.[127] Founded by such wealthy Philadelphia feminist women as Sarah Hallowell, Anne McDowall, and Grace Anne Lewis in 1876, the Century Club did practical and benevolent work for working-class women and really acted as a social science association. In 1880, in fact, it opened its rooms to a meeting of the Ladies' Social Science Association.[128]

The philanthropist Elizabeth Thompson founded the Humanitarian League in 1872 out of a general need, she said, to "alleviate suffering" and to create cooperative and humane "neighborhoods and communities." Believing in the complete integration of the sexes in all areas of life, the league worked for the appointment of women to school boards, poor-law boards, boards of commissioners of emigration, education, charities, and correction. Thompson and other members of the league hoped to divorce all reform institutions from the "politics and politicians of the day and to set up a separate order of merit" within these institutions "irrespective of sex" and based on the quality of "service to humanity." Thompson's league also advocated the reform of the school curriculum. Denying that classical education or learning by rote bore any concrete relation to life, the league sought to replace both with a new "how to live" approach to education. Thompson argued that

> children devote years to memorizing pages of certain text-books and solving problems of mathematics, which they do not retain and cannot apply to any practical purpose; but no effort is made to teach them common sanitary rules, or the laws which govern health and life. They have no idea of a moral use of dress or eating, and marry without a thought of human responsibility.[129]

Feminist moral education societies, the Woman's Christian Temperance Union, progressive spiritualists, and other such groups all shared the same interest in social science, although one could hardly call them social science associations.

Preventive Social Medicine

THE FEMINIST-ORIENTED Association for the Advancement of Women and the feminist state affiliates of the American Social Science Association explicitly committed themselves to the work of social science. The AAW acknowledged its ideological debt to the American Social Science Association by dispatching delegates to the parent body in the late 1870s and early 1880s.[130] Social science affiliates appeared in such places as Hopedale, Massachusetts (1865), Boston (1867), Nashville (1877), Cincinnati (1878), Philadelphia (1877), Indiana (1878), Illinois (1878), Southern Illinois (1879), Denver (1879), New York (1880), San Francisco (1877), and Kansas (1882). By 1880 these affiliates had far more members than the national association.[131] The Illinois membership alone exceeded that of the parent body, with well over four hundred members.[132] According to the Indiana feminist May Wright Sewall, women outnumbered men ten to one in the "study of the conditions of the poor and unfortunate and . . . in the organized, systematic effort to relieve them."[133] Social science clubs were also formed by undergraduates at a few universities, most notably at Cornell in 1878. Florence Kelley cut her reform teeth by joining the club in its first year. She was its secretary then, and she tried later, without success, to be elected president. The Cornell club resembled the faculty's Philosophical Society in its agenda, ranging over a variety of topics from "Rationalism in

Germany" and "The Economic Laws of Organization of Society" to "The Development of the Scientific View of Nature" and "The Protection of Forests." In March of 1881 it invited Matilda Joslyn Gage to speak on "Centralization."[134]

With the exception of the Philadelphia Social Science Association, which had no women and apparently little feminist spirit, feminist women dominated the affiliates.[135] Caroline Dall, of course, had helped organize the Boston Social Science Association with its seven departments and seven women directors; Lita Barney Sayles and Mrs. A. H. Whitman founded the New York City association, and Marietta Stow the affiliate in San Francisco.[136] Feminist Matilda Fletcher joined Dr. Alida C. Avery, one of the most energetic woman doctors of her generation and formerly on the Vassar faculty, and Dr. William R. Alger, author of *The Friendships of Women*, in organizing the Denver association in 1879.[137] The president of the Illinois Woman's Suffrage Association and member of the Association for the Advancement of Women, Elizabeth Boynton Harbert, organized the Illinois association in 1878. Married to a successful Chicago lawyer, William Harbert, who sympathized with her reform interests, Harbert also edited her own newspaper, the *New Era*, in the 1880s and, beginning in 1877, wrote a feminist column called "Woman's Kingdom" for the Chicago *Inter-Ocean*, one of the most influential regional newspapers in the country.[138] The Illinois affiliate thrived for years and published its own organ, the *Illinois Social Science Journal*, edited by Susan A. Richards. Matilda Joslyn Gage described this journal as "a grand contribution to woman's literature."[139]

The Association for the Advancement of Women and the various state social science associations reproduced the ameliorative and theoretical format of the national association. Feminists applied social science principles like organizaton, cooperation, and rationalization, and they busied themselves with the three ameliorative fields of social science—health, education, and social economy. They emphasized prophylactic institutional reform or reform intended to prevent the occurrence of social, economic, and psychological conflict and calamity and to make, in the words of one reformer, "the great forest of the future . . . symmetrical and all its groves healthy."[140] "The work of the hour," wrote Ella Giles Ruddy, social scientist, novelist, and daughter of the president of the Wisconsin State Board of Charities and Reform, "is not so much to save the fallen as to prevent the youth from falling."[141]

Preventive institutional reform implied a new attitude toward those suffering under the weight of new economic and social conditions. "The prevention of suffering," Samuel Eliot told the American Social Science Association in 1865, "is the crowning achievement of our time."[142] According to the social scientists, vast and often engulfing social forces, and not individuals, had caused the crime, poverty, and delinquency of the modern era. As Ella Giles Ruddy put it in her novel (really a social science tract made readable), *Maiden Rachel*, most criminals came "from ordinary types of humanity" and were "not naturally at war with society. . . . Society," however, "unintentionally has warred upon them. . . . Nobody's to blame for anything. Dreadful things are happening to someone all the while and yet it's nobody's fault." Or, as Jeanne Carr observed in her Congress paper for 1876, "The Genesis of Crime," "Moral imperfections result . . . from imperfect conditions, pernicious associations, sometimes epidemic, as disease is, often contagious as physical ills are."[143]

No self-made crminality existed: men and women were not inherently or willfully bad, but made so by social conditions. Society, not individuals, created crime. Because society determined the behavior of all individuals, all individuals (not just a few) became potential criminals. This new remissive angle on suffering explained, in part, the meaning of preventive reform. It went to the root, not to the symptoms, of crime and poverty.

In 1869 the first secretary of the American Social Science Association, Henry Villard, said that social scientists intended to replace older and more "impulsive" philanthropic remedies, which struck only at symptoms or at "things as they appear," with the rational, scientific, and more solid work of "prevention"—a position that the Massachusetts legislature implemented in 1872 when it cut off aid to "small charitable institutions."[144] In words like those of Villard, Mary Putnam Jacobi in 1873 begged the First Woman's Congress to discard older traditions of social reform based on historically "ineffective charitable impulses" or "almsgiving" and probe to the core of social dislocation through the "scientific study of social conditions." She urged them especially to take up "biological studies—the surest way of opening sociological studies to the larger class of women destined to occupy themselves with social reform."[145] So, too, Ella Giles Ruddy wrote that we must hit "at the root of all this evil" and forgo reliance on "poorhouses, private charitable funds" and the futile punishment of criminals.[146]

The new approach to social causation also demanded a more sophisticated and expanded web of institutions. If social danger enveloped everyone and if everyone harbored the germ of deviance, then everyone must be carefully watched, tended, and provided for. The AAW and other social science groups worked for the construction of women's reformatories, women's prisons (which they preferred to call "moral hospitals"), and women's industrial schools; for the creation of "homes for inebriate women" and schools for "vagrant girls;" and for the abolition of prostitution.[147] Many went so far in their preventive schemes as to propose that society "make the children" of unemployed workers, vagrants, and deviants "the wards of the state, even at the cost of maintaining them utterly in childhood; and to educate them compulsorily. It is a cheaper investment than to fill our cities with paupers and our jails with life prisoners." It would remove, moreover, the danger of social chaos, social "terror," and "communism."[148]

Preventive health reform also figured largely in the reform scheme. Women propagandized for the introduction into the school curriculum of hygiene, science, and sexual physiology. As Alice Stockham, the influential feminist doctor, said before the Illinois Social Science Association in 1878, we should "make it obligatory upon ourselves to see to it that every child should get special instruction in procreation and reproduction. Let us see to it that no girl should go to the altar of marriage without being instructed in the physiological function of maternity."[149] Social science women played a major role in shaping the information on health and physiology found scattered throughout the "reports of health boards, the transactions of societies, monographs, etc."[150] All through the 1870s and 1880s they published and distributed hygienic compendia, textbooks, and pamphlets.[151]

These women also looked beyond matters of "practical" social science. At congresses and social science meetings they spoke out in favor of the scientific, nonpartisan study of the "physical facts of life" as the only source for truth and as the principal foundation for social reconstruction.

"We are not convened for the purposes of contemplating a raid upon the Unknowable," Augusta Cooper Bristol said at the 1873 Congress. "On the contrary, we have spontaneously arranged ourselves within the limits of positive knowledge and have chosen no weaker basis than the impregnable platform of science."[152] At the same session Putnam Jacobi asked her audience to spend all their spare time in the "scientific study of . . . social conditions," and three years later, at the Philadelphia Congress, Antoinette Brown Blackwell communicated the same message: the future of women depends on the accurate assessment of the "admitted facts of physiology—the only criteria at present by which we can impartially test the relative mental ability of the sexes."[153] The president of the Philadelphia Congress, Maria Mitchell, the famous Vassar College astronomer, opened the proceedings with an exhortation to women to enter the scientific professions and to liberate themselves from those oppressive sentimental conditions that had formerly hampered the genius of women like the astronomer Caroline Hershel, who, Mitchell said, lived a smothered life in the shadow of her brother, reduced to "darning his socks" and doing "scullery."[154] Sarah Monk, a well-known "authority on the subject of natural science," declared: "We suffer countless ills because of ignorance, and carp at Providence, when scientific . . . knowledge would remove all trouble."[155] All through the 1870s women delivered papers on such scientific subjects as the "Germ Theory of Disease," "Bi-cellular Evolution," "The Physical Basis of Mind," "Herbert Spencer's Philosophy," "Physiological and Psychological Transmission as Factors in Evolution," and the "Evolution of Ideas."[156] The Illinois Social Science Association, in fact, had become so absorbed in theory by 1879 that several women complained of its indifference to the "more practical" questions of reform.[157]

CHAPTER 12

A UNIFIED WORLD VIEW: THE AMERICAN SOCIAL SCIENCE ASSOCIATION

THE American Social Science Association was the queen of bourgeois reformist organizations.[1] It spread its wings over lesser reform bodies and brought to life a myriad of new organizations. The ideas of the association were those of many, often superficially diverse societies and groups—reform spiritualists, moral educationists, Free Religionists, dress reformers, and others. Ideologically, it expressed at the highest level the major themes of the most important feminist organizations and showed how thoroughly post–Civil War feminism was shaped within the context of an emerging new world view. Seen in relationship to these other groups and societies, the American Social Science Association confirmed the widespread existence of a new ideological response to a new society, a response that was critical of but also, inevitably, reinforced existing social and economic relations. It was the response of men and women eager to settle social and institutional arrangements within their own class and between the sexes.

Organic Symmetry and Social Interdependence

BOTH THE ameliorative and the theoretical biases of the American Social Science Association revealed the ideological perspectives of the reformist bourgeoisie. Social scientists looked upon society and social laws as analogous to the structure of the human body and to the laws governing the organic universe. Reformers placed biology, the body, the physicality of the world, at the heart of their angle of vision. "The science of statistics," James Garfield, a future president of the United States, told the association in 1873, "has developed the truth that society is an organism whose elements and forces conform to laws as constant and pervasive as those that govern the material universe."[2] Frederick Wines had written similarly to Caroline Dall in 1868: "I have often said that there is an analogy between the human body and the body politic, and in this: that there is a Social Anatomy, a Social Psychology, a Social Hygiene, and a Social Surgery and Therapeutics; these should constitute the four Departments of Social Science."[3] Moreover, as in the study of the body, social scientists maintained that the facets of the social system needed to be examined synthetically and interrelatedly (thus the sweeping character of the departments), not as discrete entities apart from one another.[4]

The broad concepts of social equilibrium, symmetrical harmony, and interdependence in which all competing factions and classes converge into one unified, integrated, and healthy whole informed the thinking of this association. Throughout the early 1870s the *Springfield Republican* harped on the necessity of building a "more symmetrical organization of social forces" and recovering that "old equilibrium" between employer and employed, rich and poor, men and women that the Civil War had brought into question and that "rapid" economic changes had continued to threaten.[5] Henry Carey, the great American exponent of protectionism and author of numerous articles on social science ever since the 1850s, concurred. "The harmony of the world is only maintained by means of perfect balance of the opposing attractive and counter-attractive forces. . . . Throughout nature the more complete the subordination and the more perfect the interdependence of the parts, the greater the individuality of the whole and the more absolute the power of self-direction," he wrote in

1872. So it is on ships, in factories, and with governments: the "subordination of the subjects being essential to the well-being and the purposes of humanity."[6] Carey believed (and he was by no means alone in this) that, like the evolution of physical organisms, social institutions evolved naturally and inevitably from simple, atomic forms into more interdependent, complex, and organized forms within which individuals freely and naturally arranged themselves—like the cells and organs of the body.

Joseph Wharton, the Philadelphia political economist, and Francis Lieber, free-trade advocate and common sense philosopher, also expounded this position. In 1870, in a paper for the association, Wharton wrote:

> It is to be observed that the degree in which the individual voluntarily surrenders or modifies his original rude independence increases with the completeness of the organization to which he belongs; in the case of such complicated structures as the great civilized nations of modern times, he is compelled or restrained in every function and at every moment, in order that the great organism of which he is an almost imperceptible constituent may thrive.[7]

Similarly, "The further we advance," Lieber wrote in 1871, in an essay called "Interdependence," "the more intense as well as extensive becomes the civilizing law of interdependence, while the cravings of men multiply with every step of progress . . . Whole hemispheres depend on one another for food, health, comfort, safety, knowledge and skill."[8]

Neither Lieber, Wharton nor Carey let the matter rest there. All agreed that the individual gained greater advantages from the state by so "surrendering" his "rude independence." "The net balance of advantages to the individual," Wharton insisted in his 1870 paper, "certainly expands with increased perfection of organization."

All classes, according to the association, must understand this principle. They must learn that men and women live in a community of shared interests, that dynamic development comes only within harmonious conditions, and that "each man must concern himself with the welfare of his neighbor and fellow citizens."[9] Labor in particular must absorb the fact "that the interests of the operatives do not stand alone and cannot be promoted apart from the interests of others."[10]

Aware of the widening breach between labor and capital and many of the social and economic causes for such a breach, reformers nevertheless refused to believe that such disparities could be erased by making "social clefts into passionate antagonisms." Indeed, "overstimulated" and irrational "clamour for rights" endangered every part of the social system with disease.[11] Relying heavily on the body analogy that gave their arguments compelling psychological power, reformers feared that any social convulsion, any abrupt or spontaneous outbreak, would destroy the harmony so vital to the life of the social organism.

Men who had no direct relation with the association, positivists and neopositivists alike, spoke the same ideological language, including the most radical of mid-century individualists, Stephen Pearl Andrews, the utopian reformer Albert Brisbane, and the Free Religionist Francis Abbott. One quotation speaks for the rest. In 1866 Abbott, editor of the *Index* and a feminist, said:

> . . . the most profoundly philosophical view of human society is that which makes the race *an organism*—a Pure Individualism is the crudest type of human existence. Thoreau, in his last hut by Walden Pond, is a specimen on a higher level of the one-celled plant, or would be so, if he had not got into his head, before he went there, what never grew in the woods. . . . It is idle to cry out against organization; everyman . . . is born into organization like a nest of boxes. . . . From birth to death, men are dependent on each other in countless ways; there is no such thing as human independence.[12]

Modern historians—most recently Thomas Haskell in his book on the American Social Science Association—have argued that the educated middle and upper-middle classes gradually had to adjust their social thinking to the reality of an interdependent social and economic system, that this reality itself had changed and shaped ideology.[13] Long before an interdependent society had clearly existed, however, the most advanced whiggish and educated sectors of the American bourgeoisie had, as we have noted throughout, become accustomed to thinking of society, of human relations, of the human body in interdependent terms. For reasons that are still not altogether clear, this class embraced the concept of interdependence as its own: it became an ideological bridge to power; it helped make the reality that did come to pass and that so completely entered reformist consciousness by the end of the century.

This concept of organic symmetry and interdependence stands at the center of the ideological world view of the ASSA. Implicit within it was the varied content of bourgeois reformist consciousness. The well-balanced harmonious body found its place in the reform scheme, the ideal body suited for a civilized age, an ideal congruent with the ideal society itself, which demanded equal development of physical, moral, and intellectual faculties and promised to prolong life indefinitely. The ideal society excluded "excitability," excess passion, and "impulse" from behavior, and considered such virtues as "self-control" ("faculty of restraint under excitement"), masculine "vigor," "common sense," and "persistence in the wisdom of biding time" the signs of modern humanity.[14] These social scientists denied the existence of mind-body dualisms. Whatever evil lurked in the world came from the collapse of internal balance or harmony. Evil equaled a disturbed and pathological "one-sidedness," the splitting apart of mind and body functions.[15]

Reformers bewailed and studied the new acquisitive and aggressive "habits of life" of an "advanced industrial society" that had made ordinarily healthy and moderate people "nervous," "monomaniacal," "feeble," and immoderately self-indulgent human beings.[16] They promoted the creation of institutions—hospitals, prisons, and asylums—to rehabilitate in a "humane" way the lives of those seriously handicapped by change. "Social Science in all its aspects," the *Springfield Republican* explained in 1875, "stimulates us to deeper sympathy with human weakness."[17] Men should not design prisons to "punish" human beings, one reformer wrote. Limited to the futile exercise of external force, punishment merely hardened resistance, preventing reform. It suggested a repressive authoritarianism that no longer had relevance in modern, democratic society. The process had to be "wrought within the mind" and thus entailed a more therapeutic, more reinforcing, more maternal way of reconstructing deviant, broken souls into healthy "whole personalities." Such reformers saw no virtue in capital punishment and worked to have the gallows removed from the public realm. As a vestigial symbol the gallows offered occasions for public acts of barbarism and grossly exaggerated the latent violence in men and women.[18]

Prevention: Sanitation, Kindergartens, Industrial Schools

AS IN the feminist-dominated social science affiliates and feminist health societies, prevention was central to the thinking of the ASSA. Prevention would make reconstruction unnecessary in the first place. "Preventive institutions," observed Enoch Wines, president of the National Prison Association, "constitute the true field in which to labor for the repression of crime."[19] Preventive reform spanned a range of programs, but first came public education in the principles of personal and social hygiene.

Everyone, but especially young adults, should study the "facts of life," the "rules of their physical existence, knowledge of which is of the utmost practical moment to every human being." Hygiene, after all, "paves the road for all human advances, commercial, intellectual and moral." Obedience to hygienic rules of balance and harmony prevented physical "degeneration"—the root cause of all chaos and crime.[20] One university hygienist declared in 1865 that "students should know the human body and its relation to natural and unnatural forces" before they bothered with mathematics and languages. They should know the "four branches in the study of disease; first, those preventable by means now known; second, those preventable by proper personal hygiene; third, those preventable by proper domestic arrangements; and fourth, those which are affected by the action of communities."[21]

There was another dimension to this physiological project. By the 1860s a generation of hygienic reformers had come to accept the fact that body and mind, psychology and physiology, worked together in some kind of reciprocal relation. Some theorized that the mind acted upon the body while others defended the reverse position. Everybody appeared to agree, however, that a relationship existed between them, and for that reason concluded that medicine and education, teachers and doctors, hospitals and schools should become allies in the struggle for optimal human health. In 1875 Anna Brackett, the school administrator and enthusiastic pioneer of social science methods, argued in her paper, "The Relation of the Medical and Educational Professions," that mind and body "touch" each other constantly in "reciprocal action" and often to injurious effect. A disordered brain, for example, might bring on "depression" or an inflamed "inspiration." An "overstimulated imagination," on the other hand, might lead to "despair" or to severe physical debility. To prevent these irregularities, Brackett proposed that the "medical and educational professions should be working hand in hand, in the most perfect cooperation." Indeed, she already had them "hand in hand" in her own school, where, she remarked, her students enjoyed "regular and persistent mental activity, judiciously stimulated and controlled."[22] The noted psychologist and physician George Beard spun out his own vision of this symbiosis before one of the association's many brain children, the National Association for the Protection of the Insane and Prevention of Insanity. "For every schoolhouse," he suggested, "there should be built an asylum for those who shall hereafter trace the beginning of their insanity in endless recitings and competitions."[23] In the manner of most psychologists of his day, Beard worried about students, male and female, who suffered from overwork and competitive strain. Surely, he said, these young people should have recourse to readily accessible treatment. They also needed to be observed, their illnesses better understood. In Beard's ambitious scheme, and to a lesser extent in Brackett's, the college and high-school clinics of the twentieth century found their earliest proponents.

The desire to prevent disease, disharmony, and disorder stirred the reformist mind to imagine more than explicit measures to educate the public in matters of hygiene and physiology. It knew that knowledge would not suffice in the battle against crowding, congestion, and crime that pressed upon the life of the industrial city. More precisely, reformers knew that knowledge alone would not master the "population crisis" that the association was concerned about in the late 1860s and that would occupy a special place in reform thinking for the rest of the century. This "crisis" had two major facets: the declining fertility of the native population and the mobility of population generally from country to city. In 1866 Nathan Allen, the first association member to specialize in population studies, raised the spectre of native infertility in a speech that made headlines. He announced that the foreign population was reproducing at a rate two and three times faster than the "Yankee stock," and he attributed this in part to the inability or unwillingness of American women to bear children.[24] The following year the association debated the issue. Some members blamed the declining birth rate on female idleness, fashion, and "degeneracy." Others, notably Dall, indicted the "excesses of men." Everyone agreed, however, at this session and later ones, that the "social status of our middle class has undergone fundamental alteration," manifesting itself "in untidy homes, in discouraged housewives, in untutored children . . . in chronic indebtedness, in the desertion of the family hearth and the family altar for the theatre and the salon, in the reign of selfishness and greed which is eating out the nobility of the American heart."[25]

Unlike such contemporary reform groups as the moral education societies, the association did not deal directly with native infertility and the rationalization of marital behavior. It did support the reconstruction of the lives of women, and, indeed, practically every program it promoted was intended to add new vigor, flexibility, and strength to the "status of the middle class." And it grappled with fertility indirectly by focusing on the second facet of the "population crisis." At the 1867 session population shifts, not fertility, consumed attention.

The influential statistician Edward Jarvis presented the relevant statistics: "In twenty years, in Massachusetts . . . cities and large towns gained 109.9% and the rural districts gained 47.6%. In the United States, the gain of cities was 186.5% and rural gain was 72.5% in the same period from 1840 to 1860."[26] Jarvis saw danger written all through these figures, a danger already visible in urban crowding and poverty. To meet the danger, he and others proposed a preventive, sanitary assault—better housing and architecture, better sewers and ventilation, organized precautions against the spread of disease, quarantine regulations, vaccination programs, careful registration of births, marriages, and deaths.[27]

Reformist concern about rational, sanitary burial practices clearly showed the antisentimental tendency of the association and its kindred organizations. As one sanitarian put it in 1874, Americans yielded too easily to "sentimental deference to structural tissues" and forgot the sanitary virtues of cremation. Sentiment, of course, had its rightful place—who would deny that?—but the dead body could be "an object of loving caress" for only a "brief period of time at the most" and then, regrettably, it would begin to rot in a very unsanitary way.[28]

To help these various reforms reach their targets, Jarvis and the other social science sanitarians organized the American Public Health Association in 1872. The APHA brought "sanitary engineers," physicians, and legal counselors into one unified lobby for public health and convened in major cities from New York to Nashville. "Its first object," the *New York World* reported, "is to secure reasonable and practicable methods of sanitary administration."[29]

Reformers also wished to construct a general educational system that would educate "rich and poor equally," reform the "dangerous classes," end "idleness," and socialize children and young adults according to the principles of harmonious, positive, and utilitarian living. They early took up the cudgels, for example, in defense of the kindergarten, the institution that embodied reformist intentions at all educational levels.

The advocacy of the kindergarten system by the American Social Science Association, the Association for the Advancement of Women, and the social science affiliates illustrates the extent of reliance on education, and therefore on the state, for the implementation of preventive reform. First proposed before the war by such pioneers as Elizabeth Peabody and Mrs. Horace Mann, the kindergarten was taken up by the New England Woman's Club in the late 1860s and discussed in such periodicals as the *Woman's Advocate*, the *Woman's Journal*, *Alpha*, and the *Journal of Social Science*. It became something of a craze, a *cause célèbre*, in the middle and late 1870s.

By this time kindergartens had been established in St. Louis, Boston, Washington, and New York City.[30] Perhaps the high-water mark came in 1876, at the Centennial Exposition in Philadelphia, when the Woman's Department made space in the Woman's Pavilion for a model kindergarten where women could come to study the basic principles of kindergarten instruction.[31] A front-page story in the organ of the Woman's Department, the *New Century*, celebrated the kindergarten as one of the most fruitful new sources of female employment: "As the Kindergarten is the ground from which the schools of the future will grow, and as its comprehensive methods underly not only the intellectual but the physical and moral training of children, it will receive much attention in the *New Century*."[32]

For social science men and women, as for the progressive spiritualists, the kindergarten disclosed in microcosm both the ideals of prophylactic reform and a vision of the new society reformers wanted for themselves and for their children. On the one hand, it supposedly "freed the self-activity" and "natural playfulness" of the child. On the other hand, however, it acted "morally . . . to make this life in community serve as a means to destroy egotism."[33] As an educational strategy, the kindergarten obviously did not serve merely to give the child the rudiments of human knowledge. Its *raison d'être* was to check the earliest bent toward "isolation," "caprice," and "vain self-sufficiency" in a child and to place "self-activity" within an institutional framework of harmony and equilibrium.

Baronet Marenholtz Bulow, an influential kindergarten pioneer in Germany, observed in an article translated for the *New Century* by Elizabeth Peabody, "The child among other children is only truly himself. In that companionship alone he learns order, subordination, the value of things, and by reaction to others is character formed."[34] Alone and "self-sufficient," he becomes mentally confused and sick. The kindergarten took as its prototypic, unreconstructed child what Dr. Mary Putnam Jacobi called the "neuropathic individual" in a paper, "New Asylum Treatment of Insanity," for the American Social Science Association in 1881.[35]

The "neuropathic individual," in this analysis that anticipated the neo-Freudian thought of the twentieth century, "reveals two conspicuous elements: a profound and often unconscious egotism, resulting from the predominance of the instincts over the faculties for external relations, and a constant ineffectiveness in the maintenance of these relations." A core of symptoms accompanied the general antisocial syndrome: "nervous excitability" and "caprice"; alternating "periods of exaltation and depression, leaving scarcely any room for healthy indifference;" "jealous suspiciousness" and "an excited imagination;" an inability to bear the "weight of responsibility" and "pressure" or to "terminate the conflict in consciousness of opposing ideas by the conception of mass representing the Ego, which assimilates part of the ideas and represses the rest;" and a tendency to "submit passively" to "exciting doctrines, as in various religions or political manias" because of "the ineffectiveness of will" and a "shrunken individuality."

Putnam Jacobi did not believe that people suffering from this pathology needed hospitalization. She labored to remove the incubus by theorizing that the disorder preexisted in varying degrees of intensity in the "health" of everybody. But she warned that "egotism . . . determines the form of all delirium, which, whether primary, or engendered from the emotional insanity, invariably centre on the depression or extreme exaltation of the self," and she strongly recommended what she called a "duplex therapeutics" or "prophylaxis," based on the notion that physical and moral elements interlock inextricably within the mind of every human being. Thus her therapy combined "physical" with "educational prophylaxis" intended to treat every neuropathic case and to prevent the worsening of the disease. Physical prophylaxis entailed eating healthy food, doing gymnastic exercises, breathing pure air, and taking cold baths; "educational prophylaxis" set out to activate the "social sympathies" of the neuropath, to educate these individuals to appreciate their "insignificance relative . . . to the vastness and importance of the interests of the world," to expose them to "practical activities," "practical concepts," and "practical facts," and to direct them toward the "creation of a permanent career." Educational prophylaxis trained the neuropath "early in a sound and simple philosophy which shall provide a firm basis for thought and life without inviting to speculative thinking." Because the neuropathic condition preexisted in the "health" of everyone, in fact, such prophylaxis might commence "with the dawn of consciousness, and be extended if possible, throughout life."

Putnam Jacobi was analyzing mental sickness and preventive treatment, not the kindergarten candidate and the kindergarten. Her view of mental illness, for the most part, was compassionate and the illness she examined a very real consequence of social change. Still, it takes little imagination to see how readily these categories can be made interchangeable. Thus the neuropath becomes the unreconstructed child and the clinical situation the kindergarten. Kindergarten reformers often wrote as if every child were potentially pathological and the kindergarten were a therapeutic clinic constructed to prevent the outbreak of disease. The ease with which these two sets of categories fused shows once again how easily the conception of disease and treatment devolved into an ideological statement having little or nothing to do with disease. Disease displayed something more than the very products of social change: it was a way of life that threatened to invade the consciousness of every man and woman. In both the clinical and the kindergarten theories we find projected the harmonious, cooperative, and interdependent society of the reformist bourgeoisie, a positivist society that placed individual "growth and development" completely within structured, institutional arrangements, that looked upon idiosyncrasy, imaginative speculation, and extroverted, individualistic behavior with unrelenting suspicion and that transmuted "egotism" and "self-sufficiency" into a species of mental illness.

So, too, social scientists advanced a new "industrial system of technical education" for poor city boys unable to go to college. The great reformer Wendell Phillips, in a "Report on a Developing School and School Shops" prepared for the association in 1876 and delivered at MIT, said, "One of the great problems which confront republican statesmanship is how to manage the population of the cities. The tendency of our time is to gather men into cities" and to spawn "idleness" and "vicious classes."[36] Phillips hated the industrial city. He thought of it as a blight and a burden that compared very poorly to the purity of the rural village. Nonetheless, he faced its existence as a *fait accompli*, something odious that all Americans had to adjust to whether they liked it or not. He was prepared to attack its problems with aggressive, humane, and preventive reform. It was for this reason that, with Elizur Wright, S. P. Ruggles, Edward Hale, and others, he studied and reported on technical education for the urban poor.

These men accepted the new industrial system with its detailed division of labor, its disciplines, and its large economies of scale; they saw no feasible place left for "self-education" and apprenticeships.[37] "The conditions of society have undergone such a radical change during the last forty or fifty years that laborers must now receive a different practical education from what was required two generations ago," Ruggles wrote. "Apprenticeships having departed, never to return in their ancient form, something else must take their place, and give to our artisans practical instruction." Work had been split into "specialties." A carpenter, for example, once learned to build a whole home by himself. Now there were special establishments with complicated machines designed to make doors, sashes, and blinds of every description. The plan that Ruggles suggested and the others applauded fitted the worker for a specialized task in an interdependent system but did not at the same time turn him "into a tool." First the student entered a "school-shop" where teachers exposed him to a variety of tools: lathes, carving instruments, steam engines, jigsaws, and so forth. Next, he took instruction in the use of several tools, learning a few particularly well and getting a feel for the total mechanical process.

"We wish," as Wright put it, "to educate the boy, not into a filing tool of the highest possible perfection . . . but into a master of a large variety of tools, that he can create all the parts of some complicated and useful mechanism, so as to work and produce something." In the third step the student went into the "mechanical manufactory" itself, and "though he may be set . . . to do one thing over and over, he understands and sympathizes with all that is going on. *He catches the spirit of the place* and feels himself in some degree master of the situation."

Reformers believed that the detailed division of labor could help transform society into a fully interdependent social organism in which every class, every interest, every person would benefit. The social body was like the physical body: both flourished because both were well-balanced wholes. But this analogy was profoundly misleading. The physical interdependence of the human body, whatever its merits as a biological idea, bore no relation to the labor of men and women. After the Civil War individual labor was increasingly organized, not to make whole products but to make parts of products. The division of labor created social and economic interdependence, but it did not create whole human beings: it "dismembered the worker as well as the work."[38]

In effect and intention the industrial school resembled the kindergarten. Both sought to merge individuality, spontaneity, "non-machine-like" behavior, and balanced development with their opposites: repression of caprice, prevention of idiosyncrasy, specialized tasks, and the yielding up of the will to the good of the whole. In the one, children learned how "to combine with their fellows so as to aid and not hinder rational purposes;" in the other, they learned how to sympathize with "all that is going on" and to catch the "spirit of the place."

Even more important, social scientists conceived of the state, as well as the social institutions they supported, in both preventive and remissive terms. Dr. Henry Bowditch, in an address called "Preventive or State Medicine" which he delivered throughout the

1870s, said, "Formerly man's welfare was subordinated to that of the state. Now the theory is exactly the reverse, and the state claims to have the tenderest interest in the welfare of each and everyone, the humblest or richest of its citizens." One of Bowditch's colleagues declared in 1873 that state institutions "seek simply to save human life."[39] And Bowditch himself believed that the state existed to prevent misery; it was "caring," "non-punitive," maternal.[40]

Chairman of the Massachusetts Board of Health and one of the most powerful men in Massachusetts medicine, Bowditch insisted that the administration of public health should be completely free from the inhibiting weight of political and religious meddling. Of universally human significance, it had no place in partisan debate:

> Perfect health should be a matter of profound interest to every human being. Without it few of the relations of this life can be fully sustained. It should therefore be constantly sought after by each man and woman; by the state for its citizens; by the nation for the states and communities of which it is composed; and, finally, by every sovereign power in its relation with any and all other civilized or uncivilized sovereignties.[41]

As Joseph Wharton said in his paper on "International Industrial Competition," "Statesmen must look beyond individuals or classes, and beyond the immediate present" and "see that the body as a whole possesses vigor and symmetry; that development and robustness attend upon nutrition; that the whole organism enjoys fair play and good guidance in its strife with similar artificial bodies."[42]

To secure "perfect health" for all, Bowditch formulated a national plan that delegated extraordinary powers to the administrators of public health. In it he recommended the creation of a National Board of Health and a new cabinet post, Secretary of Health, to coordinate a system of state boards of health, "keep constant watch" over the health of the country, and "summarily abate or put under surveillance any establishment creating a nuisance, but with which the smaller . . . boards feel unable to cope." He envisioned an international system of national boards to protect the world from "violent infections" and "contagious diseases" and to "compel uncivilized states to submit to that amount of sanitary law which may be deemed necessary for the health of mankind." A former Garrisonian abolitionist and transcendentalist, faithful to the memory of John Brown, Bowditch equated disease with slavery; the oppression and invasion of the human organism by disease was like the strivings of one class or race to subdue, invade, and enslave another. As people had "combined to put down . . . the slave trade," so they must unite in a common campaign against disease.[43] For Bowditch and for his reformist generation, disease, like romantic passion, implied distance and separation, while at the same time signifying, like passion, a dangerously intimate and intrusive assault by one man or woman upon another. Disease isolated individuals, couples, communities from an interdependent society that brought men and women together without contamination and without intimacy. In a sense Bowditch saw in the nightmare of a diseased society the healthy, interdependent one he so longed to construct. Past the reality of disease, reformers saw their hygienic, cosmopolitan utopia. By overleaping and transgressing all frontiers, disease, more than anything else, underlined humanity's dependence on humanity and society. Inevitably, therefore, it required universal, state-administered solutions.

The Ideology of Unified Social Organization

"WE FELT ourselves in 1865," wrote the secretary of the American Social Science Association, Franklin Sanborn, "to be literally 'heirs of all the ages, in the foremost files of time,' and there was little that we did not fancy ourselves capable of achieving."[44] Sanborn's optimism rested on a new faith in organization, on "combined action," "action under the guidance of general laws." "The Science of Society" needs "organization" and "aims at strengthening institutions." What we need now, said Caroline Dall, "is to help the authorities instead of rendering their task impracticable."[45]

Social scientists were no longer satisfied with the usefulness of individual, unsystematic reform and private religious charities. These were not enough to cope with complex social and economic changes and interdependencies that the expansion of productive forces had brought about. Only bold, centralized reform could work: the elaboration of rational bureaucratic procedures, the diffusion of scientific knowledge throughout the population, the forging of systems of nonpartisan expertise. In 1873, for example, the association directly influenced the establishment of training schools for nurses in New York City, Boston, Philadelphia, and New Haven; in 1876 the Health Department of the association succeeded, through appeals to the governor of Massachusetts, in replacing the old system of coroners ("ignorant officials") with a new system based on a "competent force of medical examiners;" and the association worked to improve the professional status of medicine, law, and pharmacy by demanding fixed "professional traditions," systematic institutional "training," "more unity of action among the different schools," and greater "professional recognition" for university teachers.[46] The association lobbied from the 1870s into the 1880s for replacement of an "uninformed," chaotic, and fiercely partisan majoritarian system of government with minority representation; elimination of "party obstructions" to voting; enlistment of nonpartisan civil servants selected from the "best men" and employed with the prospect of "long tenure;" and a more efficient, combined system based on larger administrative units. In the 1860s Andrew Dickson White had gone so far as to advocate the abolition of state governments and boundaries and the founding of a "powerful federal government."[47] Samuel Bowles of the *Springfield Republican* made a more modest suggestion in a speech to the association in 1878, although his message matched White's:

> The changes of the century have trenched slowly and seriously . . . upon the State [Massachusetts]. . . . The multiplication and growth of cities, the transfer of our population from moderate collections of farmers and small mechanics and storekeepers into mammouth centres of manufactures and commerce, with the constant tendency to individualize the principle of local self-government, has built up a series of subsidiary representative governments within the State, that have come very much to divide, absorb and dissipate the power of the State itself.

This "imperialism" of "intimate local and individual" power had brought with it waste, ignorance, corruption, and bad government and begged to be overturned by a "common system" of political institutions under the aegis of a "flexible" and incorruptible "centralized authority."[48]

The ideology that underlay this new reform outlook mingled sentimental humanism, Hegelianism, and positivism and attempted to merge individualism with these three traditions. The thinking of two of the most influential members of the American Social Science Association, Franklin Sanborn and William T. Harris, feminists and friends, showed the impact of humanism and Hegelianism. Although Sanborn never broke with the transcendentalism of his youth and held Kantian ideas all his life, he veered decidedly toward a new kind of institutional thinking in the 1860s.[49]

Like other social science reformers, male and female, Sanborn objected to the cold "pessimism" of traditional political economy and its modern form, social Darwinism. A "sentimentalist," as he called himself, he preferred to highlight the "sympathetic" and "cooperative," not the "selfish," side of human nature:

> It is the ever advancing goal of human welfare that the students of social science aim at. Our friends of the pessimistic school dwell with grim satisfaction on their doctrine that teaches "survival of the fittest;" but if the fittest do survive, they must make the world a fitter place to live in. It is the survival and not the extinction upon which the student of social science fastens his attention.[50]

Sanborn looked forward to the end of regional and national boundaries and the growth of a truly humane, cosmopolitan "civilization . . . of no particular hemisphere or latitude."[51] As the close friend of Harris, a St. Louis Hegelian, he also adopted Hegelian ideas. In the early 1870s Sanborn read everything Harris wrote for the journals and newspapers, endorsed it, and persuaded Harris to lecture before the association.[52]

Like Sanborn and other reformers, Harris had taken up transcendentalism in his youth only to part with it in the 1860s. Later in life he called the transcendentalist period a dangerous but necessary moment of "negation," during which young men and women rejected, in healthy ways, "blind obedience . . . to arbitrary authority." But they went too far, stumbling over "illusions," renouncing "tradition" and "the aggregated thought of humanity."[53] Harris edited the *Journal of Speculative Philosophy*, the first Hegelian publication of its kind in the English-speaking world. He was superintendent of schools in St. Louis and built the first permanent public-school kindergarten system in America.[54] In the 1870s he dominated the Education Department of the American Social Science Association and contributed more articles to the association's journal than any other member except Sanborn.[55]

As Merle Curti has shown in his study, *The Social Ideas of American Educators*, Harris's Hegelianism paid homage to individualism and self-help while at the same time tying individuals to existing and emerging institutions. "As an Hegelian," Curti writes, "Harris believed that the national state was the greatest of human institutions, necessary for civilization and for the realization of true individualism."[56]

In 1885 Harris published his own retrospective on social science. In it he gloried in the technological and economic innovations—the telegraph, the express train, the morning newspaper, and international trade—that had converted a fragmentary society of "isolated and exclusive communities" into an organically interlocked, "cosmopolitan," and "spiritual" whole. "This is one of the greatest humanitarian or humanistic influences of our time," he said. "It is one of the largest counteracting forces" to the "materialism and grovelling spirit which threatens to engulf us." Technology and trade had brought men and women together, making them conscious, for the first time, of their "social" existence and of their ethical place in community life. "Coiled up in human nature is this social instinct which unfolds into human institutions; these institutions are the colossal selves of humanity—the true vocation of man is the realization of these selves."[57] Sanborn and Harris, as well as other social scientists who studied idealist thought during this time—among them Thomas Higginson, Anna Brackett, Julia Ward Howe, John Bascom, David Wasson, and Ednah Cheney—spurned the positivist mode of thinking.

Until the 1880s the majority of its members did not conceive of the ASSA as a scientific body. According to its secretary, the organization did not "satisfy the exact student of science who has been taught to divorce his wishes and aspirations from his perceptions, while he is investigating nature. . . . For the present," he went on, "we must admit that the term 'science' cannot be applied to our pursuits, in the sense that it describes the researches of the geologist, the chemist, and the astronomer." Sanborn's disclaimer notwithstanding, the association certainly had a quasipositivist, theoretical inclination.

It borrowed biological methodologies and categories of Comtian thought, with its advocacy of rational organization and symmetry, objective observation, and prediction, and its subordination of imaginative feeling and abstract theories of right to the tendencies and developments inherent in the social organism. It sought to gather all statistical and perceivable data from every field of knowledge, to synthesize such data, and to cull from it basic social principles and underlying social laws.[58]

"The object of the Social Science Association," the cofounder and secretary of the Western Social Science Association, Frederick Wines, wrote to Caroline Dall in 1868, "is to accumulate a sufficient number of facts and enlist a sufficient number of active minds in the work of digesting those facts, to enable us, after a time, to state with absolute confidence the true principles of Social Anatomy, Physiology, etc." The association was "organized for the purpose of ascertaining by an induction of facts, the fundamental laws of the social organism."[59]

In the same vein Dall herself wrote in 1865 to the newly organized Social Science Association in Hopedale that "the ASSA is not a Reform Society. . . . The people who come together in the ASSA are supposed . . . to be fully impressed with the need of reform—and they are to stimulate, not passions and prejudices, but intellectual insight, calm and cool inquiry. . . . It is not for such an Association to take up particular branches of practical work, enlist supporters and carry points," she added. "It has only to hold itself in an impartial position, to collect all facts, thoroughly discuss all theories, and test by their results all experiments. When it has reached inevitable conclusions, it may recommend modes of action, and show *how* its principles may meet the popular need."[60]

Positivism and idealism coexisted in the thought of this organization. They also coexisted in the progressive spiritualism of less influential reformers as well as in the Free Religion that most association members found so attractive.[61] Both philosophical systems, however much they conflicted, drove in the same direction—toward scientific objectivity, social hierarchy, organization, rational order, and the selfless dedication of the individual to the needs and imperatives of the social state.

As the name suggests, the American Social Science Association emphasized social reform and social institutions—the organization and rationalization of the "physical conditions" of life. Like Comte, social scientists believed that "permanent political changes cannot be effected without previous social and moral changes."[62] In his general report in 1870, Sanborn wrote with relish that "social problems"—specifically, education, sanitation, hygiene, civil service, and the reform of the sexual relation—rather than "political questions" had steadily captured the public mind and that the association had contributed to this reorientation.[63] This is not to say, however, that the ASSA ignored contemporary politics or economic matters. Indeed, it attacked—and more strongly after the demise of Radical Republican reform—the political party system for its partisanship, corruption, and indifference to the institutional needs of the country, as it bewailed efforts to absorb into the democratic franchise the most uneducated and ignorant classes. The association also fought the "powerful and corrupting centralizers of capital and controllers of legislation" and feared a new, poverty-stricken proletariat given to violent tactics to obtain its ends—both these classes spawned by a vigorous productive system and both threatening to upset the essential equilibrium of the social system. The point is that these reformers did not believe in ordinary "political" methods to bring about change, nor did they question the viability of American republicanism. They wanted a nonpolitical, harmonious republicanism, maintained by an educated, virtuous electorate and governed by an educated, virtuous elite.

These reformers were proponents of the capitalist system. Capitalism, they believed, would benefit everyone if freed to work unfettered, naturally, harmoniously. "We have in this country all the conditions that give labor its greatest productiveness and to capital its greatest rewards," the laissez faire economist David Wells wrote in 1876. Americans possessed an almost "unlimited supply of cheap, fertile land" and a great technology able to tap immeasurable industrial potential.[64] Yet the ASSA aided the working-class poor, sponsoring labor cooperatives, industrial schools, and insurance plans; and it helped formulate the cooperative ideology of the National Labor Union, which in effect defended every man's right to become a capitalist.[65] At the same time the association deplored both "selfish monopolies" for debasing the principle of equal opportunity and labor's struggle to "compel the redistribution of wealth." The ASSA wanted no alteration in the "natural" relation between labor and capital. "Suffering," "aggravation" between workers and managers, declining prices and profits, and overpopulation were the inevitable, but ephemeral, outcomes of change. They could not be legislated away without blocking the accumulation of capital and thus "diminishing the abundance" needed for further advancement and growth.[66]

The ASSA proposed little or nothing new in the way of conventional political-economic reform, but it did bring to crystallization a new social vision and channeled middle- and upper-class political energies and ambitions into a new set of social institutions that promised to minimize the insecurities and stresses of economic change; it promised to harness individualism and to make people feel a part of a larger, more caring human community. And in spite of laissez faire perspectives, it enlisted the surveillance and intervention of the state to achieve these ends.

The motives for the emergence of the association, however, went beyond the desire to meet pressing problems in new and nonpartisan ways. A powerful class motive propelled the creation of this organization, a political motive that fostered the necessity for "working in common" and that lay hidden within all pleas for equilibrium and organic symmetry. Confronted from within by doubt and disorientation, threatened from without and within by real and imagined excess, selfish abandon, and blind ambition, the reformist native bourgeoisie found itself menaced from below by new and "disturbing elements." There were new "races," "creeds," and the "spectre" of a "pauperized" working class that appeared ready, if left alone or in the hands of unscrupulous demagogues, to "hinder" the advancement of "all classes and interests," to destroy the psychological, social, and economic balance so vital to the continuing existence of bourgeois stability and identity.[67]

The American Social Science Association was an important response to these challenges. In the name of all classes, in the name of science that crossed all boundaries, in the name of both sexes, and in the name of humanity, it sought to keep or obtain for the educated middle class an invincible place in the institutional and professional centers of American society. Above all, it heralded the human "body" and its rationalization as the new frontier, the bourgeois path to sanity, order, and respectability, the new field upon which this class would struggle to reform and master its own fate and that of other men and women.

Mid-nineteenth-century feminism reflected in all ways the ideological syndrome of the American Social Science Association. It also reflected how much and how deeply feminist ideas on love and marriage, education and health, religion and science, work and fashion, had been forged within the larger quest by the educated bourgeoisie to quell the turmoil within itself, to clarify power relations within itself, and to reconstruct rationally American socal institutions. It indicated how dependent feminists were and how dependent they would remain upon the shifting needs and emerging priorities of their own class. It was no accident that feminists of the National Woman's Suffrage Association—Elizabeth Cady Stanton, Lucretia Mott, Kate Doggett, Elizabeth Boynton Harbert, and May Wright Sewall—as well as feminists of the American Woman's Suffrage Association—Julia Ward Howe, Lucy Stone, Antoinette Brown Blackwell, and Mary Livermore—attended social science congresses and conventions and occasionally chaired the meetings. There, according to the authors of the *History of Woman's Suffrage*, feminists could discuss a "wider range of subjects than could be tolerated on the platform of any specific reform."[68] There they could comfortably, and without the appearance of "political" fanaticism so often associated in the public mind with their suffrage societies, make a unified commitment to their own class interest and well-being.[69]

Male and female social science, furthermore, prefigured the Progressive movement of the early twentieth century. Although comparisons of such complex periods and groups require further study, a straight line appears to connect this comparatively small group of feminist social scientists with such later Progressives as Jane Addams, John Dewey, Walter Lippmann, and Herbert Croly. To be sure, there are differences. The earlier generation had a stronger faith in the ideal of selfless social service and in the liberating power of productive behavior in the public realm. Their assault notwithstanding, they still accepted much of the legacy of individualism—above all, economic autonomy and self-ownership—and they believed in the virtues of self-discipline and delayed gratification. Many Progressives supported a leisure-oriented and permissive ethos more in harmony with an emerging consumer-mass culture, one more committed to pleasures in the private sphere. In this later period selfless social service was partly coopted by new white-collar service industries in such areas as health, social welfare, child care, food production, and nursing.[70] Because capitalist wealth had become more concentrated and powerful, Progressives also sought government regulation more actively than the older group had done, and they followed a pragmatic, instrumentalist philosophy, repudiating the older positivist belief in fixed and regular laws.

Nevertheless, the continuities from generation to generation were many. Some of them are offered tentatively here, with full knowledge that more research must be done to prove their existence fully. Both groups came, generally, from the same native American roots and from similar middle- and upper-middle-class backgrounds. By championing mental and preventive hygiene and state intervention in such areas as health, education, and social welfare, mid-nineteenth-century reformers and Progressives displayed the same obsession with harmonious, well-adjusted health. The mental health clinics of Floyd Dell's *Love in the Machine Age*, for example, institutionalized Dr. Mary Putnam Jacobi's duplex therapeutics. So too the older feminist-reformist commitment to no secrets matured into a passion for muckraking and open government, a wider advocacy of sex education in the schools, and a more pervasive belief in the need to spread "objective" scientific knowledge throughout the population.

Both groups labored to remove conflict in the public and private spheres. On the public level, mid-century reformist conceptions of symmetry, cooperation, and communal interdependence found sophisticated and highly articulate expression in the Progressive communitarian theory of such thinkers as Charles Horton Cooley and in Herbert Croly's New Nationalism.[71] Both groups aimed to eradicate the clash of interests that formed the basis of political life, replacing this pattern with a public life dedicated to prevention of conflict and constructed by a "wise" scientific elite from "outside" the realm of politics. In the private sphere, these two groups rejected individual passion, excess, imagination, nonutilitarian behavior. Thus the Progressive sociology of marriage and parenthood inherited and made more "scientific" the earlier reformist hostility to romantic love and romantic passion.[72]

Another important feature linked both generations of reformers. Both, in large part, were feminist. We need only mention a few male Progressives who, though they differed on other matters, were feminists or had feminist sympathies, to see the continuity: Floyd Dell, Walter Lippmann, Ben Lindsay, Max Eastman, Oswald Garrison Villard, George Creel, A. Mitchell Palmer, and Herbert Croly. The extent to which feminism shaped, as much as it was shaped by, the reformist impulse is remarkable. As Gertrude Himmelfarb observed in *On Liberty and Liberalism*, the "potency" of feminism "can shape an entire ideology."[73] Health, preventive hygiene, social welfare programs, sex education, community harmony, sexual equality—all were dear to middle-class feminism for decades. All were espoused by most Progressives, male and female. All continue to determine the character of modern feminism and the ideology of liberal reform.

EPILOGUE

FEMINISM AND REFORM

EDUCATED middle-class women of the mid–nineteenth century benefited greatly from the alliance between feminism and reform. It was partly because of this alliance that many women gained access to the universities and colleges, enjoyed the results of marriage reform, and took up careers in education, journalism, charity work, and above all, in those "life-related" fields traditionally identified with the interests of women—social welfare, medicine, health, and hygiene. The foundation built in this early period prepared the way for women to enter such "life-related" academic disciplines as sociology, anthropology, psychology, and history. Ruth Benedict, Mary Beard, and especially Margaret Mead—whose intellectual roots went deep into the feminist-reformist soil of the late nineteenth century[1]—stand out among the beneficiaries of this tradition.

Feminists did not, however, achieve their goals of true love and perfect union. It would be mistaken, of course, to say that they obtained, even temporarily, complete equality for women in marriage or equal power with men within the hierarchical institutions and disciplines advocated by mid-century reformers and later by Progressives. Confronted with ethnic, religious, regional, political, and economic opposition, the reformist tradition itself lacked the strength to shape the sexual relation in American society. What power it had to do so was further diminished by the impact of World War I, the emergence of a leisure-oriented consumer-mass culture in the 1920s, and the effects of the Great Depression—all of which reestablished the sexual division of labor on new grounds.

Most important, middle-class women were unable to claim the legacy of possessive individualism effectively as their own. This male legacy, even while it became increasingly irrelevant to new economic and social conditions, continued to reinforce male dominance and power ideologically. Unlike the women of later generations, nineteenth-century feminists tried to claim this legacy for women and many of them embraced individualism with a great and liberating enthusiasm. They wanted to free women as private individuals into the "open market," where they might "sell their labor for whatever price they considered it worth," "breathe" with men the "same brisk air of healthy competition and acquisition," and benefit from the "purity of capital."[2]

This continued feminist attachment to individualism would chronically create dead ends and throw up road blocks to the advancement of bourgeois women. Committed to possessive individualism, feminists could not make genuine alliances with working-class women. That would have meant acknowledging the basic economic injustices—among them bondage to wage and monotonous work—produced under liberal capitalism. Mid-century feminism came to fruition in the great rationalist-socialist work of Charlotte Perkins Gilman, but even (or perhaps especially) she viewed the market place almost uncritically and could say of the nineteenth-century male that he "was so far the only fully human being."[3]

Equally important, the feminist attachment to individualism would leave feminists open to the charge that female emancipation would desex women, erase differences, destroy the family, cripple woman's physiology and her childbearing capacities. It was rarely argued that men too suffered from the pressures of a market society; they competed and appropriated as self-contained individuals and, as such, acted as "human beings." This ideological conception of humanity veiled the fact that most men also suffered from both the weight of economic injustice and psychological and physical repression.

Feminists' ideological inability to refute in any compelling way the charge that equality would desex women explains the tragic ease with which early feminist achievements in education, work, health, and marriage reform would be coopted, and even crushed, at the end of the century and the beginning of the next. By the 1920s the feminist demand for sexual equality in marriage had degenerated into the equal rights of both sexes to sexual pleasure and orgasmic fulfillment, transforming bourgeois women along the way into domestic playmates and little else; later, the radical feminist emphasis on self-ownership collapsed into the sober realism of planned parenthood.[4] Even in the 1890s, under the weight of "muscular" male reaction to feminization, coeducational colleges and universities had become nightmares of sexual segregation and stereotyping, women everywhere losing ground and forced to rely almost solely (and often brilliantly) on single-sex or coordinate institutions for intellectual fulfillment and leadership—a change hardly noticed by most women and easily accepted by feminist men, for they had, apparently, the least to lose by it.

When coeducation ended at Wesleyan University in 1912, one of its early male defenders, William North Rice, did nothing publicly to prevent the change, although privately he had some doubts. In a letter to a woman graduate Rice expressed pride in those who had graduated from Wesleyan, "women like yourself and many other alumnae," who represented "a far greater honor to the college than men whose chief interest is in athletics." Nevertheless, Professor Rice continued, coeducation at Wesleyan must end: the male students and faculty opposed it and other schools throughout New England and the West had ceased to support it. Rice even implied that Wesleyan could not survive as an institution if it went on admitting women, though he reminded his correspondent feebly that "coeducation in advanced studies" still existed at Harvard, Brown, and Yale, and "might continue in Wesleyan." Wesleyan might also build a separate college for women, though it might be of "inferior quality" to the male school.[5] In times of feminist crisis, when bourgeois women often turned to individualist liberal arguments or to that transparent subterfuge that had accompanied liberal polemics from the 1850s onward—feminine virtue—men like Rice became serenely ineffectual allies.

The most important declension, however, came in the area of woman's professions, and especially in the very professions that feminists had so carefully singled out from the beginning of the movement as the special provinces of women—medicine and preventive health care. By 1900 the number of American woman doctors was double that in France, England, and Germany combined. At the same time, however, female hospitals almost disappeared and a more intensive professionalization of medicine, favoring the interests of men, intruded deeply into the power base of women.[6]

Women continued to practice medicine and to participate actively in preventive health care programs until the 1920s, when the medical profession dealt a silent, decisive blow to the professional strength of women doctors. In 1921 Congress passed the Shepard-Towner Act, "the first federally funded health care program to be implemented in the United States."[7] This act made children's health, women's prenatal care, and education in hygiene an explicit function of the state, thus realizing the institutional goal of late-nineteenth-century feminist health reformers. It resulted in the use of women doctors and public-health-care nurses as the primary administrators of the program. Within the year this victory for women began to crumble: the American Medical Association attacked Shepard-Towner, claiming especially that the private family physician could just as easily and more effectively teach hygiene to children and expectant mothers. "The family physician," John M. Dodson told an AMA meeting in 1923, "who seeks to render to his patients the service which will do them the most good is bound to enter the field of preventive medicine: to become, in other words, the family health adviser as well as the family physician."[8] By the end of the 1920s health care had once again returned to the private sector, albeit into the hands of male professionals, and women professionals had suffered a defeat from which they would not begin to recover until our own time. These feminist defeats made it clear that the feminist allegiance to liberalism had little to offer bourgeois women in the way of permanent success.

It is ironic, therefore, that even before nineteenth-century feminists successfully took the liberal legacy for women, they had proposed a new cooperative world view hostile to it and more congenial to the interests of women. Even more ironic, feminists joined other reformers in attacking the romantic tradition, calling into question the right of the individual to "disharmonious" forms of self-expression. Their ideological scheme of prevention, sanitary balance, elitist centralization, and rationalization was ultimately antilibertarian, and it meshed very nicely with an emerging hierarchical corporate capitalist system dominated by men.

Nineteenth-century feminists would have been wise to abandon possessive individualism and, in their critique of individualism, to reduce their emphasis on prevention and rationalization. They might well have embraced more completely the humane and democratic character of their cooperative vision. Feminists, however, were caught within the limitations and strengths of a reform tradition that they, in many ways, helped to create.

Bibliographical Note

IN THE COURSE of this study significant use has been made of the following manuscript collections and periodicals:

Manuscript Collections

BOSTON PUBLIC LIBRARY

- Chapman Family Papers
- Thomas Wentworth Higginson Papers
- Samuel May Papers
- Weston Family Papers

COLUMBIA UNIVERSITY LIBRARY

- Sidney Howard Gay Papers
- Josephine Griffing Papers

CONCORD FREE PUBLIC LIBRARY

- Franklin Sanborn Papers

CORNELL UNIVERSITY LIBRARY

- Burt Green Wilder Papers
- Andrew Dickson White Papers

HOUGHTON LIBRARY, HARVARD UNIVERSITY

- Thomas Wentworth Higginson Papers
- Julia Ward Howe Papers
- Parton Family Papers

LIBRARY OF CONGRESS

- Susan B. Anthony Papers
- Blackwell Family Papers
- Anna Dickinson Papers
- Julia Ward Howe Papers

Louise Chandler Moulton Papers
File of the National American Woman's Suffrage Association
Elizabeth Cady Stanton Papers

MASSACHUSETTS HISTORICAL SOCIETY, BOSTON, MASS.
Caroline Dall Papers

NEW-YORK HISTORICAL SOCIETY
Theodore Tilton Papers

NEW YORK PUBLIC LIBRARY
Horace Greeley Papers
Smith Family Papers

SCHLESINGER LIBRARY, RADCLIFFE COLLEGE
Olympia Brown Papers
Blackwell Family Papers
Caroline Dall Papers
Mary Earhart Dillon Papers
(including Elizabeth Boynton Harbert Papers)
Mary Putnam Jacobi Papers

UNIVERSITY OF ROCHESTER LIBRARY
Susan B. Anthony Papers
Post Family Papers

SOPHIA SMITH LIBRARY, SMITH COLLEGE
Garrison Family Papers (including Martha Wright Papers)
Parton Family Papers

UNIVERSITY OF SOUTHERN ILLINOIS LIBRARY
Victoria Woodhull (Martin) Papers

STOWE-DAY MEMORIAL LIBRARY, HARTFORD, CONN.
Isabella Beecher Hooker Papers

SYRACUSE UNIVERSITY LIBRARY
Gerrit Smith Papers

VASSAR COLLEGE LIBRARY
Elizabeth Cady Stanton Papers

WESLEYAN UNIVERSITY LIBRARY
Wilbur Atwater Papers
William North Rice Papers
John Vleck Papers

WISCONSIN HISTORICAL SOCIETY, MADISON, WIS.
R. B. Anderson Papers

WORCESTER ANTIQUARIAN SOCIETY, WORCESTER, MASS.
Stephen Foster–Abby Kelly Papers

YALE UNIVERSITY LIBRARY
George Beard Papers
Beecher Family Papers
Theodore Tilton Papers

Periodicals

Aldine
Alpha
American Phrenological Journal
Atlantic Monthly
Ballot Box
Banner of Light
Boston Medical and Surgical Journal
Chicago Inter-Ocean
Cincinnati Commercial
Cooperative News
Cooperator
Demorest's Monthly
Godey's Lady's Book
Golden Age
Harper's Bazaar
Harper's Magazine
Hearth and Home
Herald of Health
Home Journal
Independent
Index
Journal of Social Science
Laws of Life
Liberator
Lily
Modern Thinker
Nation
New Age
New Century
New Northwest
New York Times
New York Tribune
New York World
Radical
Revolution
Sanitarian
Sibyl
Springfield Republican
Una
Water-Cure Journal
Woman's Advocate
Woman's Journal
Woodhull and Claflin's Weekly

Notes

Part 1 / Feminism and Private Lives

PROLOGUE: *True Love and Perfect Union*

1. Blake McKelvey, *The Urbanization of America* (New Brunswick, N.J. 1963), p. 4; and David Potter, *The Impending Crisis* (New York, 1973), p. 241.

2. For the Constitutional period see Eric Foner, *Tom Paine and Revolutionary America* (New York, 1976), pp. 27–44 *passim*; and for the Jacksonian period see Douglas Miller, *The Birth of Modern America* (New York, 1970), pp. 42–66, 116–139.

3. See David Rothman, *The Discovery of the Asylum* (Boston, 1971), pp. 1–77; and Michael Katz, *The Irony of Early School Reform* (Boston, 1968), pp. 1–112.

4. "The Battle of the Sexes," *Woman's Advocate*, February 1869, pp. 69–72.

5. "Woman as Companion," *Woodhull and Claflin's Weekly*, June 18, 1870.

6. *Hearth and Home*, May 6, 1871. See also, for similar statements, "Why Don't They Marry?" *Golden Age*, February 10, 1871; "Passion for Fine Houses," *Harper's Bazaar*, January 16, 1869; "Appearing Rich," *Harper's Bazaar*, May 1, 1869; "Bounty on Marriage," *American Phrenological Journal*, February 1866; *Revolution*, January 26, 1871; and *Independent*, January 3, 1867. *Harper's Bazarr* was a feminist paper at the time, edited by Mary L. Booth.

7. "Divorce," *New Northwest*, June 2, 1871. For a sample of the outpouring of articles dealing with domestic murders, see "The Case of Mrs. Wharton," *Springfield Republican*, December 13, 1871; "The Problem of Chronic Crime," *Springfield Republican*, April 25, 1874; "The Battle of the Sexes," *Woman's Advocate*, February 1869, pp. 69–72; and "Domestic Despotism and Assassination," *Woman's Journal*, July 26, 1873. For feminist commentary on the growth of divorce see "Secret Divorces," *Hearth and Home*, March 4, 1871; "A Divorce Lawyer Shot," *Inter-Ocean*, July 5, 1876; "Do You Want a Divorce?" *Inter-Ocean*, March 10, 1877; "Another Disgusting Divorce Case in High Life," *Springfield Republican*, November 28, 1865; and "Divorces in Connecticut and Elsewhere," *Springfield Republican*, March 16, 1871.

8. "Divorce," *Springfield Republican*, April 8, 1872; and "Carlsfried," *Springfield Republican*, July 1, 1872. On Field's adaptation see Melinda Jones to Anna Dickinson, September 4, 1874, Dickinson Papers, Library of Congress.

9. See, for example, Elizabeth Cady Stanton, speech on marriage and divorce at the 1870 Decade Meeting, in Paulina Wright Davis, *A History of the National Women's Rights Movement* (New York, 1871), pp. 66–81; and *The History of Woman's Suffrage* (Rochester, N.Y., 1886), Volume III, pp. 61, 116. On the liberalization of the law, see George E. Howard, *A History of Matrimonial Institutions* (Chicago, 1904), Volume III, pp. 9–31; Nathan Allen, "Divorce in New England," *North American Review*, Winter 1880, pp. 517–564; F. Saunders, *About Woman, Love and Marriage* (New York, 1868), "Frequent Divorces in New England, pp. 296–297; and James Barnett, *Divorce and the American Novel* (New Haven, 1939), pp. 15–32. Several states favored common-law marriages, while most simply passed statutes designed to facilitate marriages. See Howard, *A History*, pp. 170–191; and, on the age of consent, see David Pivar, *Purity Crusade* (Westport, Conn., 1973), pp. 104–108.

10. *Radical*, December 1869, pp. 441–454, unsigned.

11. Crocker to Isabella Beecher Hooker, May 31, 1871, Isabella Beecher Hooker Papers, Stowe-Day Memorial Library. For examples of this very widespread position see *New Northwest*, October 17, 1871, and January 26, 1871; Stanton, "The Kernel of the Question," *Revolution*, November 4, 1869; and Henry Blackwell, "Era of Monopolies," *Woman's Journal*, January 11, 1873.

12. "The History of Woman's Suffrage, 1848–1877," *Ballot Box*, September 1978; and *The History of Woman's Suffrage* (Rochester, N.Y., 1881), Volume I, p. 25.

13. *Woman's Advocate*, January 1868, p. 31.

14. The quotation in this sentence comes from Jesse Jones, *Woman's Journal*, December 23, 1871.

15. Stone to Antoinette Brown Blackwell, c. 1850, Blackwell Family Papers, Library of Congress.

16. Stanton to Antoinette Brown Blackwell, April 27, 1870, Blackwell Family Papers, Library of Congress.

17. *New York World*, July 9, 1871. See also, for an equally strong attack, "Housekeeping and Invalidism," *Woman's Journal*, August 19, 1871.

18. Sarah Perkins, "No Creed," *Woman's Journal*, January 17, 1873.

19. Degler, "Revolution Without Ideology: The Changing Place of Women in America," in *The Woman in America* (Boston, 1965), edited by Robert J. Lifton, pp. 193–210. See also William O'Neill, "Feminism as a Radical Ideology," in *Dissent* (Dekalb, Ill., 1968), edited by Alfred Young, pp. 273–301; and Aileen Kraditor, *The Ideas of the Woman's Suffrage Movement* (New York, 1965).

20. Douglas, *The Feminization of American Culture* (New York, 1977), p. 353.

21. MacPherson, *The Political Theory of Possessive Individualism* (London, 1962), p. 3.

22. In one of his letters to Lucy Stone about marriage, Henry Blackwell referred to "our new theory." The contemporary press also commented on the development of a "new theory of equal rights of married couples." See Blackwell to Stone, June 13, 1853, Blackwell Family Papers, Library of Congress; and quotation from *New York Observer* in *American Phrenological Journal*, June 1866. This "theory" emerged gradually out of the political, economic, and social revolutions of the eighteenth century.

23. Richard Hume, "The Battle of the Sexes," *Woman's Advocate*, February 1869, pp. 69–72.

24. "The Woman Question—Sex in Politics," *Springfield Republican*, June 30, 1871.

25. "Both Together," *Golden Age*, August 8, 1874; and "Cause and Outcome," *Golden Age*, April 17, 1875.

26. *New York World*, January 13, 1871.

27. "Divorces in Connecticut and Elsewhere," *Springfield Republican*, March 16, 1871. See also *New Northwest*, "Old and New," August 21, 1874, and especially "Men and Women," September 25, 1874.

28. *Woodhull and Claflin's Weekly*, March 7, 1873.

29. Croly, *For Better or Worse: A Book for Some Men and All Women* (Boston, 1874), p. 38.

30. W. R. Brock, *An American Crisis: Congress and Reconstruction, 1865–1867* (London, 1963), p. 9.

31. Fredrickson, *The Inner Civil War* (New York, 1965), p. 216.

32. Almost any feminist journal of the nineteenth century would prove this point. See, by way of illustration, the following sample of articles in *Revolution*: "Women in Russia" and "German Women in Congress," December 24, 1868; "Women's Rights in Europe," April 15, 1869; "The Position of Women in Ancient India," December 16, 1869; "From Europe," October 20, 1870; and "Peace and Liberty in Switzerland," December 17, 1868.

33. Quoted by Frederick Hinckley in a speech delivered before the Boston NWSA Convention, *Ballot Box*, July 1881.

34. "Principles and Parties," *Woman's Advocate*, March 1869, pp. 155–160.

35. *Revolution*, August 6, 1868.

36. See Lawrence Goodwyn, *The Populist Moment* (New York, 1978), pp. 4–7.

37. I have been influenced in this interpretation by Sheldon Wolin, *Politics and Vision* (Boston, 1960), pp. 1–27 *passim*.

38. "Marriage," *Radical*, December 1871, p. 343.

39. "Editorial Correspondence," *Revolution*, January 28, 1869.

40. *Woman's Journal*, July 22, 1871.

41. *Inter-Ocean*, April 14, 1877.

42. Speech at the Boston NWSA Convention, *Ballot Box*, July 1881.

43. "The *Nation* Again," Woman's Journal, August 15, 1874.

CHAPTER 1. *Hygiene and the Symmetrical Body*

1. John Scudder, *Reproductive Organs* (Cincinnati, 1874) p. 20; and Blackwell, "Dr. Clarke and Women," *Springfield Republican*, December 31, 1873. "Only familiarity with scientific methods of thought and inquiry," another feminist told a Woman's Congress in 1876, "will emancipate the sex from the tyranny of custom and superstition" (*New Northwest*, September 29, 1876).

2. On evangelical religion and health reform, see Carroll Smith-Rosenberg and Charles Rosenberg, "Pietism and the Origins of the American Public Health Movement," *Journal of the History of Medicine* 23 (January 1968), pp. 16–35; and on the early secularization of health reform, see Ronald Walters, *American Reformers, 1815–1860* (New York, 1978), pp. 145–172.

3. *The Great Harmonia* (Boston, 1868), Volume I, pp. 18–19, first published in 1850. Davis's ideas on health reflect the unmistakable imprint of Jeffersonian rationalism.

4. Stanton's words on "sickness" appear in a biographical sketch by Laura Curtis Bullard written in 1899, *Stanton Scrapbook*, III, Stanton Papers, Vassar College Library. See also Stanton to Smith, March 10, 1865, Smith Papers, Syracuse University Library. For material on Smith's extreme naturalistic position, see Octavius Frothingham, *Gerrit Smith* (New York, 1878), especially chapter 4, "Religion," pp. 44–95.

5. Mary Studley, "Our Girls," *New Northwest*, September 18, 1874; and Lois Waisbroker to Amy Post, 1867, Post Family Papers, University of Rochester Library.

6. I have been influenced in my interpretation here by Daniel Boorstin, *The Lost World of Thomas Jefferson* (Boston, 1960); Donald Meyer, *The Positive Thinkers* (New York, 1966); and Michael Foucault, *The Birth of the Clinic* (New York, 1975).

7. *Woman's Journal*, December 23, 1871; and "Social Revolution," *New Northwest*, August 14, 1874.

8. "The Sickness of the Times," *New Age*, December 4, 1875. See also, for Chadwick's position on female doctors, "The Study and Practice of Medicine by Women" (New York, 1879).

9. "Where Are We Drifting?" *Ballot Box*, August 1877. See also A. W. Stevens, "Patient and Doctors," *New Age*, first issue in 1875.

10. For examples of this Manichean thinking see Andrew Jackson Davis, *Harbinger of Health* (New York, 1867), p. 3; "The Struggle of Disease and Health in Local Politics," *Springfield Republican*, November 8, 1866; and Francis Bacon, "Civilization and Health," *Journal of Social Science* 3 (1871), pp. 58–77.

11. Tilton, *Golden Age*, February 10, 1872; and Putnam Jacobi, *The Value of Life* (New York, 1879), pp. 180–181.

12. Russell Trall, "Health and Disease of Women" (New York, 1862), p. 10. For similar statements see Jonathan Stainback Wilson, *Woman's Book of Health*, (New York, 1860), pp. 208–209; George H. Taylor, *Diseases of Women* (New York, 1871), pp. 140–141; *Sibyl*, August 15, 1856; and *Woman's Journal*, January 22, 1872. On the wretched state of the medical profession in the antebellum period, see William Rothstein, *American Physicians in the Nineteenth Century* (Baltimore, 1972); Richard Skyrock, *Medicine and Society* (Ithaca, N.Y. 1975); and Charles Rosenberg, *The Cholera Years* (New York, 1974).

13. Hunt, speech at the Worcester Woman's Convention, October 1851, *Woman's Rights Pamphlets*, NAWSA File, Library of Congress. For similar statements see Arethusa Hall, *Radical*, January 1872; and the speech of A. J. Curtis before the Cincinnati Ladies' Health Association in 1867, *History of Woman's Suffrage* (Rochester, N.Y., 1886), Volume III, p. 509.

14. Thayer to Post, May 10, 1857, Post Family Papers, University of Rochester Library.

15. *New Northwest*, November 3, 1871.

16. *New Northwest*, October 14, 1876.

17. "The Training That Girls Need," *Inter-Ocean*, April 24, 1880. See also *Woman's Journal*, September 13, 1873: "All who undertake to write on women should keep them posted in all the latest results of sanitary study."

18. Mary Putnam Jacobi, paper given at the 1876 Woman's Congress, *Papers and Letters Presented at the Third Woman's Congress of the American Association for the Advancement of Women* (New York, 1877), pp. 12–17.

19. Tacitus, "The Woman Question," *New Northwest*, February 9, 1872.

20. *Ballot Box*, October 1881.

21. "Lectures of Dr. Ludwig Buechner," *New York World*, December 31, 1872; *Woman's Journal*, January 18, 1873; and *Woodhull and Claflin's Weekly*, June 14, 1874. On Huxley see *Revolution*, December 22, 1870; *Springfield Republican*, July 22, 1874; on Buckle, *Woman's Journal*, June 17, 1871; and on Spencer, *New Northwest*, February 13, 1874, and *Woman's Journal*, December 30, 1871.

22. The feminist role in the development of social science will be examined at greater length in chapters 10–12. For references to the social scientists see Luther and Jesse Bernard, *Origins of Sociology* (New York, 1943).

23. The best discussion of this hygiene is in Steven Nissenbaum's unpublished doctoral dissertation, "Careful Love: Sylvester Graham and the Emergence of Victorian Sexual Theory, 1830–1840," University of Wisconsin, 1968. See also Richard Shyrock, "Sylvester Graham and the Popular Health Movement," *Mississippi Valley Historical Review* 29 (July 1931), pp. 172–183.

24. On the French bourgeoisie see William Coleman, "Health and Hygiene in the Encyclopédie: A Medical Treatise for the Bourgeoisie," *Journal of the History of Medicine* 29 (December 1974), pp. 399–421.

25. *Woman's Advocate*, April 1869, pp. 255–258. The terms "natural reciprocity" and "balance of power" appear respectively in Taylor, *Diseases of Women*, p. 75; and Elizabeth Blackwell, "The Religion of Health," essay published in 1871 and republished in *Essays in Medical Sociology* (New York, 1902), Volume II, p. 220. "Every faculty" and "proper cultivation" appear in a letter from Sarah Dolley, a woman doctor to Elizabeth Pennbacker, February 3, 1850, courtesy of Regina Morantz.

26. Andrew Jackson Davis, *The Great Harmonia* (Boston, 1868), Volume IV, pp. 16–17.

27. See Georgiana Bruce Kirby, *Years of Experience: An Autobiographical Narrative* (New York, 1887), pp. 4–5, 46–78, 154–166.

28. For information on Kirby's life in California see Charles Ames to Caroline Dall, c. 1866, Dall Papers, Massachusetts Historical Society.

29. "Transmission; or Variation of Character Through the Mother" (New York, 1877), pp. 16–17. For similar thoughts from another feminist Unitarian see James Freeman Clarke, *Autobiography, Diary and Correspondence* (Boston, 1891). Clarke wrote that "organic tendencies themselves have no moral quality but become virtues and vices" through "irrational abuse" (pp. 48–49).

30. *Woman's Journal*, January 22, 1870; and *Golden Age*, June 29, 1874. Such ideas appeared in *Woodhull and Claflin's Weekly*: "The healthiest persons physically are they who most evenly exercise all parts of their body" (August 30, 1873). See also Paulina Wright Davis's report to the Syracuse Woman's Rights Convention: "Symmetry, harmony, balance, are the conditions of beauty, energy and integrity," *Woman's Rights Pamphlets*, NAWSA File, Library of Congress; Russell Trall, *Sexual Physiology* (New York, 1868), p. 278; Mary Studley, *What Our Girls Ought to Know* (New York, 1878), p. 164; and Sarah Hackett Stevenson, *The Physiology of Women* (New York, 1881), p. 43.

31. "The Shame of Nudity," *Alpha*, May 1, 1878, unsigned.

32. This conception of the body appears in feminist literature over a thirty-year period from 1850 to 1880. See Andrew Jackson Davis, *The Great Harmonia*, (Boston, 1868), Volume I, pp. 138–139, 149, 310; Mrs. T. H. Keckeler, *Thaleia* (New York, 1869), pp. 13–21; *Woman's Journal*, June 20, 1874; Elizabeth Osgood Willard, *Sexology* (Chicago, 1867), p. 180; and Caroline Winslow, "Sexual Continence," *Alpha*, October 1881.

33. "Dr. Van de Walker on Sex," *Woman's Journal*, January 30, 1875.

34. Stevenson, *Physiology of Women*, (Chicago, 1881), p. 77.

35. Child, *Woman's Journal*, March 15, 1873; and Duniway, *New Northwest*, February 13, 1874. See also Studley, *What Our Girls Ought to Know*, (New York,

1878), p. 128; Charles Morrill, *The Physiology of Women* (New York, 1870), p. 409; Antoinette Brown Blackwell, *Woman's Journal*, April 25, 1874; and Anna Brackett, *The Education of American Girls* (Boston, 1874), pp. 117–118.

36. Ward, *Dynamic Sociology* (New York, 1883), Volume I, p. 23.

37. On the early Puritans see Michael Walzer, "Puritanism as a Revolutionary Ideology," in *New Perspectives on the American Past* (Boston, 1969), edited by Stanley Katz, pp. 3–35.

38. *Thaleia*, preface, pp. i–vi.

39. "Hereditary Transmission," *Herald of Health*, October 1869. See also Laura Curtis Bullard, "The Moral Code as Applied to Men and Women," *Revolution*, February 9, 1871. "The office of the educator should be to develop the body, the intellect and the heart symmetrically; not to cultivate a few of the natural gifts of either to excess, and leave untrained the qualities which are less pronounced."

40. *Revolution*, May 13, 1869.

41. Blackwell, "Sex and Evolution," in *The Feminist Papers* (New York, 1973), edited by Alice Rossi, p. 368.

42. Cridge to Dall, August 27, 1871, Dall Papers, Massachusetts Historical Society. On Cridge's free-love radicalism see "Individual Sovereignty Politically Considered, Representation of Minorities," *Woodhull and Claflin's Weekly*, December 8, 1870. See also Hal D. Sears, *The Sex Radicals, Free Love in High Victorian America* (Lawrence, Kan., 1978), pp. 162–205.

43. David Croly, *The Truth About Love* (New York, 1872), pp. 66–67. Croly's book is an extreme speculative statement (but proleptic) and therefore unrepresentative in that respect, although his hygiene is characteristic of the feminist position.

44. *Woman Under Socialism* (New York, 1904), pp. 80, 82. Bebel also quoted from contemporary sexologists that "the sexual impulse is neither moral nor immoral; it is merely natural, like hunger or thirst: Nature knows nothing of morals" (p. 82). See also, for a later American socialist position, Charlotte Perkins Gilman, *Women and Economics* (New York, 1966), edited by Carl Degler and originally published in 1898, pp. 72–73 and *passim*.

45. Following Antonio Gramsci, I believe that the concepts of symmetry, equilibrium, and interdependence emerge in consciousness when a ruling class wishes to achieve or consolidate its power. See Gramsci, "Americanism and Fordism," in *Selections from the Prison Notebooks* (New York, 1978), edited by Quentin Hoare and Geoffrey Nowell Smith, pp. 285–303. See also James Madison, Federalist 14, in *The Federalist Papers* (New York, 1961), edited by Clinton Rossiter, pp. 99–105; and in Richard Hofstadter's essay on John C. Calhoun, "John C. Calhoun: The Marx of the Master Class," *The American Political Tradition* (New York, 1974), especially pp. 105–107. I also disagree with Cynthia Russett, who has written a book given over entirely to the study of the concept of equilibrium. "Unlike evolution," she writes, "equilibrium was never a popular concept and perhaps because it had no discernibly revolutionary social and philosophical implications" (*The Concept of Equilibrium in American Social Thought* [New Haven, 1966], pp. 2–3). I think it would be more exact to say that American reformers sought to harmonize the concepts of equilibrium and evolution into a difficult, occasionally unhappy, relationship. On the concept of interdependence in common sense thinking, see Kathryn Sklar, *Catherine Beecher* (New Haven, 1973), p. 84. On Scottish common sense thought itself, see Gladys Bryson, *Man and Society: The Scottish Inquiry of the Eighteenth Century* (Princeton, 1945); Gary Wills, *Inventing America* (New York, 1978); and Jurgen Herbst, "Nineteenth-Century German Scholarship in America," unpublished doctoral dissertation, 1955, Harvard University, pp. 1–55. For the relationship between common sense and health, see John Davies, *Phrenology, Fad and Science* (New Haven, 1955), especially chapter 13, "Phrenology and Religion," pp. 149–158; and Madeleine Stern, *Heads and Headlines* (Norman, Okla., 1973).

46. Daniel C. Howe, *The Unitarian Conscience* (Cambridge, Mass., 1970), pp. 60–61. Howe also writes that the "Unitarian Conscience was not a repressive but an expressive faculty;" it resembled the "integrated personality" of the twentieth century (pp. 60–61). The analogy is provocative and probably correct, although rendered

uncritically. As I see it, the "Unitarian Conscience" tended to circumscribe—if not repress—human expression. Such a "conscience" favored the "average" over the exceptional, reason over passion and bold invention.

47. *Woman's Journal*, October 5, 1871.

48. *Physiology of Women*, p. 151.

49. *New York World*, December 3, 1871.

50. *New York World*, May 10, 1872.

51. Russell Trall, *Hydropathic Encyclopedia* (New York, 1852), Volume I, pp. 493–494.

52. Anthony, *Lily*, July 1852.

53. "Hereditary Transmission," *Alpha*, September 1, 1878.

54. Mrs. J. A. Jones, "Love, Honor, and Obey," *New Northwest*, December 20, 1873; and *Ballot Box*, August 1881. See also "Marriage Morals," *Modern Thinker* 2 (1873), pp. 129–141; and Emma Hardinge, "What Our Women Can Do," *Banner of Light*, June 22, 1862.

55. Nichols, *Experience in Water-Cure* (New York, 1849), pp. 51–52; and Preston, speech before the Westchester, Pa., Woman's Convention, June 2, 1853, *History of Woman's Suffrage* (Rochester, N.Y., 1881), Volume I, p. 657.

56. For the twentieth century see, for example, Max Eastman, *Woman Suffrage and Sentiment* (New York, 1914), p. 3. Eastman distinguishes a feminist from "a girl of either the drawing-room or kitchen variety." For the nineteenth century see Elizabeth O. Smith, *Woman and Her Needs* (New York, 1851), pp. 12–13; Jane Croly, *For Better or Worse: A Book for Some Men and All Women* (Boston, 1874); Thomas Wentworth Higginson, *Malbone* (Boston, 1869), pp. 138–139; Fuller Walker, *Aldine* (a feminist-inclined art journal), August 1872: "Woman was the toy of man when she was not his slave; and slave or toy, she was alike the victim of his caprices and his brutality;" and "Waste of Woman Power," *Harper's Bazaar*, September 25, 1869.

57. Davis, *Una*, May 2, 1853.

58. Gordon, *Woman's Body, Woman's Right* (New York, 1976), pp. 120–21.

59. On environmentalism in antebellum American thought, see Winthrop Jordan, *White over Black* (New York, 1971), pp. 287–289, 308–310, and *passim*; and Ruth Schwartz Cowan, "Nature and Nurture: The Interplay of Biology and Politics in the Work of John Galton," in *Studies in the History of Biology* (Baltimore, 1977), edited by William Coleman and Camille Lemoges, pp. 136–138.

60. Cowan, "Nature and Nurture," p. 141.

61. Stanton, "Hereditary Genius," *Ballot Box*, September 1879; "Heredity," *New Northwest*, April 10, 1879; and Stanton's letter to the 1882 NAWSA Convention, September 1, 1882, in *History of Woman's Suffrage*, Volume III, p. 245. From the late 1870s on, Stanton lectured publicly on three subjects: moral education, heredity, and Free Religion (*History of Woman's Suffrage*, Volume III, p. 194).

62. For attacks on the theory of *tabula rasa* see Mrs. T. H. Keckeler, *Thaleia*, p. 69; and Kirby, *Years of Experience*, p. 154.

63. "Maternity," *Revolution*, February 10, 1870.

64. "Hereditary Genius," *Ballot Box*, September 1879.

65. *Alpha*, March 1879.

66. *Ballot Box*, November 1877, December 1877. See also the five papers on "Enlightened Motherhood" given at the First Woman's Congress by Charlotte Wilbur, Lucinda Chandler, Caroline Corbin, Augusta Cooper Bristol, and H. M. Tracy Cutler, published in *Papers and Letters Presented at the First Woman's Congress* (New York, 1874), pp. 1–35; Georgiana Bruce Kirby, "Transmission; or Variation of Character Through the Mother" (New York, 1877); Edward Dixon, *Woman and Her Diseases* (New York, 1870), p. 313; Stephen Pearl Andrews, *Woodhull and Claflin's Weekly*, August 20, 1870, September 10, 1870; Stanton, *Revolution*, April 8, 1869, and *History of Woman's Suffrage*, Volume I, pp. 481, 495; Dr. Mary Safford Blake, "Pre-Natal Influence," *Woman's Journal*, October 31, 1874; Lucinda Chandler, "Moral Education Society," *Woman's Journal*, November 23, 1871; and *Inter-Ocean*, March 17, 1877.

67. Stanton, "Heredity," *New Northwest*, April 10, 1879.

68. Feminists tried to remain faithful to Margaret Fuller's progressive position on marriage, which is outlined in *Woman in the Nineteenth Century* (New York, 1971), pp. 72–82.

69. Quoted in John Cowan, *The Science of a New Life* (New York, 1874).

70. See Cowan, "Notices," at the back of his *Science of a New Life;* "Sanitary Science," *Inter-Ocean*, October 26, 1878. For a typical review of an earlier edition see *Banner of Light*, June 5, 1869. For similar statements see "Personal Liberty," *Alpha*, April 1875; Caroline Winslow and E. B. Foote, *Alpha*, July 1881, October, 1881, November 1881; Lucinda Chandler, "A Definition of Freedom," *Woodhull and Claflin's Weekly*, December 23, 1871; "The Pathology of Passions," *Woodhull and Claflin's Weekly*, May 30, 1874; Stevenson, *Physiology of Women*, pp. 155, 187; Studley, *What Our Girls Ought to Know*, p. 164; Trall, *Sexual Physiology*, pp. 67–69; Taylor, *Diseases of Women*, pp. 148, 307; Keckeler, *Thaleia*, pp. 34–35; Willard, *Sexology*, pp. 180, 199–200; Eliza Bisbee Duffey, *The Relations of the Sexes* (New York, 1876), pp. 230–231; and *Woman's Journal*, August 22, 1874, January 30, 1875.

71. Resolutions read by Celia Burleigh and passed by the New York City Sorosis, May 16, 1870, "Minutes," Sophia Smith Library, Smith College.

72. A. S. Graves, "Hereditary Laws," *Alpha*, June 1879.

73. See Alice Stone Blackwell, "Antoinette Brown Blackwell," Blackwell Family Papers, Library of Congress, p. 8.

74. See Stone Blackwell, pp. 3–5; Barbara Solomon, in *Notable American Women* (Cambridge, Mass., 1971), edited by Edward T. James, Volume I, pp. 158–161; and Alice Rossi's excellent biographical sketch of Blackwell in *The Feminist Papers* (New York, 1974), edited by Alice Rossi, pp. 340–346.

75. Because I have synthesized many quotations from Blackwell's *The Sexes Throughout Nature* (New York, 1875), I will not list page numbers.

76. Coleman, "Health and Hygiene," pp. 407–415. For the 1830s and 1840s see Nissenbaum, "Careful Love," pp. 219–223.

77. Mary Hibbard, *New Northwest*, November 7, 1873. See also Antoinette Brown Blackwell, *Woman's Journal*, May 30, 1874; Stevenson, *Physiology of Women*, p. 77; and Russell Trall, *Hydropathic Encyclopedia*, Volume I, p. 388. The terms "symmetrical action" and "nature's regimen" of work appear in Blackwell. "Irrational" and "civilization" appear in Stevenson and Trall. See also *Revolution*, December 15, 1870; "Nervous Women," *Revolution*, December 15, 1871; Thomas Higginson, *Women's Journal*, October 4, 1873; Matilda J. Gage, *Ballot Box*, August 1878; and Eliza Bisbee Duffey, *No Sex in Education* (Philadelphia, 1873), pp. 30–66.

78. *The Value of Life*, p. 101.

79. On Ellis, Kinsey, and Masters and Johnson see Paul Robinson, *The Modernization of Sex* (New York, 1976); and on Money see John Money and Patricia Tucker, *Sexual Signatures* (Boston, 1975). In spite of Robinson's undocumented position to the contrary, these sexologists (including Money) represent not so much a reaction to the bourgeois sexual thinking of the nineteenth century as its most complete rationalist fulfillment. Notwithstanding claims to liberate sexuality in new ways, these thinkers are applying a positivist model to sexual behavior. Indeed, their underlying direction is to "moderate" the nineteenth-century rationalist concepts of symmetry, order, and harmony, and therefore to deepen the attack on the instinctual basis of human life.

80. Beecher Hooker to Mill, August 9, 1869, Isabella Beecher Hooker Papers Stowe-Day Memorial Library.

81. Willard, *Sexology*, pp. 8–9.

82. Augusta Cooper Bristol, "Enlightened Motherhood," *Papers and Letters Presented at the First Woman's Congress*, p. 12.

CHAPTER 2. *Preparation for Marriage: Scientific Knowledge and the Cult of No Secrets*

1. *Woman's Journal*, July 11, 1874.

2. See Martha Clarke to Caroline Dall, May 27, 1878, Dall Papers, Massachusetts Historical Society.

3. Peirce, *New York World*, October 22, 1869; and Hanaford, "The Byron Controversy," *Woman's Advocate*, November 1869, pp. 287–288.

4. Beecher Hooker to Livermore, March 15, 1871, Isabella Beecher Hooker Papers, Stowe-Day Memorial Library; and Howe to Conway, 1872, Julia Ward Howe Papers, Library of Congress. See also *Golden Age*, August 14, 1875: "The only dangerous evils are the secret ones. The scandal that travels by stabs in the dark is more demoralizing than that which . . . becomes deodorized by publicity;" Octavius Frothingham, *Elective Affinities* (New York, 1870), p. 11; and Abigail Scott Duniway, "Mesmerism and Clairvoyance," *New Northwest*, June 7, 1878. Duniway defended the virtues of medical clairvoyance. "It is useful," she wrote, "because the clairvoyant can read the unspoken thought, thus lessening *crimes* because *unconcealed*."

5. "Comte's Birthday," *New York World*, January 23, 1872; and "From New York," *Springfield Republican*, January 23, 1872. Public announcements of their own marriages by feminists with Quaker backgrounds during the antebellum period helped prepare the way for the emergence of this principle. The marriages of Theodore Weld and Angelina Grimke, Stephen Foster and Abby Kelly, and Lucy Stone and Henry Blackwell are the most famous examples. Feminists used their own private marriages—their own private lives—to protest existing marriage laws. Some feminists, however, objected to this practice. "The bad taste," Elizabeth Blackwell wrote to her brother in 1855, "seems to me to result from the dragging of one's private personal affairs into public notice. . . . What need of this advertising your union?" (February 22, 1855), Blackwell Family Papers, Library of Congress) Similarly, Henrietta Ingersoll wrote Lucy Stone in 1856: "Let me tell you that I have rebelled as bitterly in my heart at the customs of civilization, in making the marriage of a virgin *public*, as at almost anything else in 'woman's lot'! From the bottom of my heart I poured silent imprecations at this thing, and have inwardly vowed that it were more modest to be seduced when no one knew it, than to live through open marrying. . . . The veil of secrecy is needed over sacred things" (March 15, 1856, Blackwell Family Papers, Library of Congress).

6. "Christianity Not for Reformers, But for Come-Outers," unpublished lecture, 1873, Mary Putnam Jacobi Papers, Schlesinger Library, Radcliffe College.

7. Putnam Jacobi, *The Value of Life* (New York, 1879), pp. 249–250.

8. Conway, *Demonology and Devil-Lore* (New York, 1879), Volume II, p. 90.

9. Ward, *Dynamic Sociology* (New York, 1883), Volume I p. 624.

10. For a discussion of these people see Hal D. Sears, *The Sex Radicals: Free Love in High Victorian America* (Lawrence, Kans., 1978), pp. 159–162; and for a selection of their writings see Taylor Stoehr, *Free Love in America: A Documentary History* (New York, 1979), pp. 319–425.

11. *Index*, November 29, 1877, quoted in Sears, *The Sex Radicals*, p. 160.

12. *Woodhull and Claflin's Weekly*, November 2, 1872. The McFarland-Richardson case was perhaps the most famous scandal until the Beecher-Tilton case of 1875 (see chapter 5 of this study). Abby McFarland obtained a divorce from her husband, Daniel, who in turn shot and killed Abby's journalist lover, Albert Richardson, in the office of the *New York Tribune*. The affair culminated in the deathbed nuptials of Richardson and Mrs. McFarland—performed by Henry Ward Beecher—and McFarland's subsequent acquittal on grounds of insane jealousy. McFarland felt so exonerated that he began to deliver public lectures on the virtues of traditional marriage. See Nelson Blake, *Road to Reno* (New York, 1962), and, in regard to McFarland's lecture career, *New York World*, February 3, 1871.

13. "Free Religionists," *New York World*, October 17, 1873.

14. Journal entry, January 11, 1869, Dall Papers, Massachusetts Historical Society.

15. See Mill, *Autobiography of John Stuart Mill* (New York, 1948), p. 96; and Foucault, *The Birth of the Clinic* (New York, 1973), p. 39.

16. Paterson, *Radical*, March-April 1870, pp. 169–185, 287–300. The specific page references are pp. 179–181. See also Tennessee Claflin's 1872 speech, "Seduction: What It Is and What It Is Not," for a similar statement, Woodhull Papers, University of Southern Illinois Library. "The human passions," she wrote, "have been found to be terrible forces, like steam and fire; and instead of studying them, in order to reg-

ulate them in accordance with their true laws and their adaptation to the world's well being, they have been feared merely, and fought down and repressed." See also Lester Ward, *Dynamic Sociology*, Volume I, pp. 11–12, 32–33, 60.

17. Burt Green Wilder, "Biographical Reminiscences," undated, pp. 1–2, Wilder Papers, Cornell University Library; and Mrs. James A. Marshall, "History of the Deacon David Wilder House, North Leicester," *Enterprise*, February 26, 1923.

18. Wilder, "Biographical Reminiscences," pp. 1–5; obituary, *Ithaca Journal*, January 23, 25, 1925; and Morris Bishop, *Early Cornell, 1865–1900* (Ithaca, N.Y., 1962), pp. 110–111. On the new curriculum and the introduction of the laboratory see Bishop, *Early Cornell*, p. 87; Richard Hofstadter, *The Development and Scope of Higher Education in the United States* (New York, 1952); Louis Snow, *The College Curriculum* (New York, 1907), p. 171; and William Coleman, *Biology in the Nineteenth Century: Problems of Form and Function* (New York, 1971), p. 6.

19. Putnam Jacobi, *The Value of Life*, p. 188.

20. Coleman, *Biology in the Nineteenth Century*, pp. 14, 100–114, 154–157.

21. Gage, "Professor Burt Green Wilder," *The Guide to Nature* 12 (September 1919), pp. 35–37.

22. Ibid., p. 37; *Ithaca Journal*, January 23, 1925, May 8, 1929; Bishop, *Early Cornell*, p. 110; and Burt Green Wilder, "The Anatomical Uses of the Cat," *New York Medical Journal*, October 1879, pp. 53, 56–58.

23. Wilder, "Two kinds of Vivisection—Sentisection and Callisection," *Nature*, September 30, 1880, p. 120. On the substitution of choloroform for the gallows see Wilder, *Boston Advertiser*, May 8, 1875; *New York Tribune*, March 21, 1875; and Henry James, Sr., *Boston Advertiser*, April 7, 1875.

24. Holmes to Wilder, November 6, 1879, Wilder Papers, Cornell University Library.

25. Wilder, *Anatomical Technology as Applied to the Domestic Cat* (New York, 1882), pp. 55–57.

26. Wilder, "On Morphology and Teleology, Especially in the Limbs of Mammalia," *Memoirs of the Boston Society of Natural History (1866–1869)*, Volume I, pp. 46–52; "Right and Left," *Atlantic Monthly*, April 1870, pp. 455–466; "Six Fingers and Toes," *Old and New*, February 1870, pp. 156–162; and "The Hand as an Unruly Member," *American Naturalist* 1 (October 1867), pp. 414–423. As the last three articles indicate, Wilder devoted special scientific study to the extremities—an illustration of his general preoccupation with order and rational behavior. "We may assume," he wrote in "The Hand as an Unruly Member," "that the tongue and the hand . . . in the functional and teleological sense, are the really characteristic organs of man, corresponding with his peculiar endowments of rationality, of thought and freedom of action; and so it is not a little significant that to these same organs alone, which, being the most capable of good, are, by perversion, the most potent for evil, can the term *unruly* properly be applied. . . . They are the agents of the individual in becoming an unruly member of society."

27. Agassiz to Wilder, November 25, 1873, Wilder Papers, Cornell University Library; and *Ithaca Journal*, January 22, 1925.

28. "The New and Old Method," *New Century*, November 4, 1876.

29. On Milne-Edwards and Cuvier see E. S. Russell, *Form and Function* (New York, 1917), pp. 32–35, 175, 197–199; and William Coleman's introductory essay to *The Interpretation of Animal Form* (New York, 1967), pp. xi–xvii.

30. *Anatomical Technology*, pp. 12–17; and "Rational Nomenclature," *Nation*, April 2, 1881. On the pursuit of rational nomenclature in other fields see Franklin Sanborn, "Prison Science versus Prison Discipline," in *Concord Pamphlets* (Concord, Mass., 1900), pp. 1–7; *Sanitarian*, April 1873; and George Beard, "Hysteria and Allied Affections," lecture at Long Island Hospital, June 1872, George Beard Papers, Yale University Library.

31. *Anatomical Technology*, p. 13.

32. Ibid., p. 16.

33. Wilder, *What Young People Should Know* (New York, 1875).

34. Ibid., p. 173.

35. Undated draft of a lecture on sexual physiology, Wilder Papers, Cornell University Library. Feminists, by the way, relied heavily on the physiological views of William Carpenter.

36. This description appears in the first petition (1878) to Wilder submitted by the Cornell senior class. See also Wilder, *Notes of Lectures on Physiology and Hygiene* (Ithaca, N.Y., 1876); and *Health Notes for Students* (New York, 1893).

37. Wilder, *Notes of Lectures on Physiology and Hygiene*, p. 13.

38. White, "Sanitary Science in Its Relations to Public Education," *New York World*, November 14, 1873. See also Andrew D. White, *Autobiography* (New York, 1907), p. 363.

39. Benedict to Wilder, February 12, 1888, Wilder Papers, Cornell University Library.

40. Lydia Marie Child, "The Intermingling of Religions," *Atlantic Monthly*, October 1871, pp. 384–495; and Warren Chase, *Banner of Light*, April 20, 1867.

41. See Bishop, *Early Cornell*, pp. 60–87, 154–169; and for a general consideration of these transformations see David F. Noble, *America By Design* (New York, 1977).

42. Burt G. Wilder, "Secret Societies," *New York Herald Tribune*, October 22, 1873; "Secret Societies," *Springfield Republican*, October 22, 1873; and George W. Curtis, *Harper's Magazine*, January 1874, pp. 293–294.

43. Wilder, "Secret Societies;" and Bishop, *Early Cornell*, pp. 132–133.

44. Burt G. Wilder, "Equal Yet Diverse," *Atlantic Monthly*, July 1870, pp. 30–40.

45. Wilder, *Notes of Lectures on Physiology and Hygiene*, p. 20; and "Secret Societies."

46. On Edward B. Foote see Sears, *The Sex Radicals*, p. 185.

47. "Scientific Culture for Women," *Woman's Journal*, January 13, 1872.

48. These three quotations appear respectively in speeches made by Mary Putnam Jacobi, Lucinda Chandler, and Stephen Smith. For Jacobi see "Social Aspects of the Readmission of Women into the Medical Profession," *Papers and Letters Presented at the First Woman's Congress of the American Association for the Advancement of Women* (New York, 1873), pp. 168–178; Chandler, "Woman's Congress," *New York World*, October 18, 1873; and Smith, "The Recent American Public Health Convention," *Sanitarian* 1 (June 1873), pp. 111–112.

49. D. P. Kayner, "The Healing Confession," *Banner of Light*, December 3, 1870.

50. "Interview with Andrew Jackson Davis," *New York World*, October 6, 1870; and Davis, *The Great Harmonia* (Boston, 1868), Volume I, pp. 206–207. See also *New Northwest*, May 11, 1877, and June 8, 1877; "A Medical Clairvoyant," *Springfield Republican*, April 27, 1875; and "Boston Reform Conference," *Banner of Light*, February 12, 1859.

51. Sidney Ditzion, Madeleine Stern, David Pivar, Bryan Strong, Kathyrn Sklar, and Linda Gordon have all pointed to the middle-class concern with sex education in the nineteenth century. See Ditzion, *Marriage, Morals and Sex in America* (New York, 1975), especially chapter 9, "Science Looks at Life," pp. 317–355; Stern, *Heads and Headlines* (Norman, Okla., 1971); Pivar, *Purity Crusade* (New York, 1973), pp. 34–39, 78–82, and *passim;* Strong, "Ideas of the Early Sex Education Movement in America, 1890–1920," *History of Education Quarterly*, Summer 1971, pp. 129–161; Sklar, *Catherine Beecher* (New Haven, 1973); and Gordon, *Woman's Body, Woman's Right* (New York, 1976). I think Gordon is wrong, however, when she writes that "prudishness . . . was an integral part of feminist thought until at least the 1880's" (p. 164). On prudery see, for example, Milton Rugoff, *Prudery and Passion* (New York, 1971) and Ronald Walters, *Primers for Prudery* (New York, 1974).

52. *Woman's Journal*, September 23, 1871.

53. *New Age*, June 1, 1876.

54. *Herald of Health* commented in September 1869 on the wave of books on health and physiology, both feminist and nonfeminist, that flooded the market during this period: "If the multiplication of books telling how to live were sufficient to redeem the race from sin and crime and aimless, worthless lives, the world would soon be a paradise, and Eden once more restored."

55. *Radical*, April 1866, pp. 452–453.

56. Dall Journal, February 1858, Dall Papers, Massachusetts Historical Society.
57. *Woman's Journal*, May 29, 1875.
58. "Books, Authors and Art," *Springfield Republican*, August 3, 1876.
59. Wilder, *What Young People Should Know*, p. 172.
60. *Inter-Ocean*, April 4, 1876.
61. Oakes Smith, review of Rachel Gleason's *Talks to My Patients*, *Herald of Health*, June 1870; and Davis, *The Great Harmonia*, Volume IV, p. 64. See also Alice Williams, "Mrs. Grundy," *Golden Age*, September 16, 1871; Jane Croly, "The Physical Life of Women," *Woman's Journal*, January 14, 1871; *Revolution*, May 20, 1870; *Woodhull and Claflin's Weekly*, March 15, 1873, September 23, 1871, April 5, 1873; *Sibyl*, February 1, 1857, August 15, 1856, October 1, 1855; and Amelia Bloomer, "Mock Modesty," *Lily*, May 1851. The *Liberator* also objected to the prudery of some reformers and regularly published articles on "licentiousness," marriage, and sexual hygiene. See Lewis Perry, *Radical Abolitionism* (Ithaca, N.Y., 1973), p. 227.
62. Warren Chase, *The Fugitive Wife* (Boston, 1866), p. 20.
63. Charles Merrill, *The Physiology of Women* (New York, 1870), pp. 4, 410.
64. Brackett, *Education of American Girls* (New York, 1874), pp. 61–63.
65. *New Century*, October 7, 1876.
66. Review of Jane Croly's *For Better or Worse*, in *New Century*, September 2, 1876. See also, on the need for sexual and physiological education in the schools, Caroline Corbin, speech at the First Woman's Congress, *Woman's Journal*, October 25, 1873; Lucinda Chandler, *Woman's Journal*, November 23, 1872; J. Noble, *Woman's Journal*, January 4, 1873; Harriet Brooks, *Woman's Journal*, April 7, 1870; and C. F. Butler, *Alpha*, June 1, 1878. See also, for other feminist statements dealing with the necessity for openness in regard to sexual matters both before and after marriage, Lydia Fowler, *History of Woman's Suffrage* (Rochester, N.Y., 1881), Volume I, p. 664; John Scudder, *Reproductive Organs* (Cincinnati, 1874), p. vi; Victoria Woodhull, "Social Freedom," *Springfield Republican*, February 5, 1872; May Wright Sewall, "Honesty in the Family," *Alpha*, July 1, 1882; John Hooker, *Some Reminiscences of a Long Life* (Hartford, Conn., 1899), p. 310; "Mrs. Swisshelm," *Springfield Republican*, April 29, 1875; Caroline Corbin, "Hiding Places of Wrongs," *Woman's Journal*, October 25, 1873; Rachel Gleason, *Talks to My Patients: Hints on Getting Well and Keeping Well* (New York, 1880), p. 151; Robert Dale Owen, *Moral Physiology* (London, 1859), p. 31; and Isabella Beecher Hooker, *Womanhood: Its Sanctities and Fidelities* (Boston, 1873). "The welfare of the race," Hooker wrote, "requires that moral and physical truths be understood; and language that plainly conveys its meaning is far better in its moral effects than that which deals with its subject in covert and ambiguous expression and thus suggests concealed indelicacy and stimulates unwholesome curiosity" (pp. 9–10). Hooker wrote of her book to anti-feminist Madeleine Dahlgren that "every woman in our movement is in sympathy with the spirit of this little book" (January 15, 1878, *History of Woman's Suffrage* [Rochester, N.Y., 1886], Volume III, p. 100).
67. "The Secret Vices of Children," *Laws of Life*, August 1868, a journal edited by James C. Jackson. The terms "impulsive action" and "rational desire" appear in Ward, *Dynamic Sociology*, Volume I, p. 616.
68. Wilder, *What Young People Should Know*, p. 163.
69. Information on the "physiological" careers of Paulina Wright Davis, Mary Gove Nichols, Caroline Dall, Harriet Hunt, Fidelia Harris, Mrs. Markham Wheeler, Lydia Fowler, and Caroline Severance can be found in *History of Woman's Suffrage*, Volumes I–III; for Mary Studley see *What Our Girls Ought to Know* (New York, 1878), introductory biographical essay; and for Mary W. Johnson see Johnson to Amy Post, April 10, 1852, Post Family Papers, University of Rochester Library. On women's physiological societies, see Caroline Dall to Caroline Severance, May 30, 1859, Dall Papers, Massachusetts Historical Society.
70. On the various feminist physiological societies, see Phoebe Hanaford to Isabella Beecher Hooker, August 9, 1871, Isabella Beecher Hooker Papers, Stowe-Day Memorial Library; *Ballot Box*, May 1881; *Woman's Journal*, July 18, 1874; *Cincinnati Commercial*, May 11, 1869; and *Inter-Ocean*, August 2, 1879.
71. "Female Physicians," *Springfield Republican*, July 15, 1865.

72. *New York World*, January 15, March 12, 1871.

73. For Boston see Bowditch, "The Medical Education of Women," *Boston Medical and Surgical Journal* 15 (July–December 1881), pp. 289–292; for California and Kansas see *History of Women's Suffrage*, Volume III, pp. 706, 757.

74. On the various universities see "Coeducation at Michigan University," *Woman's Journal*, June 6, 1874, July 17, 1875; "Notes" on the University of Wisconsin, *Inter-Ocean*, June 22, 1876; James R. Chadwick, "The Study and Practice of Medicine by Women" (New York, 1879); "Minutes" of the New York City Sorosis, December 1, 1872, Sophia Smith Library, Smith College; and *New York World*, April 12, 1874.

75. "Michigan University," *Springfield Republican*, February 5, 1875; Chadwick, "The Study and Practice of Medicine by Women;" and *New Northwest*, May 28, 1875.

76. Emily Talbot to Caroline Dall, June 14, 1877, Dall Papers, Massachusetts Historical Society.

77. Sarah Hackett Stevenson, speech to the International Council of Women, in Henry Blackwell, *Report of the International Council of Women* (Washington, 1888), p. 171. See also, on regulars, "Notes," *Ballot Box*, October 1881; and "Feminine M.D.'s," *Inter-Ocean*, February 27, 1877. On irregulars see *Herald of Health*, April 1864; *Woman's Journal*, January 22, 1870, December 16, 1871, September 25, 1873; and *Sibyl*, December 15, 1859.

78. "The Cause of Women in Italy," *Revolution*, February 23, 1871.

79. For biographical material on Lozier see *New Northwest*, June 29, 1877; and "Bright Haven," *New Northwest*, April 9, 1875, and November 17, 1876.

80. "Charlotte Denman Lozier," *Revolution*, January 13, 1870.

81. "People Worth Knowing," *Woman's Journal*, March 4, 1871.

82. Studley, *What Our Girls Ought to Know*, p. 9.

83. "Social Aspects of the Readmission of Women into the Medical Profession," *Papers and Letters Presented at the First Woman's Congress of the American Association for the Advancement of Women* (New York, 1874), p. 173. See also, for the women doctors who gave public lectures, Ann Preston, Frances White, and Elizabeth Keller (Philadelphia), *Woman's Journal*, March 27, 1875; Caroline Winslow and Lucy Abell (Washington), *Inter-Ocean*, August 2, 1879; Marie Zakrzewska (Boston), "Hospital Lectures for Women," *Woman's Journal*, February 15, 1878; Alice Stockham (Chicago), "Conversations upon Health," *Inter-Ocean*, August 2, 1879; and Mary Putnam Jacobi (New York), "The Medical Education of Women," *Woman's Journal*, June 26, 1875.

84. "Lectures of Dr. Densmore," *Revolution*, March 19, 1868.

85. Ibid. See also, for Densmore's work in Sorosis, "Minutes" of the New York City Sorosis, Sophia Smith Library, Smith College.

86. "Free Religious Association," *Springfield Republican*, October 31, 1874; and Howe, diary entry, December 21, 1874, Houghton Library, Harvard University.

87. *Woodhull and Claflin's Weekly*, February 17, 1872.

88. Davis to Woodhull, May 1872, Woodhull Papers, University of Southern Illinois Library.

89. *Woodhull and Claflin's Weekly*, August 9, 1873.

90. Stowe, *Lady Byron Vindicated* (Boston, 1870), p. 327; and Stowe to Sarah Parton, June 1, 1869, Parton Family Papers, Sophia Smith Library, Smith College.

91. Stanton to Mott, April 1, 1874, Wright Family Papers, Sophia Smith Library, Smith College. See also Stanton to Paulina W. Davis, April 1, 1872, Stanton Papers, Vassar College Library; Stanton to Isabella B. Hooker, April 1, 1871, and Stanton to Martha Wright, January 19, 1871, Stowe-Day Memorial Library. "I think it would be wiser and kinder," Stanton wrote Beecher Hooker, "if you cut short all this raking up of Victoria Woodhull's antecedents. . . . A man's home is his castle . . . where he pulls up the drawing bridge and closes the gate none must follow: so there are . . . limits to every individual personality that none have the right to invade."

92. *Woodhull and Claflin's Weekly*, July 15, 1871.

93. "Minutes," New York City Sorosis, Sophia Smith Library, Smith College.

94. Ibid., April 15, 1872.

95. Howe, *Woman's Journal*, November 12, 1872; Bowles, *Springfield Republican*, June 27, 1873; and Warner, "The American Newspaper," *Journal of Social Science* 14 (1881), pp. 52–70. See also, for similar contemporary attacks on newspaper reporting "Paul Pry," "Marital Infelicities," *Inter-Ocean*, January 27, 1877: "It is unfortunately a prominent feature of our modern social system that domestic privacy is no longer the private and sacred thing it used to be when newspapers were in their infancy;" and "A New Development in Journalism," *New York World*, March 24, 1872.

96. "Industrial and Social Revolution," *Revolution*, August 27, 1868.

97. "Marriage Laws," *Banner of Light*, December 10, 1859.

98. Channing to Dall, April 2, 1872, Dall Papers, Massachusetts Historical Society.

99. "Minutes" of the New York City Sorosis, May 16, 1870; June 19, December 4, 1871; February 5, April 15, May 20, June 3, 1872; and January 19, 1874.

100. Higginson, "The Fact of Sex," *Woman's Journal*, March 23, 1870.

101. Stanton, *Revolution*, May 21, 1868; and Chase, *Banner of Light*, September 30, 1871.

102. Tilton, "Secret Societies," *Independent*, October 24, 1867; Bowles, "Secret Societies," *Springfield Republican*, October 24, 1874; Charles Warner, "Women in Men's Colleges," *Springfield Republican*, August 9, 1871; Curtis, *Harper's Magazine*, January 1874, pp. 132–133 See also "Secret Societies," *Woman's Journal*, March 5, 1874; and Virginia Penny, "Secret Societies," in *Think and Act* (New York, 1869), pp. 220–221.

103. Channing to Dall, April 2, 1872, Dall Papers, Massachusetts Historical Society.

104. *Cincinnati Commercial*, January 8, January 23, February 6, February 15, 1878.

105. *Revolution*, April 9, 1868.

106. Scudder, *Reproductive Organs*, p. 91; and "Eclectic Medical Institute," *Cincinnati Commercial*, January 23, 1878. In the very month of the uproar over Wilder's book, Scudder called the role of graduates at an Eclectic Medical Institute convocation.

107. On Woodhull and Tennessee Claflin see *Woodhull and Claflin's Weekly*, November 2, 1872, December 28, 1872, and Emanie Sachs, *The Terrible Siren* (New York, 1928), pp. 168–212; on the Fowlers and Whitman see Madeleine Stern, *Heads and Headlines*, pp. 105–117, 246–266; on Noyes see Robert Parker, *A Yankee Saint* (New York, 1935), pp. 267–283. See also, on suppression of periodicals, *Ballot Box*, July 1881; *New Northwest*, November 6, 1879; *Alpha*, March 1, 1878; Ditzion, *Marriage, Morals and Sex*, pp. 169–180; and Sears, *The Sex Radicals*, pp. 61–107 *passim*.

108. On feminist organization against suppression see *Ballot Box*, July 1881; and Olive Smith to Amy Post, May 5, 1878, Post Family Papers, University of Rochester Library; and on reformist protest against suppression of *Truth Seeker*, see *New Northwest*, November 6, 1879, and Ditzion, *Marriage, Morals and Sex*, pp. 169–180.

109. See, on the National Grange, Solon Buck, *The Granger Movement* (Cambridge, Mass., 1933), pp. 40–43; on the Knights of Labor, Norman Ware, *The Labor Movement in the United States, 1860–1895* (New York, 1929), pp. 26–27, 91–94 *passim*; and on the Ku Klux Klan, Kenneth Stampp, *The Era of Reconstruction* (New York, 1965).

CHAPTER 3. *A Healthy Mind in a Healthy Body: Equal Physical Development and the Coeducational Experience*

1. For information on Gleason's water cure see Caroline Dall's journal, July 27 and August 30, 1871, Dall Papers, Massachusetts Historical Society.

2. See Stanton, quoted in Russell Trall, *Mothers' Hygienic Handbook* (New York, 1869), p. 7; Elizabeth Stuart Phelps, "Where It Goes," *New Northwest*, September 1871; "Men and Muscle," *Women's Journal*, September 16, 1871; "Girls, Like Boys, Are Human," *Laws of Life*, October 1868; and Caroline Dall, *College, Market and Court* (Memorial Edition, New York, 1914), p. 157.

3. Rochester (pseudonym of Frances Russell), "Dot and I—Boy and Girl," *Revolution*, March 17, 1870.

4. Duffey, *No Sex in Education* (Philadelphia, 1874), pp. 38–39, 40–41.

5. *Harper's Bazaar*, October 10, 1869. A popular fashion magazine, *Harper's Bazaar* had a strong feminist bias during its first twenty years. On Booth's plans to edit a feminist newspaper see "Women and the Press," *Liberator*, June 27, 1862; and for a long sketch of Booth's early career as a translator and essayist, see Susan Conrad, *Perish the Thought* (New York, 1977).

6. "Commencement Exercises," *Liberator*, September 13, 1862; and "Physical Education," *Sibyl*, June 1863.

7. "Russell Trall, Portrait, Character and Biography," *Herald of Health*, July 1864.

8. On Trall's conception of disease see *Herald of Health*, February 1864; and on support of women's rights see Trall, *Sexual Physiology* (New York, 1868), "College Department," *Herald of Health*, April 1864, and "Dr. Trall on Female M.D.'s" *Sibyl*, February 15, 1861.

9. "College Department," *Herald of Health*, April 1864.

10. Jackson, "Reminiscences and Incidents in My Life," *Laws of Life*, January 1870.

11. Ibid. For material on other feminist men, see chapter 13 of this study. The rest of the information on Jackson's life comes from his serialized biography in *Laws of Life*, February, May, and July 1870.

12. For differing views on health and women see Ann Douglas, *The Feminization of American Culture* (New York, 1977) and G. J. Barker-Benfield, *The Horrors of Half-Known Life* (New York, 1976). In her analysis of a declining male ministry and a "rising" female literary "influence" Douglas equates male "invalidism" with powerlessness; Barker-Benfield, however, establishes a relationship between female sickness and powerlessness and dependence on male authority. Both analyses suffer from "phallic" thinking (for want of a better term), linking the acquisition or quest for power with health. Such thinking certainly enhances the rhetorical effects of an argument, but it does little to illuminate the social complexities of the lives of nineteenth-century men and women. Worse, it fixes our attention on a relationship between physical health and power which may not exist.

13. "Anniversary Report of Our Home," *Laws of Life*, December 1873.

14. Hooker, *Some Reminiscences of a Long Life* (Hartford, 1899), pp. 101–103; and *Laws of Life*, December 1873, January 1875.

15. "Anniversary Report of Our Home," *Laws of Life*, December 1873.

16. *Laws of Life*, July 1868.

17. "Psycho-Hygienic Health," *Laws of Life*, July 1875.

18. *Laws of Life*, October 1868, September 1873.

19. For accounts of the introduction and evolution of coeducation see Thomas Woody, *A History of Woman's Education in the United States* (New York, 1929), Volume II, pp. 224–278; Albert Hazen Wright, *New York Historical Source Studies* numbers 15–19, 20–21, 23–24, 35 (Ithaca, N.Y., 1953–1966), Nancy Cott, *The Bonds of Womanhood* (New Haven, 1977), pp. 101–125; "Worcester High School," *Woman's Journal*, January 13, 1871; "The High Schools of Boston," *Woman's Journal*, July 13, 1871; "Progress of Coeducation," *Woman's Journal*, April 10, 1875; Lebanon Normal School," *Woman's Journal*, June 15, 1872; and "Coeducation of the Sexes," *Woman's Journal*, September 21, 1872. The *Woman's Journal* offers the best source for feminist ideas on coeducation.

20. Albert Hazen Wright, *New York Historical Source Studies* 20–21 (Ithaca, N.Y., 1960), pp. 1–5.

21. Ibid., pp. 13–14. See also, on feminist backing, Harrison Howard (organizer of the People's College) to Amy Post, September 6, 1852, and November 25, 1852, and Sarah Thayer to Amy Post, March 9, 1853, Post Family Papers, University of Rochester Library

22. Quoted in John Humphrey Noyes, *History of American Socialisms* (Philadelphia, 1870), pp. 455–56. For information on Weld's school see Gerda Lerner, *The Grimke Sisters* (New York, 1973), pp. 316–339. On the Fourieristic utopian com-

munities see Noyes, *History of American Socialism,* pp. 307–460, and Raymond L. Muncey, *Sex and Marriage in Utopian Communities* (Bloomington, Ind., 1973), pp. 64–79.

23. Prospectus of Eagleswood School, Garrison Papers, Sophia Smith Library, Smith College.

24. Ibid., p. 1.

25. Ibid., p. 2. See also Weld to Martha Wright, December 2, 1856, Garrison Family Papers, Sophia Smith Library, Smith College.

26. Ibid., p. 2.

27. Martha Wright to David Wright, December 2, 1856, Garrison Family Papers, Sophia Smith Library, Smith College. On Elizabeth Smith Miller's children, see Elizabeth to her father, April 19, 1858, and March 15, 1861, Smith Family Papers, Syracuse University Library.

28. Albert Hazen Wright, *New York Historical Source Studies* pp. 20–21, 56–59; and Lerner, *The Grimke Sisters,* pp. 340–347.

29. *Woman's Journal,* September 30, 1874.

30. May Wright Sewall, speech to the International Council of Women, *Report of the International Council of Women* (Washington, D.C., 1888), pp. 51–63.

31. On these changes see Richard Hofstadter, *The Development and Scope of Higher Education in the United States* (New York, 1952), pp. 11–42; Louis Snow, *The College Curriculum* (New York, 1907), pp. 171–180; and Morris Bishop, *Early Cornell, 1865–1900* (Ithaca, N.Y., 1962), pp. 54–140 *passim.*

32. See *Tributes to Professor William North Rice* (New York, 1915), pp. 21–36.

33. "College New Departures," *Springfield Republican,* September 20, 1871.

34. Rice, in *Tributes to William North Rice,* pp. 36–37.

35. On White see Bishop, pp. 146–147, and White to Gerrit Smith, July 2, 1874, Smith Family Papers, Syracuse University Library; and on Vleck see Rice's reminiscences in *Tributes to William North Rice,* p. 36.

36. *Tributes to William North Rice,* pp. 21–22.

37. Ibid., p. 31. For Rice's most famous attempt to fuse Darwinism with religion see *Twenty-Five Years of Scientific Progress* (Boston, 1894). For biographical material on Rice see *Tributes to William North Rice.*

38. Atwater to Rice, December 2, 1872, Atwater Papers, Wesleyan University Library. For biographical material on Atwater see *Dictionary of American Biography* (New York, 1943), Volume I, pp. 417–418; and Charles Rosenberg, *Dictionary of Scientific Biography* (1972), pp. 325–327. It was through Atwater's lobbying that in 1887 Congress passed the Hatch Act, which funded every state to maintain at least one experimental station. From 1887 to 1888 Atwater headed the office of Experimental Stations of the United States Department of Agriculture while still holding his position at Wesleyan. During this twenty-year period, he also wrote books on the quantitative analysis of foods and became world famous as an expert on food science. With E. B. Rosa, a professor of physics at Wesleyan, he invented the calorimeter, an instrument that measured the amount of heat produced by food when eaten and oxidized and that provided the basis for still reliable caloric tables. For a good history of agricultural science see Margaret Rossiter, *The Emergence of Agricultural Science* (New York, 1976).

39. Atwater, "Late Advances in the Application of Science, Farm Experimenting, and the Experiment Station," unpublished and undated lecture, p. 1, Atwater Papers, Wesleyan University Library.

40. Atwater, "Suggestions Regarding the Nomenclature of Nitrogenous Compounds of Animal and Vegetable Substances," unpublished and undated lecture, Atwater Papers; Atwater, "They Believe in the Department of Agriculture and Endorse the Policy of the Present Commissioner. More help to Agriculture, More Science, Less Politics," unpublished and undated lecture, Atwater Papers, Wesleyan University Library; Rosenberg, *Dictionary of Scientific Biography,* pp. 325–327.

41. Louise Wilby Knight, "The 'Quails': The History of Wesleyan University's First Period of Coeducation, 1872–1912," Senior Honors Thesis, Wesleyan University, 1972, pp. 8–9.

42. Wilby Knight, "The 'Quails,'" p. 73. On Cornell see Rachel Leedom Moore,

"Sage College," *New Century*, August 12, 1876; and on Michigan, *New Century*, October 7, 1876.

43. *Argus*, July 20, 1871.

44. *Olla Podrida*, March 1873.

45. Wilby Knight, pp. "The 'Quails,'" pp. 50–60.

46. See Woody, *A History of Woman's Education in the United States*, Volume II, pp. 147–192; and Mabel Newcomer, *A Century of Higher Education for American Women* (New York, 1959), pp. 1–53.

47. On feminist efforts at Harvard see "Coeducation of the Sexes," *Woman's Journal*, September 21, 1872, and "Harvard Examinations for Women," *New Century*, July 22, 1876; at Cornell see Bishop, pp. 144–145, and Stanton, *Revolution*, August 13, 1868; at Swarthmore see Helene Baer, *The Heart Is Like Heaven* (Philadelphia, 1964), pp. 135–140; at Columbia see Katherine Blake, *Champion of Women* (New York, 1943), pp. 110–111; and at Michigan see "The University of the Future," *Independent*, February 21, 1867. On Tilton's further efforts at New York University see "Our Cherished Mother," *Independent*, April 18, 1869. See also, in defense of coeducation, Mary Putnam Jacobi, *The Question of Rest for Women During Menustration* (New York, 1877), which criticizes E. H. Clarke's argument against the higher education of women, *Sex in Education* (Boston, 1874); Julia Ward Howe, editor, *A Reply to E. H. Clarke* (Boston, 1874); Alexander Wilder, *Plea for the Collegiate Education of Women* (New York, 1874); Eliza Bisbee Duffey, *No Sex in Education* (New York, 1876); Anna Brackett, "Coeducation at Cornell," *New England Journal of Education* 7 (1878), p. 137; Sarah Hackett Stevenson, "Coeducation of the Sexes in Medicine," a paper for the Chicago Woman's Club in 1867, republished in Stevenson, *The Physiology of Women* (Chicago, 1881), pp. 146–228; Dall, *College, Market and Court*, pp. 6–7; and Zadel Barnes Buddington, "Medical Coeducation of the Sexes," *Golden Age*, June 20, 1874.

48. "Coeducation of the Sexes," *Woman's Journal*, September 21, 1872. See also Jane Croly, *For Better or Worse: A Book for Some Men and All Women* (New York, 1876), p. 38; and *Revolution*, November 11, 1869. To my knowledge, feminists during this period never supported single-sex colleges and universities for women with the same vigor they supported coeducational ones. On single sex institutions see Sheila Rothman's excellent discussion of Vassar, Smith, and Wellesley, in *Woman's Proper Place* (New York, 1978), pp. 26–42.

49. See S. W. Dodds, *Woman's Journal*, March 7, 1874; Lydia Maria Child, "Physical Strength of Women," *Woman's Journal*, March 1, 1873; and Antoinette Brown Blackwell, "Sex and Work," Woman's Journal, March 28, 1874.

50. Fuller Walker, M.D., "Woman Fit for Suffrage." *Golden Age*, September 19, 1874.

51. See "Sex in Education," *Golden Age*, August 15, 1874; "Smith College," *Golden Age*, June 5, 1875; Antoinette Brown Blackwell, "Sex and Work," *Woman's Journal*, March 7, 1874; Lester Ward, *Dynamic Sociology* (New York, 1883), Volume I, p. 652; and *Woman's Journal*, October 4, 1873.

52. "Sex in Education," *Golden Age*, August 15, 1874.

53. *Woman's Journal*, October 4, 1873.

54. Stanton, on coeducation at Cornell, *Revolution*, August 13, 1868, and at Northwestern, *Woodhull and Claflin's Weekly*, December 20, 1873. For Stanton's continued commitment to coeducation see "Coeducation Will Not Repel but Attract the Sexes to Each Other," *Boston Investigator*, 1899, Stanton Papers Scrapbook, Volume III, Vassar College Library.

55. Susan B. Anthony to Isabella Beecher Hooker, September 4, 1876, Isabella Beecher Hooker Papers, Stowe-Day Memorial Library. It is important to note that Stanton sent her two youngest daughters to Vassar, probably on the basis of Vassar's excellent curriculum, although her reasons were almost certainly more complex.

56. Blackwell, "Sex and Work," *Woman's Journal*, June 22, 1872.

57. *Alpha*, January 1, 1878.

58. Higginson, "Fact of Sex," *Woman's Journal*, March 23, 1870. Educational reformers used these same arguments to justify public school education. See Michael Katz, *The Irony of Early School Reform* (Boston, 1968), pp. 41–68, 122–129. "As

populations increase," declared Horace Mann, "and especially as artificial wants multiply, the temptations increase and the guards and securities must increase also or society will deteriorate" (Katz, p. 41). Educationists (and feminists) argued that public, free-school education would create harmony between groups and classes, "refine" the people, and eliminate the pernicious influence of the private schools.

59. See "Humanitarian League," *New York World*, June 6, 1872; "About Women," *New York World*, October 1, 1871; *Springfield Republican*, August 13, 1874; Clemence Lozier, "Lecture," *New Northwest*, June 22, 1877; resolutions of President Mrs. C. M. Palmer of the California State Woman's Suffrage Association, *New Northwest*, May 22, 1874; and *Inter-Ocean*, November 23, 1874.

60. Andrew Dickson White to Gerrit Smith, September 1, 1862, quoted in Albert Hazen Wright, *New York Historical Source Studies* 23, pp. 4–5.

61. Professor J. K. Hosmer of the University of Wisconsin, address on "Coeducation" to the Education Association of Detroit, *Springfield Republican*, August 13, 1874. Hosmer did not name the school in question.

62. "Friendship Between the Sexes," *Revolution*, October 20, 27, 1870.

63. See, for example, "Minutes," New York City Sorosis, March 21, 1870, Sophia Smith Library, Smith College; and Stanton's and Antoinette Blackwell's comments on cooperative homes, *New York World*, October 18, 1873.

64. *Woman's Journal*, December 12, 1874; and Weiss, *Springfield Republican*, April 22, 1872. The 1874 piece also appeared in *New Northwest*, April 9, 1875. See also Junus Henri Browne, "Companionship Should Precede Marriage," *Woman's Journal*, September 16, 1873; Abigail Scott Duniway, *New Northwest*, March 16, 1872; and "Humanitarian League," *New York World*, June 6, 1872.

65. *Alpha*, January 1, 1878.

66. Published in Paulina Wright Davis, *A History of the National Woman's Rights Movement* (New York, 1871), p. 62.

67. *Journal of Social Science* 5 (1874), pp. 36–45.

68. Wilby Knight, "The 'Quails,'" p. 42; and "Women in the University," *New Century*, October 17, 1876.

CHAPTER 4. *Sexual Ownership and the Rationalization of Sexual Desire*

1. "Motherhood," *Woodhull and Claflin's Weekly*, May 13, 1871.

2. "Matilda Joslyn Gage on the Right of Habeus Corpus," *Ballot Box*, November 1876.

3. *Radical*, April 1868, p. 492.

4. The quotation is from an article by Dr. Mrs. Smith, "The Fact Divested of Romance," *New Northwest*, November 7, 1878. See also "Objections to Large Families," *Woman's Journal*, January 30, 1875; "Mrs. Hooker's New Book," *Springfield Republican*, October 11, 1873; Sara Spencer, "Crime and Reform," *Ballot Box*, November 1877; "Personal Liberty," *Alpha*, April 1, 1879; Elizabeth Boynton Harbert, "Woman's Kingdom," *Inter-Ocean*, March 17, 1877; Abram W. Stevens, "Some Suggestions to Working People," *New Age*, May 27, 1876; S. B. Butler, "Man and His Relations," *Banner of Light*, September 3, 1859; "Ante-Natal Murder," *New Northwest*, December 27, 1872; "Unwilling Maternity," *New Northwest*, March 27, 1879; "Unwelcome Children," *New Northwest*, September 18, 1879; Russell Trall, *The Hydropathic Encyclopedia* (New York, 1853), Volume II, pp. 493–494; Edward Dixon, *Woman and Her Diseases* (New York, 1855), pp. 313–316; Eliza Bisbee Duffey, *The Relations of the Sexes* (New York, 1876), pp. 228–229; and Isabella Beecher Hooker, *Womanhood, Its Sanctities and Fidelities* (Boston, 1873), pp. 269–270.

5. See Stone's speech before the 1853 New York Convention, in *History of Woman's Suffrage* (Rochester, N.Y., 1881), Volume I, p. 576. The quotations are from Stone's letter to Susan B. Anthony, November 25, 1856, Blackwell's Family Papers, Library of Congress.

6. Wright to Stanton, June 5, 1858, Garrison Family Papers, Sophia Smith Library, Smith College.

7. Stanton to Nichols, August 21, 1852, in *Elizabeth Cady Stanton as Revealed in Her Letters, Diary and Reminiscences* (New York, 1922), edited by Theodore Stanton and Harriet S. Blatch, pp. 42–43.

8. See Wilson Grabill, Clyde V. Kiser, and Pascal K. Whelpton, "A Long View," in *The American Family in Social-Historical Perspective* (New York, 1973), edited by Michael Gordon, pp. 374–396; Daniel Scott Smith, "Family Limitation, Sexual Control and Domestic Feminism in Victorian America," in *Clio's Consciousness Raised* (New York, 1974), edited by Lois Banner and Mary Hartman, pp. 119–136; Linda Gordon, *Woman's Body, Woman's Right* (New York, 1976), pp. 48–49; and Carl Degler, *At Odds: Women and the Family in America, 1776 to the Present* (New York, 1980), especially chapters 2 and 4.

9. *Journal of Social Science* 5 (1873), pp. 71–97. Walker offered statistics to prove the existence of a lower birth rate in the industrialized states (p. 78).

10. "Social Aspects of the Readmission of Woman into the Medical Profession," *Papers and Letters Presented at the First Woman's Congress of the Association for the Advancement of Women* (New York, 1874), pp. 172–173; and "The Baby Crusade," *Inter-Ocean*, May 20, 1877. See also "Startling Statistics," *New York World*, March 24, 1872. Feminist commentary along these lines, both sensationalist and "scientific," abounded during this period.

11. On aspects of this change see Paula Fass, *The Damned and the Beautiful* (New York, 1977); Christopher Lasch, *Haven in a Heartless World* (New York, 1977); and Sheila Rothman, *A Woman's Proper Place* (New York, 1978).

12. Unpublished speech on the assassination of President Lincoln, c. 1866, Dickinson Papers, Library of Congress.

13. "Woman's Adversity," *Revolution*, February 11, 1869. See also Barbara Berg, *The Remembered Gate: Origins of American Feminism, the Woman and the City* (New York, 1978), pp. 174–220.

14. See A., "Marriage and Maternity," *Revolution*, July 8, 1869. For similar sympathetic portraits of working-class women, see Ella Giles, *Maiden Rachel* (Boston, 1879); Antoinette Brown Blackwell, *Island Neighbors* (Boston, 1874); Mary Clemmer Ames, *Eirene; or a Woman's Right* (New York, 1871); Lillie D. Blake, *Fettered for Life* (New York, 1874); Eleanor Kirk, *Up Broadway* (New York, 1870); and Marie Howland, *Papa's Own Girl* (New York, 1874).

15. Quoted by Julia Ward Howe, diary entry, March 18, 1871, Howe Papers, Houghton Library, Harvard University.

16. On Stanton see *History of Woman's Suffrage* (Rochester, N.Y., 1886), Volume III, p. 594; and on Gage see *Alpha*, December 1881, and *Ballot Box*, September 1881. Gage left the editorship of *Ballot Box* to become a regular columnist for *Alpha*.

17. On Blake see *Alpha*, December 1877, *Laws of Life*, January 1875, and *New Age* February 10, 1876; on Severance see *Report of the International Council of Women* (New York, 1888), p. 272; and on Woolson see "About Women," *New York World*, December 24, 1871, and *Woman's Journal*, November 18, 1871.

18. On Severance see *Woman's Journal*, July 22, 1871; on Woolson and Blake see *Woman's Journal*, November 8, 1873.

19. On these women (with the exception of Bristol) see "The Washington Conference," *Ballot Box*, June 1877 and January 1880. See also Augusta Cooper Bristol, "The Relation of the Maternal Function to the Woman Intellect," published by the Washington Moral Education Society (Washington, 1883).

20. *New York World*, December 24, 1871; "Moral Science Association," *Springfield Republican*, November 10, 1871; *Report of the International Council of Women*, p. 272; and David Pivar, *The Purity Crusade* (New York, 1974), pp. 80–81. Pivar has written an enormously useful and informative volume, but he focuses largely on the moral educationists' interest in prostitution and its regulation and not their marital ideology.

21. Pivar, *Purity Crusade*, pp. 82–83.

22. Wood to Hooker, November 27, 1874, Isabella Beecher Hooker Papers, Stowe-Day Memorial Library. See also Lucinda Chandler, *Woman's Journal*, September 13, 1873; *Woman's Journal*, July 31, 1875; and Frederick Hinckley, "The Relation of Sex," *New Age*, June 1, 1876.

23. "Anti-Fashion Convention," *Woman's Journal*, January 3, 1874; and "Woman's Greenback Club," *Alpha*, September 1, 1878. On the women of Vineland, see "Vineland, New Jersey," *Banner of Light*, April 27, 1867; Elizabeth Stanton, editorial correspondence, *Revolution*, September 17, 1868; "American Free Dress League," *Woman's Journal*, April 25, 1874; "George Eliot Is Dead," *Ballot Box*, January 1881; and Eliza Duffey, *Relations of the Sexes* (New York, 1876), pp. 309–310.

24. "Lucinda Chandler," biographical sketch in *A Woman of the Century* (1893), edited by Frances Willard and Mary Livermore, pp. 165–166.

25. *Papers and Letters*, pp. 15–19; and Pivar, *Purity Crusade*, p. 80.

26. Willard and Livermore, p. 165; Lucinda Chandler, "Margaret Fuller Society," *Inter-Ocean*, October 18, 1880; and Elizabeth B. Harbert, "Margaret Fuller Society and Its Scope and Objects," *Inter-Ocean*, August 16, 1880. Chandler is quoted in Linda Gordon, *Woman's Body, Woman's Right*, p. 113. For Willard's position see "Obituary," *Revolution*, January 26, 1871. "What is human freedom," she wrote, "but a right to one's own time and vital energy—a right to the fruits of one's labor? And what is slavery but the unjust appropriation of one's time and labor and energy by another? The laborer is still the slave of capital, whereas capital should be the servant of the laborer. . . . The woman's movement is a labor movement. It hinges and turns on the right of labor. The ballot is merely a protection to these and other individual rights."

27. *Ballot Box*, June 1877, January 1880, and July 1881.

28. *Proceedings of the First Margaret Fuller Society, 1880–1881* (Chicago, 1881), pp. 3–12, Schlesinger Library, Radcliffe College.

29. *Woodhull and Claflin's Weekly*, April 29, May 6, May 13, September 30, 1871; "Marital Equality," *Revolution*, September 7, 1871; and "Motherhood and Its Duties," *Revolution*, August 30, 1871. See also "Free Thought: An Appeal to the Women of America," *Banner of Light*, September 20, 1871; "An Open Letter to Dr. Thomas," *Alpha*, August 1, 1878; "Woman's Sphere of Motherhood," *Woman's Journal*, September 6, 1873; and "Motherhood," *Woman's Journal*, November 1, 1873.

30. See Cowan's *Science of a New Life* (New York, 1874), Eliza Bisbee Duffey, *Relations of the Sexes*; and Georgiana Bruce Kirby, "Transmission; or Variation of Character Through the Mother" (New York, 1877). For their importance to moral education see *Alpha*, June 1878 and February 1882.

31. "Marital Equality," *Revolution*, July 20, 1871; and "Enlightened Motherhood," *Papers and Letters*, p. 15.

32. *Proceedings of the First Margaret Fuller Society*, p. 3.

33. "Marital Equality," *Revolution*, July 20, 1871.

34. "Motherhood," *Woodhull and Claflin's Weekly*, May 13, 1871.

35. "Marital Equality," *Revolution*, July 20, 1871.

36. Ibid.

37. Ibid.

38. "Woman's Sphere of Motherhood," *Woman's Journal*, September 6, 1873.

39. Ibid.

40. "Motherhood and Its Duties," *Woodhull and Claflin's Weekly*, August 30, 1871.

41. "Motherhood," *Woodhull and Claflin's Weekly*, April 29, 1871.

42. "Motherhood and Its Duties," *Revolution*, August 31, 1871; "Moral Education Society," *Woman's Journal*, November 23, 1872; and "Motherhood," *Woodhull and Claflin's Weekly*, April 29, 1871. The term "premeditating desire" comes from a lecture by Frederick Hinckley at the Boston Moral Education Society, *New Age*, June 17, 1876. Hinckley's ideas exactly reproduce Chandler's.

43. "Moral Education Society," *Woman's Journal*; and "Enlightened Motherhood," *Papers and Letters*, p. 15.

44. "Enlightened Motherhood," *Papers and Letters*, p. 18.

45. "Moral Education Society," *Woman's Journal*; and Enlightened Motherhood," *Papers and Letters*, p. 21.

46. "Moral Education Society," *Woman's Journal*, November 23, 1872.

47. "The Relation of Sex," *New Age*, June 1, 1876.

48. "Marital Equality," *Revolution*, July 20, 1871; "Enlightened Motherhood,"

Papers and Letters; and speech before the International Council of Women, *Report of the International Council of Women*, pp. 284–285.

49. See also chapter 1 of this study, "Hygiene and the Symmetrical Body."

50. "An Appeal to Women," *Revolution*, September 7, 1871.

51. Ibid.

52. Gordon, *Woman's Right, Woman's Body*, p. 109.

53. The quotation comes from a speech by Sara Spencer at the Cleveland Woman's Congress, October 13, 1877, *Ballot Box*, November 1877. See also, for a sampling of views, Elizabeth Cady Stanton, "Maternity," *Revolution*, February 10, 1870; Henry C. Wright, *Marriage and Parentage* (New York, 1855), pp. 74–75; Tennessee Claflin, *Tried as by Fire* (New York, 1874), Woodhull Papers, Southern Illinois University Library; Laura Curtis Bullard, *Revolution*, August 24, 1871; and Russell Trall, *Hydropathic Encyclopedia* (New York, 1852), Volume II, pp. 493–499.

54. *Banner of Light*, September 3, 1859. For material on Britten see R. Lawrence Moore, *In Search of White Crows* (New York, 1978), p. 225.

55. "Enlightened Motherhood," *Papers and Letters*, pp. 10–14.

56. "Pre-Natal Culture," *Alpha*, June 1, 1879.

57. "The Woman Question—'Sex in Politics,'" *Springfield Republican*, June 30, 1871.

58. See Winslow, "Sexual Continence," *Alpha*, October 1, 1881; and Duffey, *Relations of the Sexes*, pp. 96–97, 177, 221–222.

59. For Napheys's view on the limitation of births, see "Personality and Liberty," *Alpha*, April 1, 1879; and for Carpenter see "The Beauty and Freedom of Love," *Banner of Light*, May 25, 1867: "It seems the rule, which scientific research corroborates, that on the ascending scale of life the more individuals or nations become intellectualized and spiritualized, the less in numerical ratio, the progeny produced."

60. Napheys, *The Physical Life of Women*, pp. 11, 74–75.

61. Stevenson, *Physiology of Women* (New York, 1881), pp. 17–18; and Frothingham, "A Psychological Glance at the Woman Question," *Herald of Health*, January 1870.

62. On Carpenter see "From Boston," *Springfield Republican*, May 15, 1873; and on Mary Carpenter see W. L. Burn, *The Age of Equipoise* (London, 1965), pp. 151–152.

63. Carpenter, *Principles of Human Physiology* (London, 1876), 8th ed. pp. 893–895, 899. In this book Carpenter also asserted that the "intellectual powers of Woman are inferior to those of Man" and that this was due in large part to the absence in women of "volitional power" (p. 965). Carpenter, however, claimed that this disparity resulted from environmental, not natural, causes. In another book, *Mental Physiology*, he wrote that "dependence on the bounty and therefore the will of others" produces not only the "paralysis of volition" in those who submit, but also "hysteria" and "morbid imaginings." Quoted by Laura Goodell, "Submission or Equality?" *New Northwest*, November 6, 1875.

64. "Notes on Neurasthenia," *Detroit Lancet*, April 1880. For similar views on the importance of sexual pleasure in marriage, see Russell Trall, *Sexual Physiology* (New York, 1880), pp. 69–70; and Edward Foote, *Medical Common Sense* (New York, 1858), p. 196.

65. Quoted in an article by Dr. Alexander Widler, "Studies upon a Forbidden Topic," *Woodhull and Claflin's Weekly*, October 15, 1870. For Storer's view see Storer to Burt Green Wilder, December 18, 1872, Wilder Papers, Cornell University Library. "Whatever may be said to the contrary by superficial observers," Storer wrote, "there can be no doubt that intercourse, unless *complete*, is prejudicial to the health of both parties."

66. See Antoinette Brown Blackwell, *Sexes Throughout Nature*, pp. 30–33; Henry C. Wright, *Marriage and Parentage*, pp. 36–37; Sarah H. Stevenson, *Physiology of Women*, p. 78; John Scudder, *Diseases of Women* (Cincinnati, 1857), pp. 37–38; John Cowan, *The Science of a New Life* (New York, 1874), p. 55; and Burt G. Wilder, undated essay on "sexual desire," Wilder Papers. According to the *Springfield*

Republican, Carpenter's work was "perhaps more read in this country than any other author's (*Springfield Republican*, May 15, 1873)."

67. Ward, *Dynamic Sociology* (New York, 1882), Volume I.

68. Ward's attack on secrecy and the imagination contradicted the romantic facts of his own sexual history. At the age of nineteen he met his first wife, Lizzie, loved her passionately, and, before they married, slept with her. "Have I ever been so happy?," he confided to his diary. "We lay with our faces together. I unfastened my shirt and put her tender little hands on my bare breast, and there we counted the beatings of our hearts like the whispering of angels. She gave me her heart and her body" (*Young Ward's Diary, 1860–1870* [New York, 1935], edited by Bernhard Stern, pp. 14–15). A year later he wrote: "That evening and night we tasted the joys of love and happiness which only belong to a married life" (pp. 75–76). Ward kept their affair secret; all his thoughts of Lizzie he confided to a diary written in French; and after their marriage in 1862 he kept that secret for a while too. Such privacy deepened the intensity of their relationship. "My heart's darling," he recorded on August 13, 1862, "whom I have loved so long, so constantly, so frantically, is mine! We are keeping it a secret, but it has been guessed but not yet discovered. How sweet it is to sleep with her!" (p. 90). On November 6, 1863, he wrote: "I have just enjoyed a sweet, glorious, sublime season with my darling, my beautiful, the life of my heart, my existence. . . . Without her, I should not wish to live longer" (p. 125). And on November 5, 1865, the year they moved to Washington, D.C., where Ward first had contact with a large metropolis and with feminists: "How happy we are to live all alone together and to enjoy . . . our ravishing intellectual creations" (p. 184).

69. In this context Ward wrote: "The system of education which makes art take precedence over science, and culture over knowledge, is a perverted system which, at best, only leaves the world where it finds it" (p. 71).

70. Unpublished speech on the recruitment of fashionable women, 1870, Stanton Papers, Library of Congress.

71. Cowan, *Science of a New Life*, p. 3.

72. "Woman's Kingdom," *Inter-Ocean*, March 3, 1877.

73. *Alpha*, October and November 1881.

74. "Report of the Sixth Woman's Congress," *Inter-Ocean*, October 11, 1879.

75. Duffy, *Relations of the Sexes*, pp. 229–231.

CHAPTER 5. *The Vindication of Love*

1. Conway, *Earthward Pilgrimage* (London, 1874), p. 290; and Emerson, *The Conduct of Life* (Boston, 1860), quoted in *The American Transcendentalists* (New York, 1957), edited by Perry Miller, p. 177. "Poets have no right," wrote Pascal, "to picture love as blind: its bandage must be pulled off and henceforth it must be given the use of its eyes" (quoted in Ortega y Gasset, *On Love*, (New York, 1957), p. 190.

2. On the transformation of American Protestantism see Sidney Mead, *Nathaniel Taylor* (New York, 1967); William McLoughlin, *Modern Revivalism* (New York, 1959); Donald Mathews, "The Second Great Awakening as an Organizing Process, 1780–1830: An Hypothesis," *American Quarterly*, June 1969, pp. 22–43; H. Richard Niebuhr, *The Kingdom of God* (New York, 1959); Perry Miller, *The Life of the Mind in America* (New York, 1965); and Daniel W. Howe, *The Unitarian Conscience* (Cambridge, 1970). For an excellent recent discussion of the relationship between the sexual division of labor and evangelicalism, see Nancy Cott, *The Bonds of Womanhood* (New Haven, 1977).

3. For accounts of the emergence of this sexual differentiation see Christopher Lasch and William Taylor, "Two Kindred Spirits: Sorority and Family in New England, 1839–1845," *New England Quarterly* 36 (March 1963), pp. 154–174; Keith Melder, *Beginnings of Sisterhood: The American Woman's Rights Movement* (New York, 1977); Kathryn Sklar, *Catherine Beecher* (New Haven, 1973); and G. J. Barker-Benfield, *The Horrors of the Half-Known Life* (New York, 1976).

4. Andrew Jackson Davis, *The Magic Staff* (New York, 1857), p. 55.

5. *New York World*, "Easter Fashions," March 31, 1872.

6. For a provocative discussion along these lines see Michael Rogin, *Fathers and Sons* (New York, 1975), pp. 20–73.

7. *Independent*, April 20, 1865.

8. See Ruth H. Block, "Untangling the Roots of Modern Sex Roles: A Survey of Four Centuries of Change," *Signs* 4 (Winter 1978), pp. 237–252; and Carl Degler, *At Odds: Women and the Family in America, 1776 to the Present* (New York, 1980), especially chapters 2 and 4.

9. "The Administration of Wealth," *Herald of Health*, November 1867, p. 227. My analysis here has been influenced by William McLoughlin's *The Meaning of Henry Ward Beecher* (New York, 1970), pp. 110–116. On sentimentalization see also Ann Douglas, *The Feminization of American Culture;* John Tomsich, *A General Endeavor: American Culture and Politics in the Gilded Age* (Stanford, 1971); and Carl Bode, *The Anatomy of American Popular Culture, 1840–1861* (Berkeley, 1959).

10. "Minutes," New York City Sorosis, May 16, 1870, Sophia Smith Library, Smith College.

11. *Dictionary of American Biography* (New York, 1936), pp. 551–553; *Golden Age*, April 10, 1871; *Theodore Tilton vs. Henry Ward Beecher* (New York, 1875), Volume I, pp. 389–412; Leon Oliver, *The Great Sensation, History of the Beecher-Tilton-Woodhull Scandal* (Chicago, 1873), pp. 16–21; Robert Shaplen, *Free Love and Heavenly Sinners* (New York, 1954); and newspaper scrapbook, Long Island Historical Society.

12. "One Blood All Nations," *Independent*, February 25, 1864, republished in Tilton, *Sanctum Sanctorum* (New York, 1870); "Lone Woman's Bed and Board," *Golden Age*, April 8, 1871. See also Tilton, *Independent*, November 1, 1869. On various aspects of Tilton's reform program see "Woman in Society," *Golden Age*, March 18, 1871; "The Rights of Women," *Tract I* (New York, 1871); "The Health Life," *Golden Age*, April 1, 1871; reviews of the physiological books of James Hinton, George Taylor, and George Napheys, *Golden Age*, July 18, 1874, and June 3, 1871; "A Law Against Women," *Independent*, January 18, 1866; "The Political Rights of Women," *Independent*, February 21, 1867; "The University of the Future," *Independent*, February 21, 1867; and "The Municipal Crisis," *Golden Age*, November 11, 1871.

13. See James McPherson, *The Struggle For Equality* (1964), pp. 145–153.

14. Tilton to Parton, February 1874, Parton Papers, Harvard University Library.

15. See, for example, these *Golden Age* articles: Octavius Frothingham, "Free Religious Movement," June 15, 1872; Tilton, review of the American scholar Samuel Johnson's *Oriental Religions and Their Relation to Universal Religion* (Boston, 1872), October 19, 1872, and review of Darwin's *Descent of Man*, March 18, 1871; and Lillie D. Blake, "Free Churches for Free Christians," April 22, 1871.

16. Edward Pessen, *Riches, Class and Power Before the Civil War* (New York, 1973), pp. 3–4, 16, 38; and Shaplen, *Free Love and Heavenly Sinners*, pp. 16–24.

17. "Gymnastics," *Herald of Health*, September 1866. See also Tilton to Anne Dickinson: "Come nibble at some late viands tonight with me—say at Delmonico's" (October 4, 1864) and "I have just received a basket of grapes. Come and you shall have some!" (undated, but probably 1870), both Dickinson Papers, Library of Congress).

18. Tilton to Beecher, November 30, 1865, August 7, September 18, September 24, 1863, Beecher Family Papers, Yale University Library. See also Shaplen, *Free Love and Heavenly Sinners*, pp. 27, 30, and *Theodore Tilton vs. Henry Ward Beecher*, Volume II, p. 11.

19. Memoir of Elizabeth Tilton, May 20, 1872, Beecher Family Papers, Yale University Library. Both men enjoyed experimenting with unconventional forms of social behavior. According to one observer close to the Tilton family, Tilton "liked the oriental style of kissing; he liked to see gentlemen kiss, and he should be very sorry to see his wife with the fastidious notions that some ladies had on the subject of kissing." Beecher once entertained the pleasures of woman's dress. "It seems," Eliza-

beth Cady Stanton observed in a letter to the *Revolution,* "that once on a visit to Mrs. Stowe . . . Mr. Beecher was imbued with the spirit of decoration and urged his nieces to curl and friz his hair . . . With the addition of becoming bonnet, skirt, mantilla and fan, he presented so lady-like an appearance that he was quite unwilling to return to the somber garb of manhood." See Sarah Putnam's testimony in *Theodore Tilton vs. Henry Ward Beecher*, Volume II, pp. 152–154; Stanton to editor, *Revolution*, July 14, 1870; and Beecher's "Notes on Rupture With Tilton," in which he describes his "familiarity" with both men and women and says that "not always, but usually, it included in it such salutations as they were accustomed to interchange among themselves. Also kissing *men*" (Beecher Family Papers, Yale University Library).

20. "Chromo-Civilization," *Nation*, September 24, 1874.

21. For a discussion of Beecher's *Norwood*, see McLoughlin, *The Meaning of Henry Ward Beecher*, pp. 84–97.

22. Tilton, *Tempest Tossed* (New York, 1873), pp. 375–376.

23. Tilton to his wife, February 6, 1867, *Theodore Tilton vs. Henry Ward Beecher*, Volume I, pp. 497–498; *Sanctum Sanctorum*, p. 56; and Tilton to Dickinson, April 9, 1870, Dickinson Papers, Library of Congress.

24. For examples of Tilton's appeals to famous men, see Tilton to Gerrit Smith, July 14, 1868, Smith Papers, Syracuse University Library; Tilton to Garrison, March 31, 1870, Garrison Papers, Boston Public Library; Tilton to Greeley, April 9, 1861, Greeley Papers, New York Public Library; and Tilton to Judge Brand, February 3, 1865, Tilton Papers, New York Historical Society.

25. For biographical material on the Beecher-Tilton affair, see *Dictionary of American Biography*, pp. 551–553; *Golden Age*, April 10, 1871; Oliver, *The Great Sensation*; and Shaplen, *Free Love and Heavenly Sinners.*

26. Tilton to R. B. Anderson, November 15, 1881, Anderson Papers, Wisconsin Historical Society; and Tilton, "The Phantom Queen of Sparta," unpublished poem, Yale University Library.

27. Tilton, "Thou and I," *Complete Poetical Works* (New York, 1897), pp. 15–66; written in 1870.

28. "Confessions of a Pyramid," 1903, unpublished manuscript, Yale University Library.

29. On romanticism see Henry May, *The Enlightenment in America* (New York, 1976) p. 190, pp. 359–360; Morse Peckham, *The Triumph of Romanticism* (New York, 1970); Raymond Williams, *Culture and Society* (New York, 1966), especially "The Romantic Artist," pp. 30–49; R. A. Yoder, "The Equilibrist Perspective: Toward a Theory of American Romanticism," *Studies in Romanticism* 12 (Fall 1973), pp. 705–740; and Susan Conrad, *Perish the Thought* (New York, 1976).

30. For an analysis of the emergence of romantic love through the medium of the magazine, see Herman R. Lantz et al., "Pre-industrial Patterns in the Colonial Family in America: A Content Analysis of Colonial Magazines," *American Sociological Review* 33 (February 1968), pp. 413–426. See also Nancy Cott, "Eighteenth-Century Family and Social Life Revealed in Massachusetts Divorce Records," *Journal of Social History* 10 (Fall 1976), pp. 20–43; and Cott, "Divorce and the Changing Status of Women in Eighteenth-Century Massachusetts," *William and Mary Quarterly* 33 (October 1976), pp. 586–614. On England, for the same development, see Lawrence Stowe, *The Family, Sex, and Marriage in England, 1500–1800* (New York, 1977), pp. 284, 318.

31. This interpretation owes a great deal to Norman O. Brown, *Life Against Death* (New York, 1959), especially chapter 4, "The Self and the Other: Narcissus," pp. 40–54.

32. I have borrowed the term "crystallization" from Stendhal's *On Love* (London, 1916): "I call crystallization the operation of the mind which, from everything which is presented to it, draws the conclusion that there are new perfections in the object of love. . . . Crystallization scarcely ceases at all during love. This is its history: so long as all is well between the lover and the loved, there is crystallization by imaginary solution; it is only imagination which makes him sure that such and such per-

fection exists in the woman he loves. But after intimate intercourse, fears are continually coming to life, to be allayed only by more real solutions" (pp. 23, 31). I would like to thank Donald Meyer of Wesleyan University for directing me to this source.

33. De Rougement, *Love in the Western World* (New York, 1956), p. 52. De Rougement's book represents a modern version of the nineteenth-century feminist position on love. In a later book, *Love Declared: Essays on the Myths of Love* (New York, 1963), De Rougement tempered his attack on romanticism while still maintaining his allegiance to an equilibrist perspective. Of the Tristan and Don Juan myths he wrote: "Yet without them, what would our love be?" (p. 162).

34. For an account of the degeneration of the romantic vision, see Mario Paz, *The Romantic Agony* (London, 1933).

35. Tocqueville, *Democracy in America* (New York, 1966), edited by J. P. Mayer, Volume II, p. 597.

36. Blackwell writes in her preface that "the pictures drawn in the following pages are only broken reflections of the real objects as I saw them mirrored in the little pools among the rocks, where there was almost always a disturbing ripple from the sea-breezes; and where the tides swept over often enough to break into fragments many of the veritable images which I should like very much to have preserved" (p. v.). Marie Howland's *Papa's Own Girl* (New York, 1874) and Louisa May Alcott's *Moods* (Boston, 1864) also approach romantic love from sympathetic, but very critical, points of view and might be analyzed along the same lines that I analyze Blackwell's book.

37. "The Poetic Endeavor," anthologized in Miller, *The American Transcendentalists*, pp. 201, 203.

38. Stendhal, *On Love*, p. 55. It is appropriate here to remark that Stendhal had strong feminist sympathies. See, by way of illustration, chapter 54, "Of the Education of Women," pp. 222–226, and Simone de Beauvoir, *The Second Sex* (New York, 1970), pp. 223–233.

39. Stendhal, *On Love*, pp. 25–27.

40. De Rougement, *Love Declared*, pp. 59–63.

41. "Work in Relation to Home," *Woman's Journal*, May 30, 1874.

42. For a good, but still inadequate, treatment, of Stanton's early romantic career, see Alma Lutz, *Created Equal: A Biography of Elizabeth Cady Stanton* (New York, 1974) chapter 5, "Peterboro and Henry B. Stanton," pp. 13–23. Elizabeth Cady married Henry Stanton against her father's wishes. See also, on Mary Putnam Jacobi, Ruth Putnam, *The Life and Letters of Mary Putnam Jacobi* (New York, 1925) chapters 7–9; and on Caroline Dall see chapter 10 of this study.

43. I disagree with De Rougement's contention that romantic passion is the principal culprit by abolishing "the world and society" in order to exist (*Love Declared*, pp. 59–63). De Rougement has the mistaken notion that the "logic" of the "myths of love" alone, apart from the historical conditions in which they express themselves, determine the character of man's amorous behavior (p. 55). He writes of Andre Gide as a "helpless victim of the tyranny" of the Don Juan and Tristan myths (p. 181).

44. "Literary Notices," *Una*, April 1853.

45. "The Other Side," published in *The Education of American Girls* (New York, 1874), edited by Anna Brackett, pp. 156–157.

46. Much of the information about Blake comes from Katherine Blake, *Champion of Women: The Life of Lillie Devereux Blake* (New York, 1943).

47. Letter to editor, *Revolution*, July 27, 1871

48. Celia Burleigh, *New York World*, May 28, 1871

49. Stanton to Davis, April 1, 1872, Stanton Papers, Vassar College Library

50. Gay to Dall, October 29, 1859, and February 24, 1859, Dall Papers, Massachusetts Historical Society; and Dall's journal entry, September 9, 1859. It must be noted here, however, that some feminists—notably Laura Curtis Bullard and Elizabeth Cady Stanton—admired Michelet for insights into female sensibility. See Bullard, "Michelet," *Revolution*, April 13, 1871; and Stanton, "Book Notices," *Radical* (1867–1868), pp. 347–348. Of Michelet's *L'Amour* and *Le Femme* Bullard wrote, "They

revealed a knowledge of the feminine nature that was surprising in its extent and comprehension."

51. *Una*, January 1, 1855.

52. *College, Market and Court* (Memorial Edition, New York, 1914), pp. 50–51.

53. "Woman in Literature," *Papers Read Before the Fourth Woman's Congress* (Washington, 1877), pp. 47–51. For similar statements see Marie Brown, "Parasites," *Radical* (March 1869), pp. 207–208; and Marie Howland, *Papa's Own Girl* (New York, 1874), pp. 213–14.

54. For a fine discussion of the novel as a bourgeois form, see Ian Watt, *The Rise of the Modern Novel* (Berkeley, 1974).

55. Jane Croly, *For Better or Worse* (Boston, 1874), pp. 27–28; and Thomas Nichols, *Esoteric Anthropology* (New York, 1873), p. 85.

56. Ezra Heywood, "Cupid's Yoke and the Holy Scriptures" (Boston, 1878), p. 4.

57. Stearns, *Love and Mock Love* (New York, 1860), pp. 38–39. See also "The Rationale of True Love," *Liberator*, October 19, 1860.

58. "What Justifies Marriage?" *Revolution*, August 18, 1870.

59. "Happy Marriages," *Woman's Journal*, February 18, 1871.

60. "Idols and Iconoclasts," *Concord Lectures on Philosophy* (Boston, 1883), edited by Raymond L. Bridginan, pp. 49–50. See also "The American Heroine," *Springfield Republican*, November 5, 1875, an article dealing with the heroines of Louisa May Alcott's novels: "The security of the American girl's position and the conditions of society in which she lives are not romantic. Her very independence and freedom of action cut her off from those situations of trial and danger which have served to make the heroines of the old World . . ." For similar attacks on romantic love see "Second Marriage," *New Northwest*, January 12, 1877; "Matrimonial Quarrels," *New Northwest*, March 13, 1879; Mary Studley, *What Our Girls Ought to Know* (New York, 1878), p. 202; Edward B. Foote, *Medical Common Sense* (New York, 1859), p. 216; John Cowan, *The Science of a New Life* (Boston, 1874), pp. 46–48; "The Decline of Marriage," *Golden Age*, October 10, 1874; John Scudder, *Reproductive Organs* (Cincinnati, 1874), pp. 22–35; Eliza B. Duffey, *The Relations of the Sexes* (New York, 1874) pp. 58–100; and Mark Lazurus, *Love vs. Marriage* (New York, 1852), pp. 68–69. These attacks on romantic love also characterized the marital polemic of the Progressive period. See, for example, George E. Howard, *A History of Matrimonial Institutions* (New York, 1904), Volume III, p. 258. "Through ignorance and defiance of the rules of health," Howard wrote, "we are destroying our physical constitutions. Under the plea of 'romantic love' we blindly yield to sexual attractions in choosing our mates, selfishly ignoring the welfare of the race."

61. "A Woman Who Dared," *Revolution*, November 11, 1869. I would also cite Eleanor Kirk's *Up Broadway* (New York, 1870), Caroline Corbin's *A Woman's Secret* (Boston, 1867), Mary Clemmer Ames's *Eirene*; *or A Woman's Right* (Philadelphia, 1871), and Marie Howland's *Papa's Own Girl.*

62. Sargent, *The Woman Who Dared* (New York, 1869). For biographical material on Sargent see *The National Cyclopedia of American Biography* (New York, 1897), Volume III, pp. 243–44.

63. Stowe, *Lady Byron Vindicated* (Boston, 1870). The Byron controversy was to "high culture" what the McFarland-Richardson "tragedy" was to popular culture." Feminists responded the same way to both.

64. Howe, of course, represents the most famous example of a woman raised almost overnight to a position of eminence within the feminist movement; her evolution—as well as Child's, Lozier's, Beecher Hooker's and that of innumerable other women—requires more historical study. On Lozier see *New Northwest*, June 29, 1872; and for the most recent, but still flawed, assessment of Child, see Kirk Jeffrey, "Marriage, Career, and Feminine Ideology in Nineteenth-Century America: Reconstructing the Marital Experience of Lydia Maria Child," *Feminist Studies*, 1975, pp. 113–130. Unfortunately, Jeffrey fails to mention Child's feminism or the feminist significance of her marriage, which can be better understood in feminist, not "feminine," terms. For an indication of Child's evolving feminist views see "Woman and

Suffrage," *Independent*, January 10, 17, 1867; "Physical Strength of Women," *Woman's Journal*, March 1, 1873; and Child to Stanton, May 24, 1863: "How many times when I have seen articles from your pen, have I said to my husband, 'There is a woman after my own heart!' and he responded, 'Amen!'" (Stanton Papers, Library of Congress). See also, on the beginnings of Beecher Hooker's feminist career, "Woman Suffrage," *Springfield Republican*, May 16, 1871.

65. Stowe to Parton, July 25, 1869, Parton Papers, Sophia Smith Library, Smith College. In the same letter she wrote: "Yes, I do believe in Female Suffrage—the more I think of it the more absurd this whole government of men over women looks . . ." In another letter to Parton, February 15, 1869, she declared: "I hold to woman's rights to the extent that a woman's own nature never ought to die out and be merged in the name of any man whatever."

66. See "The Woman Question," *Hearth and Home*, August 17, 1869, in which Stowe briefly compared Mill's book with Horace Bushnell's *Woman's Suffrage*, making clear throughout her contempt for Bushnell's position. Indeed, she praised the "woman's rights party" for being responsible for changing unjust marriage laws and suggested that women read the tracts published by the AWSA, "most of all, the report of the first Worcester Convention in 1851." Stowe also published a fierce attack on Bushnell in an earlier number of her journal; she had wanted to write the piece herself, but, as she explained to Parton in another letter, she could not because of family connections. "You see," she wrote on July 25, 1869, "as I am on friendly terms with the family, I couldn't have done it and so I am immensely tickled that it's done and well done" (Parton Papers, Sophia Smith Library, Smith College). See also, for an indication that Stowe contemplated taking a leadership position in the feminist movement, John Hooker to Isabella Beecher Hooker, July 1869, Isabella Beecher Hooker Papers, Stowe-Day Memorial Library. Stanton expressed pleasure at plans to appoint Stowe editor of the *Revolution* and for Hooker to become president of the NWSA. See Stanton to Paulina Wright Davis, July 1869, Isabella Beecher Hooker Papers. The best assessments of Stowe's life can be found in Constance Rourke, *Trumpets of Jubilee* (New York, 1927) and Edmund Wilson, *Patriotic Gore* (New York, 1966).

67. For an interpretation of Stowe's *Lady Byron Vindicated* that differs from mine, see Ann Douglas, *The Feminization of American Culture* (New York, 1977), pp. 245–247.

68. Stowe to Greeley, December 19, 1869, Greeley Papers, New York Public Library.

69. See Leon Oliver, *The Great Sensation* and Francis Williamson, *Beecher and His Accusers* (Philadelphia, 1874). Douglas comments on Byron's popularity with the public in *The Feminization of American Culture*, p. 246.

70. "The Moral of the Byron Case," *Independent*, September 9, 1869.

71. More, "The Byron Scandal," *Revolution*, October 28, 1869; and Livermore, "Mrs. Stowe's Vindication," *Woman's Journal*, January 22, 1870. See also, for favorable reception of *Lady Byron Vindicated*, John Neal, "Lady Byron," *Revolution*, January 20, 1870; and H., letter to the editor, *Independent*, August 26, 1869. Some feminists, however, disagreed with Stowe's argument. See in particular Theodore Tilton's hot retort, "The Byron Revolution," *Independent*, August 29, 1869; and Julia Ward Howe's gentle rebuke, "Lady Byron Vindicated," *Woman's Journal*, January 22, 1870. Howe's and Tilton's reservations sprang from political "nostalgia," not from any affection for Byron's sexual ethics. They fondly remembered Byron as the liberator of the Greek people (which he was not) and hated to have his image tarnished. Moreover, Samuel Gridley Howe, Julia's husband, had fought with Byron in Greece and possessed much Byron memorabilia. See Harold Schwartz, *Samuel Gridley Howe* (Cambridge, 1956), pp. 18–24.

72. Conway, *Earthward Pilgrimage* (London, 1870), pp. 271–272. See also George Curtis, "Editor's Easy Chair," *Harper's New Monthly Magazine*, October 1869.

73. For biographical material on Conway see his *Autobiography* (New York, 1904), Volume I, pp. 1–30, 99–121; Mary Elizabeth Burtic, *Moncure Conway* (New Brunswick, N.J., 1952); and Loyd Easton, *Hegel's First American Followers* (Athens, Ohio, 1966), pp. 134–145.

74. Conway, *Demonology and Devil-Lore* (New York, 1879), Volume I (originally a series of lectures delivered in 1872), pp. 233, 244. Conway's study represents an early and impressive study of comparative religions and mythologies. See also Volume II, especially the feminist chapters on Eve and Lilith (whom Conway called the "prototype of female independence"), pp. 85–106. The work of the American scholar Samuel Johnson, is similar to Conway's both conceptually and in its feminist orientation. See Johnson, *Oriental Religions and Their Relations to Universal Religion* (Boston, 1877), especially Part IV, chapter 1, "Patriarchialism," pp. 670–685.

75. *Demonology and Devil-Lore*, Volume I, pp. 210–211.

76. Emerson, "The Poetic Endeavor," in Miller, *The American Transcendentalists*, p. 211.

77. *Demonology and Devil-Lore*, Volume I, pp. 418–419.

78. For an extended discussion of these changes see chapter 7 of this study.

79. "Sex and Work," *Woman's Journal*, June 27, 1874.

80. It is appropriate here to remark on the meaning of "free love" within this context. Free love, in many respects, was the most rational democratic expression of true love; not only did free-lovers demand an absolute moral, intellectual, and sexual reciprocity in marriage, they utterly repudiated "bondage to the passions." They differed from the more moderate true-lovers only in their overt and strenuous refusal to accept external legal sanctions and in their affirmation of the most "well-developed conscience." Even here, however, true love or the more pervasive variant of feminist rational love approached free love in character. See *Woodhull and Claflin's Weekly*, March 15, 1873, and May 9, 1874.

81. *Revolution*, August 13, 1868.

82. "An Extinct Type," *New Century*, November 11, 1876.

83. *New Northwest*, March 6, 1877.

84. The terms "educated," "intelligent," and "organized" love appear respectively in Woodhull (quoted from an updated essay on free love, Woodhull Papers, University of Southern Illinois Library); Russell Trall, *Sexual Physiology* (1881), p. 267; and David Croly, *The Truth About Love* (New York, 1872), p. 38.

85. Stevenson, *The Physiology of Women* (Chicago, 1881), p. 79.

86. On the various facets of this progress, see, on law, Morton Horowitz, *The Transformation of American Law* (Cambridge, Mass., 1977); on labor, David Montgomery, *Beyond Equality* (New York, 1967); on technology, David Noble, *America By Design* (New York, 1977); and on education, Burton Bledstein, *The Culture of Professionalism* (New York, 1976).

87. Davis to Parsons, c. 1850, NAWSA Papers, Library of Congress.

88. "Dot and I," *Revolution*, May 5, 1870.

89. "Marriage," *Woodhull and Claflin's Weekly*, December 13, 1873. For similar statements see also Charles Morrill, *The Physiology of Women* (1870), p. 410; Henry C. Wright, *Marriage and Parentage* (1855), p. 129; and Elizabeth Cady Stanton's attack on "heterogeneous" marriages, *History of Woman's Suffrage* (Rochester, N.Y., 1881), Volume I, p. 719.

90. See Edmund Morgan, *The Puritan Family* (New York, 1966), pp. 29–65; and Keith Thomas, "Women and the Civil War Sects," in *Crisis in Europe, 1560–1660* (New York, 1964), edited by Trevor Aston, pp. 318–319.

91. Morgan, pp. 48–50; and Richard L. Bushman, *From Puritan to Yankee* (New York, 1967), pp. 189–190.

92. On Edwards's view of the affections, see Perry Miller's introduction to Edwards, *Images or Shadows of Divine Things* (New York, 1977), pp. 1–41.

93. *Minima Moralia* (London, 1978), p. 76.

94. Lawrence, *Studies in Classic American Literature* (New York, 1970), p. 19.

95. Channing to Stanton, March 9, 1899, Stanton Papers, Vassar College Library. For a similar statement see Channing's letter to his friend Caroline Dall, which he wrote on January 10, 1859, soon after divorcing his first wife, Susan. "I wish to study and bring out in due proportion and harmony the laws of social life with marriage at its centre," he wrote, "which can be treated properly only in relation with other things. . . . A thought has taken hold of my mind, revealing a law of unity which presides over all God's universe." Channing also sought to reassure Dall that

he still "strove to keep the balance of all my faculties" and that his decision to divorce his wife had not been influenced "by passion or illusion" but only by a process of "calm deliberation" (Dall Papers). Channing's struggle to justify his act bore the mark of modernity: for him, as for Dall, divorce represented the most irrational moment in an otherwise rational existence, so powerfully irrational, in fact, that it shaped the character of Channing's reform career for the rest of his life. Today, however, the irrational has become increasingly domesticated.

96. Nathan Allen, "The Act of Securing Longevity," *Sanitarian*, November 12, 1873.

97. See Raymond Williams, *Culture and Society* (London, 1958), pp. 66–69.

98. Nietzsche, quoted in De Rougement, *Love Declared*, p. 134. Nietzsche continues: "The images of myth must be invisible and omnipresent tutelary spirits favoring the development of the adolescent soul, and whose signs foretell and explain the grown man's life and struggles."

Part 2 / Feminism in the Public Realm

CHAPTER 6. *From the Private to the Public Realm*

1. Howe, "Idols and Iconoclasts," in *Concord Lectures on Philosophy* (Boston, 1883), edited by R. L. Bridginan, pp. 49–50.

2. On Fruitlands see Raymond Lee Muncey, *Sex and Marriage in Utopian Communities* (Bloomington, Ind. 1973), pp. 79–92.

3. "Individualism," in *Concord Lectures on Philosophy*, pp. 30–31. For an account of Alcott's evolution see Franklin Sanborn and William T. Harris, *A. Bronson Alcott, His Life and Philosophy* (Boston, 1893), Volume II, pp. 593–637. Louisa May Alcott, by the way, declared in her novel *Silver Pitcher* (New York, 1876) that only fashionable women refused to read Buckle, Mill, and the *Social Science Reports*. See Luther and Jesse Bernard, *American Sociology* (New York, 1943), p. 545.

4. "A New Party," *Revolution*, February 10, 1870.

5. "Noisy Women and Gentle Women," August 12, 1869; and "Mrs. Holloway on Charlotte Brontë," *Revolution*, February 10, 1870.

6. Pillsbury to Samuel May, January 25, 1859, May Family Papers, Boston Public Library. For material on Pillsbury's early religious views and on his abolitionism, see David B. Pillsbury and Emily A. Getshall, *The Pillsbury Family* (New York, 1898), pp. 70–71, 131; Louis Filler, "Parker Pillsbury: Anti-Slavery Apostle," *New England Quarterly* 19 (September 1946), pp. 315–337; and Pillsbury, *Anti-Slavery Bugle*, March 12, 1853.

7. The quotation appears in a letter from Pillsbury to Ellen Wright, January 14, 1875, Garrison Family Papers, Sophia Smith Library, Smith College.

8. On Pillsbury's feminist views see the following articles in the *Revolution*: "Women in History," July 8, 1869; "A Wife's Debts and Earnings," August 5, 1869; and "Disability of Sex and Marriage," March 3, 1870.

9. Charles Griffing to Samuel May, March 12, 1863, May Papers, Boston Public Library.

10. "Free Public Bathing," *Revolution*, July 20, 1868. See also in *Revolution* "Turkish Baths," February 11, 1869, and "The Healing Art," April 15, 1869.

11. Pillsbury to the editor, *Alpha*, January 1882 and June 1882.

12. Pillsbury, "The Popular Religions and What Shall Be Instead" (Concord, N.H., 1891), pp. 17–25. See also "Free Religious Conventions," *Revolution*, June 18, 1868; Pillsbury to Gerrit Smith on Free Religion, April 1, 1871, Smith Papers, Syracuse University Library; editorial, *Index*, August 17, 1872; and "Notes from the Field," *Index*, July 15, 1871.

13. The quotation comes from an article by Gertrude Lenzer, "Mind-forged Manacles: Auguste Comte and the Future," *Marxist Perspectives* (Fall 1979), pp. 62–88. See also *Auguste Comte and Positivism* (New York, 1975), edited by Lenzer, especially her excellent introduction.

14. "The Positivist Problem," *Modern Thinker*, 1 (1870), pp. 49–72.

15. See, for example, *Woodhull and Claflin's Weekly*, *Revolution*, *New Northwest*, *Woman's Journal*, *New Age*, *Radical*, *New Century*, *Golden Age*, *Banner of Light*, and *Index*.

16. For an excellent recent discussion of the swift transformation of Unitarianism from a religion still rooted in Congregationalism during the 1830s to one deeply influenced by Comtian positivism in the 1860s, see Charles D. Cashdollar, "European Positivism and the American Unitarians," *Church History* 45 (December 1976), pp. 490–506. Cashdollar observes, however, that American Unitarians did not completely align themselves with Comtian thought; rather, "they fused Comte's thought with parts of the British empirical tradition as represented especially by John Stuart Mill and Herbert Spencer" (p. 491). Julia Ward Howe read Comte's *Cours de Philosophie Positive* in 1850 and much preferred his scientific analysis to his discussions of theology and metaphysics. She also praised Comte in her parlor lectures. See her *Reminiscences, 1819–1899* (Boston, 1899), pp. 198, 211, 305–307. For further observations on the Frothingham school see Stanton, "Editorial Correspondence," *Revolution*, August 13, 1868.

17. These feminists included Elizabeth Cady Stanton, Antoinette Brown Blackwell, Lydia Maria Child, Julia Ward Howe, Martha Wright, Caroline Severance, Lucy Stone, Sara Underwood, Lillie Devereux Blake, Olympia Brown, Lucretia Mott, Thomas Higginson, Caroline Dall, Anna Garlin Spencer, Anna Brackett, Ednah Cheney, Caroline Putnam, Sally Holly, Henry Blackwell, Franklin Sanborn, Matilda Joslyn Gage, and Kate Hilliard. Feminists played a leadership role in the creation of Free Religion. See Stow Persons, *Free Religion* (New Haven, 1947), pp. 38–49, *passim;* and Sidney Warren, *American Freethought, 1860–1914* (New York, 1943), chapters 3 and 4. See also, for lists of the feminist members, *Springfield Republican*, December 12, 1871, and June 3, February 19, 1872. Lester Ward also enrolled himself as a "Free Religionist." On the feminist role in the ideologically similar Free Church Movement, see Ralph Harlow, *Gerrit Smith* (New York, 1939), pp. 193–218; and Ida Husted Harper, *Susan B. Anthony* (Indianapolis, 1899), Volume I, p. 167.

18. *Springfield Republican*, November 28, 1873.

19. "Modern Principles of Free Religion," *Index*, January 7, 1871.

20. "Some Radical Resolutions," *New York World*, January 29, 1874.

21. *New York World*, November 23, 1872.

22. See Caroline Dall, journal entry, April 6, 1868, Dall Papers, Massachusetts Historical Society; Julia Ward Howe's report on the Boston Radical Club, *Woman's Journal*, April 2, 1870; and *Sketches and Reminiscences of the Radical Club* (Boston, 1880), edited by Mrs. John T. Sargent, introduction.

23. Warren, *American Freethought*, p. 162; and *New Age*, July 22, 1876.

24. For a sample of spiritualist writing in a Free Religion paper, *New Age*, see "The Spiritualistic Philosophy of Life," November 6, 1876; and for the reverse see "The Religion of Positivism," *Banner of Light*, October 12, 1867.

25. Gage to Wright, February 12, 1871, Garrison Family Papers, Sophia Smith Library, Smith College.

26. *New York World*, April 16, 1871.

27. *New Northwest*, April 17, 1874.

28. *New Northwest*, November 7, 1873.

29. Willard, *Sexology* (Chicago, 1867), pp. 8–10. Willard first drew public acclaim in 1869 when she gave a lecture before a convention of western suffragists in Chicago, a lecture considered the "most able and lucid address of the convention, arguing and explaining away the disgraceful 'sphere' doctrine so much referred to by the opponents of the Woman's Suffrage Cause" (*Woman's Advocate*, October 1869). In 1870 Willard wrote a long series of articles for the *Revolution* in which she responded critically and effectively to the antifeminist posturing of E. L. Godkin's *Nation*. See *Revolution*, December 15, 1870, February 2, April 6, May 11, 1871. The *Revolution* also reviewed her book very favorably. See "New Books," February 5, 1868; and "The Author of *Sexology*," March 18, 1868.

30. Willard to Dall, February 8, 1868, Dall Papers, Massachusetts Historical Society; and Willard, *Sexology*, pp. 8–10, 59, 145–157.

31. See Hester Pendleton, *Husband and Wife; or The Science of Human Development Through Inherited Tendencies* (Boston, 1863); Antoinette Brown Blackwell, *The Sexes Throughout Nature* (New York, 1875); and Sarah Hackett Stevenson, *Boys and Girls in Biology* (Chicago, 1875) and *The Physiology of Women* (Chicago, 1881). Both very popular with feminists and both representative of other feminist women who wrote such books, Stevenson and Blackwell drew on the most advanced biological and sociological literature. Pendleton apparently attracted less attention from feminists, although the *Springfield Republican* praised her book very highly (see review, July 24, 1871). No less than the others, however, she based much of her analysis on the writings of Gorge H. Lewes (follower of Comte, companion of George Eliot), Spencer, and Darwin. An important female physician from Chicago and the first female member of the American Medical Association, Stevenson thought of herself principally as a student of Thomas Huxley.

32. For a short biographical sketch of David Croly see Elizabeth B. Schlesinger, "The Nineteenth-Century Woman's Dilemma and Jennie June," *New York History* (October 1961), pp. 367–379; and for a description of the first *Modern Thinker* see *Banner of Light*, October 15, 1870. I have also drawn material on Jane Croly (who wrote under the name Jennie June) from the Schlesinger article.

33. "People Worth Knowing," *Woman's Journal*, September 10, 1870.

34. For a history of Central College see Albert Hazen Wright, *New York Historical Source Studies* 23 (Ithaca, N.Y., 1953–1966), pp. 16–74. Jane Cunningham's name appears on a list of Central College students, p. 71.

35. See Andrew Jackson Davis's autobiography, *The Magic Staff* (New York, 1857), p. 443; Croly, letter to the editor, *New York Herald*, February 3, 1856; and Zulu Woodhull's scribbled notes on Croly, Woodhull Papers, Southern Illinois University Library.

36. *Modern Thinker* 1 (1870), pp. 185–201.

37. For an account of the business and work of the Positivist Society in New York, as well as a list of its members, see Carlsfried, "From New York," *Springfield Republican*, January 31, April 8, and April 24, 1872.

38. For information on the New York City Social Science Association, see *Cooperator* (June and July, 1881); and *Cooperative News* (October 1887); for a sample of Bristol's work for the *Revolution*, see "An Independent Life," April 20, 1871; and, on her role in the AAW, see *Report of the Association for the Advancement of Women* (1878–1879), Vassar College Library. For a biographical sketch of Bristol see *A Woman of the Century* (Boston, 1893), edited by Frances E. Willard and Mary Livermore, pp. 123–124. Bristol also wrote for the *Radical*. See, for example, "The Old Song and the New," *Radical*, February 1867.

39. *Life and Letters of Mary Putnam Jacobi* (New York, 1915), edited by Ruth Putnam, p. 16. See also, for other material on Jacobi, Barbara Sicherman, "The Paradox of Prudence: Mental Health in the Gilded Age," *Journal of American History* March 1976, pp. 890–912.

40. Putnam to E. G. Putnam, March 24, 1868, and Putnam to her mother, February 5, 1870, in Putnam, *Life and Letters*, pp. 171–172, 271. Putnam also became a close friend of the French male feminist Laboulaye. See Putnam to her mother, September 30, 1867, p. 169.

41. Quoted in Putnam, *Life and Letters*, pp. 281–282.

42. Putnam to her mother, September 15, 1870, p. 271. On the Réclus family see Putnam's commentary in *Life and Letters*, pp. 280–281.

43. Putnam to her father, May 7, 1871, pp. 277–278.

44. Ibid., p. 281

45 "Comte's Birthday," *New York World*, January 23, 1872; and "From New York," *Springfield Republican*, January 23, 1872.

46. Because I quote profusely and synthetically from Putnam Jacobi's text, I will not list all the page numbers.

47. Wright to Stanton, December 27, 1870, Garrison Family Papers, Sophia Smith Library, Smith College; and "Woman's Suffrage," *New York World*, May 16, 1874.

48. "A Tour Among Good People," *Revolution*, August 4, 1870; and Charles Miller to Gerrit Smith, February 10, 1872, Smith Family Papers, New York Public Library.

49. *New Century*, May 27, 1876 and June 3, 1876.

50. Thomas Huxley's paraphrase of Comte, "Comte, Spencer, and Huxley," *Modern Thinker* 2 (1873), pp. 37–56.

51. "Feminine Influence of Positivism," *Revolution*, April 16 and 30, 1868.

52. Stanton to Gerrit Smith, January 29, 1868, Smith Family Papers, New York Public Library.

53. Stanton to Miller, June 20, 1853, *Elizabeth Cady Stanton as Revealed in Her Letters, Diary, and Reminiscences* (New York, 1922), edited by Theodore Stanton and Harriet S. Blatch, Volume II, pp. 52–53.

54. Stanton to her husband, October 9, 1867, Stanton Papers, Library of Congress.

55. "Reasons Why Some Marriages Are Unhappy," *Revolution*, October 15, 1868.

56. *Revolution*, October 15, 1868.

57. See Ellen Dubois, "On Labor and Free Love: Two Unpublished Speeches of Elizabeth Cady Stanton," *Signs* 1 (Autumn, 1975), pp. 257–265. Dubois provides important introductory material to each of these speeches, the first delivered in 1868 and the second in 1870. She does not make clear enough, however, that Stanton's views on separation and divorce differed from (but did not contradict) her views on married life. Her free-love "radicalism" applied only to the former and not to the latter. Moreover, it can be argued that Stanton had developed a free-love perspective on divorce long before 1870.

58. Address before the New York State legislature, 1854, *History of Woman's Suffrage*, Volume I, p. 598.

59. "Book Notices," *Radical*, January 1868, p. 351. This review of Caroline Dall's book was signed "E. C." and thus has gone unnoticed as a product of Stanton's pen. See *Contributions Towards a Bibliography of the Association of Collegiate Alumnae* (Association of University Women, 1897) for the correct citation.

60. Dubois, "On Labor and Free Love," *Signs* 1 (Autumn 1975), pp. 263, 267.

61. See, for example, Lita Barney Sayles, "The Marriage Question," *Revolution*, February 17, 1870; Anna Dickinson, unpublished speech on marriage, 1870, Dickinson Papers, Library of Congress; Robert Collyer, *Springfield Republican*, April 24, 1871; Theodore Tilton, "Love, Marriage and Divorce," *Independent*, December 1, 1870, editorial, *Golden Age*, September 9, 1871; John Neal, "The Woman Who Dared," *Revolution*, November 11, 1869; Epes Sargent, *The Woman Who Dared* (Boston, 1869); Eliza Bisbee Duffey, *The Relations of the Sexes*, (New York, 1876); Mary Clemmer Ames, *Eirene; or A Woman's Right* (New York, 1871); Elizabeth Osgood Willard, "Bill of Rights for Women," *Revolution*, March 1871; Laura Curtis Bullard, "Shall We Go a Solitary Path?" *Revolution*, November 24, 1870; Eleanor Kirk, *Up Broadway and Its Sequel* (New York, 1870); Moncure Conway, "Marriage," *Radical* (December 1871), pp. 340–353; Abigail Scott Duniway, *New Northwest*, February 16, 1872; Mary L. Booth to Caroline Dall, June 5, 1860, Dall Papers, Massachusetts Historical Society; Isabella B. Hooker "Mrs. Hooker's Position," *Springfield Republican*, August 11, 1874; Gerrit Smith, "Gerrit Smith on Divorce," *Revolution*, February 10, 1870; Mary F. Davis, *Banner of Light*, July 10, 1858; Paulina Wright Davis, letter to the editor, *Revolution*, February 10, 1870, and "The Apollo Hall Resolutions," *Revolution*, June 15, 1871; Lucy Stone to Elizabeth Cady Stanton, April 16, 1860, Stanton and Blatch, editors, *Elizabeth Cady Stanton as Revealed in Her Letters;* Olympia Brown to John Hooker, September 22, 1873, Isabella Beecher Hooker Papers, Stowe-Day Memorial Library; Mary Walker, *HIT* (New York, 1871) especially chapter 6, "Divorce," pp. 137–146; "The Divorce Law," *New Northwest,* January 18, 1877; Samuel Bowles, *Springfield Republican*, May 19, 1870; and William Robinson, *Springfield Republican*, May 27, 1870, and April 20, 1871. The women of the New York City Sorosis—including Jane Croly, Celia Burleigh, and Charlotte Wilbour—also followed Stanton's view. See "Minutes," New York City Sorosis, May 16, 1870, Sophia Smith Libary, Smith College.

62. "Divorce," December 16, 1871, and December 23, 1871, *Woman's Journal.* See also *Woman's Journal* for March 11, April 19, and August 26 of that year.

63. For discussion of the anarchist position see Hal D. Sears, *The Sex Radicals, Free Love in High Victorian America* (Lawrence, Kans., 1978); James Joll, *The Anarchists* (London, 1964); and George Woodcock, *Anarchism* (New York, 1967).

64. "Book Notices," *Radical,* January 1868.

65. Ibid., p. 349. See also Croly, paper given at the Syracuse Woman's Congress in 1874, *Association for the Advancement of Women Papers, 1874–1875* (Syracuse, N.Y., 1875), pp. 41–51, Vassar College Library; and Gilman, *Women and Economics* (New York, 1966), pp. 134–136. "Monopoly and appropriation are the strongest instincts of man's nature," Croly wrote. "They had to be. They formed the basis of our civilization, and the march of civilization is not infrequently over the bleeding body of poor humanity." American Comtians first formulated this position in the late 1860s.

66. See Ellen Dubois, *Feminism and Suffrage: The Emergence of an Independent Women's Movement in America, 1848–1869* (Ithaca, N.Y., 1978), pp. 112–159. On Stanton's admiration for Favre, Mazzini, and Mill, see *New Northwest*, July 20, 1871.

67. *Golden Age*, April 27, 1872.

68. "Book Notices," *Radical*, January 1868, p. 350.

69. "I Have All the Rights I Want," *Revolution*, April 1, 1869.

70. Ibid.

71. Speech to the American Equal Rights Association, *Revolution*, May 13, 1869. See also Matilda J. Gage, "Series of Conventions of the NWSA," *Ballot Box*, June 1881 for an almost verbatim reproduction of Stanton's views. "The male element," Gage writes, "has thus far held high carnival, crushing out all the divine elements of human nature. . . . The present disorganization of society warns us that in the disenfranchisement of woman we have let loose the reins of violence and ruin which she only has the power to avert. . . . All writers recognize women as the great harmonizing elements in the 'new era.' "

72. Stanton, *Una*, January 1853. Stanton wrote privately to Anna Dickinson on August 6, 1867, that "when philosophers come to see that ideas as well as babies need the mother soul for their growth and perfection, that there is sex in the mind and spirit, as well as body, then they will appreciate the necessity of a full recognition of womanhood in every department of life" (Dickinson Papers, Library of Congress).

73. "Reverend Henry Edgar," *Revolution*, June 10, 1869; and "Maternity," *Revolution*, February 10, 1870.

74. On Wilcox see *Woodhull and Claflin's Weekly*, September 10, 1870, October 15, 1870, and *Cincinnati Commercial*, February 19, 1880; on Brisbane, Andrews, and Channing see Stanton to Theodore Tilton, October 1871, and Stanton to Isabella Beecher Hooker, February 2, 1872, Isabella Beecher Hooker Papers, Stowe-Day Memorial Library; and on Miller, "What People Say to Us," *Revolution*, January 14, 1869. For further indications of Stanton's interest in positivism and in social science, see, in the *Revolution*, "Positivism in Paris," June 4, 1868; "American Social Science Association," October 29, 1868; "Positivism," November 19, 1868; "Socal Science Convention," November 26, 1868; and "Social Science," January 15, 1868. See also Rutger B. Miller, "The Nicene Creed Contemplated From the Positive Point of View," *Radical* 4 (August 1868), pp. 101–106. I might add here that Stanton considered the *Radical* "the best and most important, after all, of the magazines." See *Revolution*, review of the *Radical*, December 16, 1869.

75. Smith to Stanton, March 23, 1868, Stanton Papers, Vassar College Library.

76. *Revolution*, August 13, November 26, December 17, 1868, and May 13, 1869. See also December 24, 1868, January 21, 1869, April 8, 1869, and April 29, 1869.

77. *History of Woman's Suffrage*, Volume I, p. 22; and Stanton, *Eighty Years or More* (New York, 1898), p. 3.

78. "The True Republic," *Woodhull and Claflin's Weekly*, May 18, 1872.

79. Quoted in *Woodhull and Claflin's Weekly*, January 27, 1872.

80. *History of Woman's Suffrage*, Volume I, p. 540.

81. *Revolution*, May 13, 1869.

82. Matilda J. Gage, speech at the Fifth NWSA Convention, January 16, 1873, *History of Woman's Suffrage*, Volume II, pp. 527–528. I have quoted Gage, but Stanton almost certainly shared these ideas. See also Gage's speech before the Social Science Club at Cornell, *Ballot Box*, May 1878, January 1880 and February 1881; and Susan B. Anthony, *New Northwest*, November 22, 1872. "What have we gained by all the blood and treasure of this four years' war," Anthony demanded, "if we have not established this one principle—the supremacy of the *national* law to protect the American citizens over that of any and every State law to the contrary?"

83. Speech on marriage and divorce at the 1870 Decade Meeting, in Paulina Wright Davis, *A History of the National Women's Rights Movement* (New York, 1870), pp. 66–81; and *History of Woman's Suffrage* (Rochester, N.Y., 1886), Volume III, pp. 61, 116.

84. "Women and the International Association," *Revolution*, October 12, 1871. See also "Woman's Rights in Europe," April 15, 1869; "Letter from Mrs. Miller," January 27, 1870; and "Foreign Correspondence," *Revolution*, May 5, 1870.

85. "Maternity," *Revolution*, February 10, 1870.

86. Letter to the 1882 NWSA Convention, September 1, 1882, *History of Woman's Suffrage*, Volume III, p. 181.

87. Report of the People's Convention, *Woodhull and Claflin's Weekly*, May 18, 25, 1872; and editorial correspondence, *Revolution*, August 20, 1868.

88. Stanton and Blatch, editors, *Elizabeth Cady Stanton as Revealed in Her Letters*, pp. 330–331.

89. Speech on marriage and divorce at the 1870 Decade Meeting, in Davis, *A History*, pp. 70–71. Davis pointed to the unique character of this speech in a letter to the *Revolution*, December 1, 1870. Apropos of the Boston feminists who objected to the bias of Stanton's argument, Davis wrote: "I inquired if Mrs. Stanton's address on the divorce bill (1860), published by the Equal Rights Society, had ever been read there? 'Certainly, but then this new lecture is something very different,' Not at all, I replied, it is the same in substance and spirit, only more finished and enlarged on the history of marriage, and the laws which attempt to regulate it."

90. See David Pivar, *Purity Crusade* (Westport, Conn., 1973), pp. 104–105, 139–146.

91. Howard, *A History of Matrimonial Institutions* (New York, 1907) Volume II, pp. 183–184.

92. "Marriage and Divorce," *Revolution*, October 22, 1868.

93. Review of Dall's book, *Radical*, January 1868; letter to *Revolution*, February 9, 1871; and speech on marriage and divorce, Dubois, *Signs* 1 (Autumn 1975), p. 268.

94. Hartz, *The Liberal Tradition in America* (New York, 1955), pp. 185–189. Hartz's view is that the southern sociological analysis developed by such men as George Fitzhugh and Henry Hughes died with the end of the Civil War, and that it found no comparable expression in the North. The thinking of Stanton and others refutes this contention, although it does not refute Hartz's basic argument that American social criticism, bordering often on socialist analysis, was overwhelmed by the historical weight of liberalism.

95. *Woman's Journal*, November 21, 1874.

96. *Woman's Journal*, November 6, 1875.

97. Comte, "The Subjection of Women," *Modern Thinker* 1 (1870), p. 171.

98. Ward, *Dynamic Sociology* (New York, 1883), p. 130.

99. *New York World*, May 5, 1872.

100. Jane Croly, "The Love Life of Auguste Comte," p. 190; "Talks With Women," *Revolution*, September 23, 1869; "Women in Journalism," *Association for the Advancement of Women Papers, 1874–1875*, pp. 41–51, Vassar College Library; and "Woman's Parliament," *New York World*, October 22, 1869.

101. Croly to Dickinson, c. 1871, Dickinson Papers, Library of Congress.

102. Brisbane, "The Civilization of the Future," *Modern Thinker* 2 (1873), pp. 225–246; and Henry Edgar, "The Social Problem," *Woodhull and Claflin's Weekly*, May 25, 1872.

103. Undated essay on women in the labor force, Putnam Jacobi Papers, Schlesinger Library, Radcliffe College.

104. "George Eliot Is Dead," *Ballot Box*, January 1881 (quoted in *History of Wom-*

an's Suffrage, Volume I). See also "George Eliot," *Inter-Ocean*, October 17, 1879; *New Century*, May 27 and June 10, 1876; and Sara Underwood, "George Eliot," in *Heroines of Free Thought* (New York, 1876), pp. 297–327. A member of the NWSA, Underwood wrote many articles for the *Revolution*. The popular English feminist Emily Faithful also distinguished Eliot from other novelists in the fact that she "instructs" (*New York World*, February 16, 1873). For an excellent modern assessment of Eliot's thought see Steven Marcus, "Human Nature, Social Orders and Nineteenth-Century Systems of Explanation: Starting in with George Eliot," *Salmagundi*, Winter 1975, pp. 20–42. Interestingly enough, Stanton, Gage, and Anthony viewed Charlotte Brontë as having "genius and ignorance." Eliot, however, had "genius and culture" (*Ballot Box*, January 1881).

105. Quoted in a favorable review of Croly's book in *New Century*, September 2, 1876.

106. *New Age*, June 17, 1876.

107. "Stirpiculture: Regulated Human Production," *Modern Thinker* 2 (1873), p. 57, unsigned.

108. Bristol, "Enlightened Motherhood," *Papers and Letters Presented at the First Woman's Congress of the Association for the Advancement of Women* (New York, 1874), pp. 10–14; and Stanton, "The True Republic," *Woodhull and Claflin's Weekly*, May 18, 1872.

109. Bristol, "The philosophy of the Woman's Era," *Papers and Letters Read Before the Fourth Woman's Congress* (Washington, 1877), p. 62. See also, from the *Modern Thinker*, Frederic Harrison, "The Positivist Problem," 2 (1873), pp. 52–65; J. D. Bell, "Religion and Science," 1 (1870), pp. 121–147; and Andre Poey, "Good and Evil—Their Origin," 1 (1870), pp. 151–162.

CHAPTER 7. *Economic Autonomy for All Women*

1. The term "social fiction" appears in Epes Sargent, *The Woman Who Dared* (New York, 1869), p. 264.

2. "Parasites," *Radical*, March 1869, p. 203.

3. "Anniversary of the American Equal Rights Association," *Revolution*, May 13, 1869.

4. These quotations appear in the following essays by feminists: Antoinette Brown Blackwell, "Caroline Herschel," *New Century*, September 2, 1876, and "Industrial Reconstruction," *Woman's Advocate*, January 1869; Matilda J. Gage, "Opportunity for Development," *Woman's Advocate*, February 1869; Emma Marwedel, "A Horticultural School for Women," *Herald of Health*, September 1869; and "Differentiation and Natural Selection," *Ballot Box*, January 1879.

5. Unpublished speech on marriage and work, 1870, Dickinson Papers, Library of Congress.

6. See, for attacks on the woman's sphere and "domestic vine theory," Harriet Beecher Stowe, *Lady Byron Vindicated* (Boston, 1870), pp. 114–119; Elizabeth S. Phelps, *Inter-Ocean*, May 17, 1873; Sargent, *The Woman Who Dared*, pp. 151–153; Jane Croly, "Woman's Work," *New Century*, July 29, 1876; and Laura Curtis Bullard, "The Moral Code as Applied to Men and Women," *Revolution*, February 9, 1871. The term "pedestal and goddess theory" appears in *New Century*, May 20, 1876.

7. "Woman Sovereignty," *Revolution*, January 5, 1871; and Stanton, speech at the 1870 Decade Meeting, in Paulina Wright Davis, *A History of the National Woman's Rights Movement* (New York, 1871), p. 63. See also Anna Dickinson, speeches on marriage and work, Dickinson Papers, Library of Congress; Stanton, "The Parable of the Ten Virgins," *Ballot Box*, December 1877; "The Coming Housewife," *Golden Age*, December 30, 1871; and "An Extinct Type," *New Century*, November 11, 1876.

8. S. C., "About Our Girls," *Woman's Journal*, August 30, 1873; and M. F. Burlingame, "Emancipation," *Woman's Advocate*, March 1869, p. 298. The feminist literature on these matters is enormous. See, by way of illustration, all the published reports of the Woman's Congresses, 1873–1880; *Inter-Ocean*, January 25, 1879, and

April 21, 1880; *Radical*, June 1869; *Alpha*, June 1879; *New Northwest*, November 11, 1871, May 24 and 27, 1872, October 6, 1871; *Woman's Advocate*, November 1869; *Woman's Journal*, July 25, 1874; *Revolution*, May 23, 1871, October 1, 1868, May 27, 1871, and May 27, 1869; *New Age*, October 21, 1876; *New Century*, May 20, June 24, October 14, 1876; and *Banner of Light*, June 22, 1867, and April 2, 1870.

9. See, for example, Thomas Higginson, "Woman and Her Needs," *Una*, May 2, 1853; Moses Coit Tyler, "Fragmenatry Manhood," *Independent*, November 22, 1869; and Stanton, *Revolution*, April 22, 1869.

10. "A Review," *Woman's Advocate*, November 1869, pp. 268–276.

11. The feminist literature on this subject is also vast. See, for example, Caroline Dall, *College, Market and Court* (Memorial Edition, New York, 1914), pp. 224–225, 245, 363, and *passim;* Mary Putnam Jacobi, *The Question of Rest For Women During Menstruation* (New York, 1877), pp. 15–62; "Ida Lewis," *Hearth and Home*, June 26, 1869; "An Independent Wife" and "How a Woman Had Her Way," *Revolution,* April 20, 1871; "Woman as Heroine" and "What Cannot a Brave Woman Do?" *Revolution*, January 29, 1870, and September 2, 1869; "People Worth Knowing," *Woman's Journal*, July 9, July 20, 1870, and March 4, 1871; and "The Exhibition of Woman's Work," *New Century*, May 13, 1876.

12. Higginson, "The Limitation of Sex," *Woman's Journal*, February 22, 1873; and Lydia Maria Child, "The Physical Strength of Women," *Woman's Journal*, March 15, 1873. See also *Woman's Journal*, January 3, March 7, 1874, and February 13, 1875.

13. C. Clark, "An Average Woman," *Woman's Advocate*, March 1869, pp. 256–301.

14. Burleigh, "Another Milestone," *Woman's Journal*, January 14, 1871; Parker Pillsbury, "The Amazons," *Revolution*, April 10, 1869; and Higginson, "Sappho," *Atlantic Monthly*, July 1871, pp. 83–93.

15. Higginson to Dall, March 22, 1854, Dall Papers, Massachusetts Historical Society.

16. "Woman's Kingdom," *Inter-Ocean*, March 27, 1880; and "Women and Medical Science," *Woman's Journal*, February 5, 1870.

17. Dall, *Historical Pictures Retouched* (Boston, 1860), pp. 10–11. See also Sara A. Underwood, *Heroines of Free Thought* (New York, 1876); and, for a discussion of this kind of reconstruction by other educated women, Susan Conrad, *Perish the Thought* (New York, 1977), pp. 108–120.

18. Stanton, "Book Notices," *Radical*, January 1868.

19. See, for example, Stanton's eloquent treatment of Esther in "The Women of the Bible," *Revolution*, September 30, 1869.

20. See, for example, "Woman as Inventor—Silk Invented by A Woman," *Revolution*, May 21, 1868; and *Woman as Inventor* (New York, 1870). Gage's findings appeared in the popular press. See, for example, "The News," *Cincinnati Commercial*, May 8, 1868.

21. Gage to Wright, c. 1871, Garrison Family Papers, Sophia Smith Library, Smith College.

22. "A Great Inquiry," *Ballot Box*, October 1878.

23. "Social Aspects of the Readmission of Women into the Medical Profession," *Papers and Letters Presented at the First Woman's Congress of the Association for the Advancement of Women* (New York, 1874), p. 169.

24. David Montgomery, *Beyond Equality* (New York, 1967), p. 33; and Edith Abbott, *Women in Industry* (New York, 1912), p. 3.

25. Barbara Wertheimer, *We Were There, The Story of Working Women in America* (New York, 1976), pp. 163–166; and Montgomery, *Beyond Equality*, pp. 33–36. See also Shirley Dare, *Herald Tribune*, March 18, 1870; and "Among the White Slaves of New York City," *Woodhull and Claflin's Weekly*, September 17, 1870.

26. Elizabeth Dexter, *Career Women in America, 1776–1840* (New York, 1950), pp. 139–169.

27. Unpublished lecture on labor, Anna Dickinson Papers, Library of Congress.

28. "Women and the Kitchen," *Springfield Republican*, September 2, 1874.

29. Mary Rose Smith, *Catalogue of Charities Conducted by Women as Reported to the Women's Centennial Executive Committee* (New York, 1876), pp. 9–70.

30. Maud Muller, letter to the editor, *Woodhull and Claflin's Weekly*, May 24, 1870. See also, for historical background on the emergence of this association, Carroll Smith-Rosenberg, *Religion and the Rise of the City* (Ithaca, N.Y., 1970), p. 180; "Female Labor," *New York World*, November 2, 1873; and *Woodhull and Claflin's Weekly*, September 17, 1870, November 2, 1873, and February 12, 1873.

31. "Protection for Poor Women," *New Century*, May 26, 1876.

32. "Female Labor," *New York World*, November 2, 1873.

33. "Working Girls in Boston," *New Century*, September 2, 1876; and "Buffin's Bower," *New Century*, September 30, 1876.

34. See Elizabeth Hoxie, biographical sketch of Collins, in *Notable American Women* (Cambridge, Mass., 1971), edited by Edward T. James, Volume I, pp. 362–363.

35. "Working Girls in Boston," *New Century*, September 2, 1876.

36. On early industrial education see Smith, *Catalogue of Charities Conducted by Women*, p. 18; Smith-Rosenberg, *Religion and the Rise of the City*, pp. 221–222; and Thomas Woody, *A History of Women's Education in the United States* (New York, 1929), Volume II, pp. 75–84.

37. S. S. Packard, letter to the editor, *Revolution*, February 23, 1871; and "Telegraphy for Women," *Revolution*, January 7, 1869.

38. On the State Industrial School at Champagne, Ill., see Mary Safford Blake, *Woman's Journal*, November 28, 1874; and on the Maine Industrial School for Girls see *Woman's Journal*, February 6, 1875. The Illinois school was coeducational, with an enrollment of ninety women and three hundred men. See also Emma Marwedel, "A Horticultural School for Girls," *Herald of Health*, July, September 1869; and on international efforts to establish industrial, scientific, and art schools for women, see a report on the World Women's Convention in Berlin, Germany, "Home Intelligence," *Woman's Advocate*, July 1869.

39. "New York City," *Cincinnati Commercial*, May 14, 1869.

40. Parker Pillsbury, "Industrial Schools," *Revolution*, August 15, 1869; and "Training School for Girls," *Revolution*, July 15, 1869.

41. "Report on a Developing School and School Shop" (Boston, 1877), pp. 1–17.

42. "Women at Home," *Herald of Health*, June 1867. For a European position identical to this one, see the *Springfield Republican*, September 21, 1875, which reported an address by Professor von Scheel of the University of Berne, in Switzerland, on "Women and Education." The professor "attributed abstinence from marriage to the fact that a woman is less able than formerly to help her husband, on account of the extent to which corporate manufacture has superseded domestic industry. Hence to exclude women from any profession for which they prove themselves qualified is to inflict serious injury upon society." The most famous early American assessment along these lines came from Horace Bushnell in 1851. See his "The Age of the Homespun," *Litchfield County Centennial Celebration* (Boston, 1851), pp. 107–130. See also Theodore Parker, *Una*, August 1, 1855.

43. On these changes see, for example, "Decrease in Marriage," *Revolution*, October 5, 1871.

44. "The Wages Question," *Springfield Republican*, July 9, 1874.

45. On these matters see, for example, "Cause and Outcome," *Golden Age*, April 17, 1875; Lillie D. Blake, "Woman as Worker," *Golden Age*, April 3, 1875; and *Woman's Journal*, August 14, 1875.

46. "Book Notices," *Radical*, January 1868, p. 347.

47. On Bushnell see John Hooker, "Woman Suffrage and the Marriage Relation," *Revolution*, March 31, 1870. See also, for comments on single women, Mary Livermore, *What Should We Do with Our Daughters?* (New York, 1875); "The Wages of Women," *Woman's Journal*, July 8, 1874; and Virginia Penny, *Think and Act* (Boston, 1869), pp. 109–113.

48. "Female Missionaries," *New York World*, November 20, 1870. See also, on foreign missions, Page Smith, *Daughters of the Promised Land* (Boston, 1970), pp.

181–201; and on domestic missions see Smith-Rosenberg, *Religion and the Rise of the City*, pp. 125–159.

49. "Women's Work in the Church," *New York World*, November 10, 1871.

50. "Female Missionaries," *New York World*, November 20, 1870.

51. Ruddy, *Maiden Rachel* (New York, 1879), pp. 222–223

52. Diary entry, *Elizabeth Cady Stanton as Revealed in Her Letters, Diary and Reminiscences* (New York, 1922), edited by Harriet S. Blatch and Theodore Stanton, p. 195. See also, on feminist fascination with the Catholic sisterhoods, Caroline Dall, introduction to her journal, 1837, Dall Papers, Massachusetts Historical Society; "Missionary Fund," *Banner of Light*, July 6, 1867; Georgiana Bruce Kirby, *Years of Experience* (New York, 1887), pp. 182–183; *New Northwest*, June 7, 1872; and *New Century*, June 10, 1876. "That the practical use of sisterhoods," declared the *New Century*, "are part of the organization of the church of Rome, is hardly a valid objection for there can be no denying its working power." See also, on Catherine Beecher's admiration for the Catholic model, Katherine Sklar, *Catherine Beecher* (New Haven, 1973), pp. 172–176.

53. For examples of such advertising see the want ads of the *Cincinnati Commercial*, January 9, 1868, to May 1, 1868.

54. Dexter, *Career Women in America*, pp. 139–169.

55. *Springfield Republican*, April 7, 1875; J. G. G., "New York Correspondence," *New Century*, May 20, 1876; and "Woman's Kingdom," *Inter-Ocean*, February 17, 1877.

56. On Demorest see *History of Woman's Suffrage* (Rochester, N.Y., 1881), Volume I, p. 48; *New York World*, October 13, 1872; and "Madame Demorest," *Woodhull and Claflin's Weekly*, August 6, 1870. See also, for a biographical sketch of Demorest, Caroline Bird, *Enterprising Women* (New York, 1976), pp. 35–57.

57. *New York World*, October 13, 1872. For material on another tea company owned by a woman, see Susan King, "A Woman's Enterprise," *Springfield Republican*, October 31, 1871.

58. *College, Market and Court* (Memorial Edition, New York, 1914), pp. 224–228, 245, 363.

59. "A Revolution in Printing," *Springfield Republican*, May 3, 1875. For historical information on the revolution in printing see Helmut Lehmann-Haupt, *The Book in America* (New York, 1951), pp. 139–191; and Bernard A. Weisberger, *The American Newspaperman* (Chicago, 1961), pp. 64–121.

60. On these women see Susan Conrad, *Perish the Thought*; and Ruth Finley, *The Lady of Godey's: Sarah Josepha Hale* (New York, 1931).

61. On these women see *History of Woman's Suffrage*, Volume I, p. 48; *New Century*, June 3, 1876; "Woman's Kingdom," *Inter-Ocean*, February 17, 1877; and Madeleine Stern, *Purple Passage, The Life and Times of Mrs. Frank Leslie* (Norman, Okla., 1953).

62. "About Women," *New York World*, July 14, 1872.

63. Howe, circular on the "New Orleans Exposition," December 1, 1884, Julia Ward Howe Papers, Houghton Library, Harvard University.

64. On diseases and preventive hygiene see George Beard, "Great Subjects—Disease," *The Congregationalist*, March 27, 1882, and "Hysteria and Allied Affections," lecture at Long Island Hospital, June 1872, Beard Papers, Yale University Library; on professionalization see Burton Bledstein, *The Culture of Professionalism* (New York, 1976), pp. 31–32; and on traditions of female nurture see Sklar, *Catherine Beecher*.

65. See Mary Roth Walsh, "*Doctors Wanted: No Women Need Apply*": *Sexual Barriers in the Medical Profession 1835–1975* (New Haven, 1977), pp. 14–15; Mary Putnam Jacobi, "Social Aspects of the Readmission of Women into the Medical Profession," in *Papers and Letters*, p. 113, and "The Law and the Lady," *New Century*, October 7, 1876.

66. Walsh, p. 181; "Notes," *Ballot Box*, October 1881; "Feminine M.D.'s," *Inter-Ocean*, February 27, 1877; and "Medical Women," *Springfield Republican*, May 7, 1870.

67. *New York World*, January 15, June 11, 1871; and *New Century*, May 13, 1876.
68. *New York World*, December 25, 1870.
69. *New York World*, March 25, 1872.
70. "About Women," *Cincinnati Commercial*, February 27, 1880.
71. "About Business Women," *New Century*, May 20, 1876.
72. Dall journal entry, August 16, 1871, Dall Papers, Massachusetts Historical Society.
73. "Women in the University," *New Century*, October 7, 1876.
74. *Ballot Box*, May 1881.
75. Walsh, "*Doctors Wanted*," p. 84.
76. Mary Elizabeth Massey, *Bonnet Brigades* (New York, 1966), pp. 131–133, 340.
77. Ibid., p. 340; and "Woman and the Government Office," *Springfield Republican*, December 28, 1866.
78. "About Women," *New York World*, September 1, 1872, March 16, 1873, and January 18, 1874; and *Ballot Box*, May 1881.
79. On Illinois see *Springfield Republican*, December 2, 1870, and *New York World*, June 11, 1871; on Michigan, Iowa, and Missouri see *New Century*, October 7, 1876, "The New Era," *New Northwest*, December 12, 1873, and *Springfield Republican*, December 2, 1870; and on Minnesota and California see *New York World*, January 11, 1874, "Woman's Kingdom, *Inter-Ocean*, April 7, 1877, and February 22, 1879.
80. "Lockwood Bill," *Ballot Box*, February 1879.
81. *Inter-Ocean*, February 22, 1879. On Fuller's contention see Ruddy, *Maiden Rachel*, p. 154. The number of women lawyers did increase to a thousand by 1900. See Bledstein, *The Culture of Professionalism*, p. 120.
82. On this relationship see Morton Horowitz, *The Transformation of American Law* (Cambridge, Mass., 1977); and William E. Nelson, *The Americanization of the Common Law* (Cambridge, Mass., 1975).
83. On this educational ideology see Sklar, *Catherine Beecher*.
84. Massey, *Bonnet Brigades*, p. 130; and Sklar, *Catherine Beecher*, p. 180.
85. On the passage of these laws see *New Northwest*, December 12, 1873; *New York World*, April 20, 1873, June 21, 1874, and January 11, 1874; and *Ballot Box*, February 1880. On response by women see *New Northwest*, January 8, 1874; *New York World*, April 27, 1873; and *New Century*, October 28, 1876.
86. *New Northwest*, December 12, 1873; and *New York World*, June 21, 1874.
87. Rothman, *Woman's Proper Place* (New York, 1978), p. 59.
88. "Properly Regulated School Teachers," *Cincinnati Commercial*, February 15, 1880.
89. "Property Rights of Women," *North American Review* 99 (1864), pp. 34–64.
90. "Property Rights of Women," *North American Review* 99 (1864), pp. 63–64; and William Cord, *A Treatise on the Legal and Equitable Rights of Women* (1861), p. 603.
91. "The Legal Status of Women," *Woodhull and Claflin's Weekly*, May 21, 1870. See also John Proffat, *Woman Before the Law* (New York, 1874), pp. 60–64.
92. "Property Rights of Women," *North American Review* 99 (1864), p. 44.
93. Bishop, *Review of Commentaries on the Law of Married Women* (New York, 1871), p. 73.
94. See Suzanne D. Lebsock, "Radical Reconstruction and the Property Rights of Southern Women," *Journal of Southern History*, May 1977, pp. 196–215.
95. Duffey, *Relations of the Sexes* (New York, 1876), p. 56. See also Harriet Robinson, in *History of Woman's Suffrage* (Rochester, N.Y., 1886), Volume III, pp. 290–291; the introduction to *History of Woman's Suffrage*, Volume I, p. 14; and "How to Succeed," *Woman's Journal*, March 20, 1875.
96. *Springfield Republican*, January 13, 1871.
97. Charles Almy, Jr., and Horace Fuller, *The Law of Married Women in Massachusetts* (Boston, 1878), pp. vi–viii; and *Woman's Journal*, March 23, 1874.
98. "The Need for Women Lawyers," *Woman's Journal*, September 12, 1874.
99. Almy and Fuller, *The Law of Married Women*, p. viii.
100. On Nebraska see *History of Woman's Suffrage*, Volume III, pp. 241–242; on

Illinois see *History of Woman's Suffrage*, Volume III, pp. 570–571, and "Married Women in Illinois," *Woman's Journal*, September 6, 1873; on California see "Property Rights of Women in California," *New Northwest*, October 31, 1873, "Woman's Rights in California," *New York World*, February 6, 1873; on Oregon see "Married Woman's Property Bill," *New Northwest*, October 31, 1878, and *Inter-Ocean*, November 23, 1878; on Connecticut see *Ballot Box*, May 1877, and *Mrs. Lucy A. Allen's Appeal to the Legislature of the State of Connecticut for the Protection of Women* (New Haven, 1878); and on New Hampshire see *New Northwest*, August 4, 1871.

101. Bishop, *New Commentaries on Marriage, Divorce and Separation* (New York, 1891), p. 519.

102. *History of Woman's Suffrage*, Volume III, pp. 241–242; and *New York World*, February 6, 1873.

103. See Willard Hurst, *Law and the Conditions of Freedom in Nineteenth-Century United States* (New York, 1956), pp. 10–22; Nelson, *Americanization of the Common Law*, pp. 103–153; Horowitz, *The Transformation of American Law*, pp. 253–268, and especially the chapter called "The Triumph of Contract," pp. 160–209; and George E. Howard, *The History of Matrimonial Institutions* (New York, 1904), Volume III.

104. *History of Woman's Suffrage*, Volume I, p. 4.

105. Horowitz, *The Transformation of American Law*, p. 160, p. 265.

106. On the unrelenting expansion of grounds for divorce, see Howard, *History of Matrimonial Institutions*, Volume III, "Divorce Legislation," pp. 3–160.

107. Hurst, *Law and the Conditions of Freedom*, pp. 14, 22–23; and Horowitz, *The Transformation of American Law*, "The Rise of Legal Formalism," pp. 253–265.

108. "Property Rights of Women," *North American Review* 99 (1864), pp. 52–53. See also Lebsock, "Radical Reconstruction and the Property Rights of Women," *Journal of Southern History*, May 1977, 193–216; Peggy Rabkin. "The Origins of Law Reform: The Social Significance of the Nineteenth-Century Codification Movement and Its Contribution to the Passage of the Early Married Women's Property Act," *Buffalo Law Review*, Spring 1975, pp. 683–760; and Mary Beard, *Women as Force* (New York, 1946).

109. *History of Woman's Suffrage*, Volume I, p. 14.

110. Horowitz, *The Transformation of American Law*, p. 265; Joel Bishop, *Commentaries on the Law of Marriage and Divorce* (New York, 1852), pp. 32–41; and Cord, *A Treatise*, pp. 8–11, 116–117.

111. "Property Rights of Women," *North American Review* 99 (1864), pp. 52–53. See also Norma Basch, "In the Eyes of the Law: Women, Marriage and Property in Antebellum America," paper delivered at Berkshire Conference, August 1978; Proffat, *Woman Before the Law*, pp. 66–67; and Cord, *A Treatise*, pp. 137–138.

112. "Mrs. Broat's Brother-in-law," *New Century*, September 23, 1876.

113. *Woman's Journal*, May 10, 1873.

114. "Are We Worse than Our Grandmothers!" *Woman's Advocate*, July 1869, pp. 1–8.

115. Dall, *College, Market and Court*, pp. 351, 367. See also, for similar claims, Richard Hildreth, "A Word for Man's Rights," *Putnam's Monthly* 7 (Spring 1856), pp. 208–213.

116. "How to Succeed," *Woman's Journal*, March 20, 1875. See also speech of Mrs. O. E. Graves before the Illinois Woman's Suffrage Association, *Inter-Ocean*, March 3, 1877: "We have secured the repeal of many unjust statutes regulating property rights of women."

117. *Woman's Journal*, October 18, 1873.

118. "The Indiana Female Suffrage Convention," *Cincinnati Commercial*, June 10, 1869. See also *History of Woman's Suffrage*, Volume III, p. 570.

119. *History of Woman's Suffrage*, Volume III, pp. 483–484.

120. *History of Woman's Suffrage*, Volume III, p. 765.

121. John Hooker, *Some Reminiscences of a Long Life* (Boston, 1899), pp. 57, 67. A feminist, Hooker drafted this law.

122. Stanton, "Anniversaries," *Revolution*, May 21, 1868.

123. Many historians have studied this involvement, so I offer here, at the risk of simplifying a complex matter, only a summary account of it, See Eleanor Flexner, *Century of Struggle* (New York, 1972), pp. 131–141; Israel Kugler, "The Trade Union Career of Susan B. Anthony," *Labor History* 2 (Winter 1961), pp. 97–111; Montgomery, *Beyond Equality* pp. 387–424; and Ellen Dubois, *Feminism and Suffrage: The Emergence of an Independent Women's Movement in America* (Ithaca, N.Y., 1978), pp. 103–151. Dubois provides the best discussion to date.

124. Report of the Working Women's Association, *Revolution*, September 24, 1868. See also Dubois *Feminism and Suffrage*, pp. 126–141.

125. "Woman's Typographical Union," *Revolution*, October 15, 1868; and "The Female Labor Question," *Revolution*, October 22, 1868.

126. "The Platform of the NLU," *Revolution*, September 24, 1868.

127. Stanton to Beecher Hooker, c. 1872, Isabella Beecher Hooker Papers, Stowe-Day Memorial Library.

128. *Woodhull and Claflin's Weekly*, April 10, 1872.

129. Stanton, letter to the editor, *Golden Age*, April 27, 1871.

130. Quoted from the platform on principles, read by Mary F. Davis before the Working Women's Association, *Revolution*, November 5, 1868.

131. For feminist views on domestic servants in these terms, see Ruddy, *Maiden Rachel*, pp. 110–135; Lillie D. Blake, *Fettered for Life* (New York, 1874), pp. 30–32; Epes Sargent, *The Woman Who Dared* (New York, 1869), pp. 215–219; and Caroline Corbin, *A Woman's Secret* (New York, 1870).

132. See, for example, "Report of the Committee in Industrial Avocations," in *Proceedings of the Woman's Rights Convention* (1851), Worcester, NAWSA Papers, Library of Congress; and report of the Woman's Rights Convention in Massedon, Ohio, in the *Liberator*, June 18, 1852.

133. Dubois, *Feminism and Suffrage*, pp. 147–148.

134. Among them were Mary Putnam Jacobi, Kate Hilliard, Matilda Joslyn Gage, Paulina Wright Davis, Laura Curtis Bullard, Mary F. Davis, Ernestine Rose, Jane Swisshelm, Celia Burleigh, Charlotte Fowler Wells, Elizabeth Oakesmith, Elizabeth Stuart Phelps, Fanny Fern, Frances Gage, Eleanor Kirk, Mary Livermore, Tracy Cutler, and Anna Brackett. For material on the Sorosis and for lists of its members, see the "Minutes" of the New York City Sorosis, 1868–1874, Sophia Smith Library, Smith College. I might add here that both the NWSA and AWSA paid close attention to the meetings of the Sorosis and supported its activities. For an indication of the space devoted to it by the *Revolution*, see Sorosis reports, January 14, 21, 1869, February 4, 8, 1869, and October 7, 1869. The *Revolution* covered the work of the Sorosis more closely than any other contemporary newspaper.

135. *Revolution*, March 31, 1870.

136. "Minutes," New York City Sorosis, November 22, 1868. Sophia Smith Library, Smith College. "We cannot," Sorosis declared, "as an association, assume the care or management of the Association for Working Women."

137. See, for example, "Minutes," March 20, 1870, February 14, 1871, December 16, 1872, and October 6, 1873.

138. See, in particular, Alger's chapters, "Friendships of Women With Women" and "Pairs of Female Friends," in *The Friendships of Women* (Boston, 1870), pp. 266–281, 282–363.

139. "Sorosis," *Revolution*, March 31, 1870; and Croly, *For Better or Worse: A Book for Some Men and All Women* (Boston, 1876), pp. 61–68, 110–111, 146–147.

140. *New York World*, October 22, 1869.

141. Davis, letter to the editor, *Revolution*, July 29, 1869.

142. Quoted in *Historical Account of the Association for the Advancement of Women, 1873–1893* (New York, 1893), edited by Julia Ward Howe, pp. 5–6.

143. *New Century*, September 30, 1876.

144 "How Can Woman Best Associate?" unpublished speech before the 1873 Woman's Congress, Howe Papers, Houghton Library, Harvard University.

145. "Woman's Congress," *New York World*, October 13, 1873.

146. The presidents, vice-presidents, and directors included such women as Mary Putnam Jacobi, Antoinette Brown Blackwell, Sara Spencer, Abby May, Augusta

Cooper Bristol, Jane Croly, Kate Hilliard, Lita Barney Sayles, Mary Livermore, Mary F. Davis, and Anna Brackett. See Lita Barney Sayles, *History and Results of the Past Ten Congresses of the Association for the Advancement of Women* (New York, 1882), pp. 2–16; and *Association for the Advancement of Women Papers, 1874–1875* (Syracuse, N.Y., 1875), pp. 1–6, Vassar College Library.

147. *Revolution*, August 18, September 30, 1869.

148. *New Century*, July 15, 1876; *Inter-Ocean*, April 4, 1876; and Julia Ward Howe, diary entry, October 17, 1874, Howe Papers, Houghton Library, Harvard University.

149. *New York World*, December 18, 1870.

150. Diary entry, September 12, 1873, Howe Papers, Houghton Library, Harvard University. See also, for her interest in social science, diary entries, September 4, 1871, June 8, 1872, June 15, March 21, 1873, May 14, 1873.

151. Diary entries, May 22, 1875 and December 31, 1877, Howe Papers, Houghton Library, Harvard University.

152. Unpublished address before the AAW, 1882, Howe Papers, Houghton Library, Harvard University.

153. "The Law and the Lady," *New Century*, October 7, 1876.

154. "Social Aspects of the Readmission of Women into the Medical Profession," in *Papers and Letters*, p. 174.

155. *Papers and Letters*, pp. 17, 168–178; *New York World*, October 15, 1873; *Papers Heard at the Fourth Congress of Women* (New York, 1877), p. 12; Howe, *Historical Account of the Association*, pp. 9, 25; Lita Barney Sayles, *History and Results of the Past Ten Congresses*, pp. 17–18; and *Association for the Advancement of Women Papers, 1874–1875*, pp. 41–51, Vassar College Library.

156. On Hallowell see *Woman's Journal*, December 2, 1871

157. *New Century*, November 4, 1876.

158. *New Century*, July 1, 1876; Laura Curtis Bullard, "The Cause of Women in Italy," *Revolution*, February 2, 1871; and Julia Ward Howe, "Industries for Women," speech before the Woman's Department of the Boston Industrial Exhibition, New England Manufacturers' and Mechanics' Institute, 1883, and circular on New Orleans Exposition, 1885, Howe Papers, Houghton Library, Harvard University.

159. Unpublished speech on the recruitment of fashionable women, 1870, Stanton Papers, Book III, Library of Congress.

160. Howe, *Historical Account of the Association*, p. 9. For Dall's observations on Chace, see Dall to her father, March 2, 1874, Dall Papers, Massachusetts Historical Society.

161. Howe, *Historical Account of the Association*, p. 9; and "Social Sciences," *Inter-Ocean*, August 30, 1879.

162. "Woman's Congress," *New Century*, October 7, 1876.

163. *College, Market, and Court*, p. 224.

164. "Domestic Legislation," *Ballot Box*, September 1881.

CHAPTER 8. *The Sexual Division of Labor and Organized Cooperation for Public Life*

1. Quoted by Alfred Cridge, *Woodhull and Claflin's Weekly*, December 8, 1870.

2. "A New Year's Greeting," *Revolution*, January 5, 1871.

3. "The Apollo Hall Resolutions," *Revolution*, June 15, 1871. For similar statements see Lita Barney Sayles, "The Marriage Question," *Revolution*, February 17, 1870, and "Friendship Between the Sexes," *Revolution*, October 20, 27, 1870; "Minutes," New York City Sorosis, May 16, 1870, Sophia Smith Library, Smith College; Eleanor Kirk, *Up Broadway and Its Sequel: A Life Story* (New York, 1870), p. 196, 246–247, 267–268; Mary D. Davis, speech before the Rutland Convention in Vermont, *Banner of Light*, July 10, 1858; and Mary Clemmer Ames, *Eirene; or A Woman's Right* (New York, 1871).

4. *History of Woman's Suffrage* (Rochester, N.Y., 1881), Volume I, p. 752.

5. Blackwell to Stone, August 21, 1853, Blackwell Family Papers, Library of Congress; and *Woman's Advocate*, February 1869, pp. 94–97.

6. "The Conservative," *Woman's Advocate*, March 1869, pp. 197–202.

7. Croly, *For Better or Worse: A Book for Some Men and All Women* (Boston, 1874), p. 82. See also Thomas Higginson, *Women and Men* (Boston, 1888), a collection of essays written over a twenty-year period, pp. 105–106; Esther Hawkes, M.D., "Marriage—Its Sacredness," *Revolution*, February 10, 1870; and "The Woman Question," *Springfield Republican*, June 10, 1871.

8. For extensive treatment of this theme see Ronald Walters, *The Antislavery Appeal* (Baltimore, 1976), pp. 20–91; and Ellen Dubois, *Feminism and Suffrage: The Emergence of an Independent Women's Movement in America* (Ithaca, N.Y., 1978), pp. 45–48, 135–137.

9. "The Moral Aspect," *Revolution*, February 10, 1870. See also *Una*, January 1854; Laura Curtis Bullard, *Revolution*, September 15, 1870, and "The Slave Women of America," *Revolution*, October 6, 1870; Francis Barry, *Woodhull and Claflin's Weekly*, November 12, 1870; Stanton's review of *Ruth Hall* in *Una*, January 1, 1855; and C. L. James, letter to the editor, *Woodhull and Claflin's Weekly*, June 24, 1871.

10. "The Third Decade Meeting," *New Northwest*, August 15, 1878. See also "Where Are We Drifting?" *Ballot Box*, August 1877; and report of the thirtieth anniversary meeting, *Ballot Box*, August 1878.

11. *Woman's Journal*, October 31, 1874; and report of the First Woman's Congress, *Woman's Journal*, November 8, 1873. For typical discussions of domestic technologies see Harriet Beecher Stowe and Catherine Beecher, *The American Home; or Principles of Domestic Science* (New York, 1869); Mrs. Horace Mann, *Hearth and Home*, October 30, 1869; Jane Croly, *For Better or Worse:* and "Kitchen Aids," *New Northwest*, December 7, 1872. On flats, apartments, and boarding, see *Woman's Journal*, October 16, 1875; *Golden Age*, December 7, 1872; and Ella Giles Ruddy, *Maiden Rachel* (New York, 1879), pp. 9–12.

12. Stanton, "Book Notices," *Radical*, January 1868.

13. "Industrial Reconstruction," *Woman's Advocate*, January 1869.

14. "Liberty for Married Women," *New Northwest*, August 15, 1873.

15. *New Northwest*, July 5, 1872.

16. *Woman's Journal*, September 7, 1872. See also "Matrimonial Partnerships," *Woman's Journal*, April 3, 1875.

17. Davis, "Pecuniary Independence For Wives," *Una*, January 1854; Smith, *Woman and Her Needs* (New York, 1851), p. 45; and Ernst, "Rights and Wrongs of Women," *Antislavery Bugle*, March 25, 1853.

18. *History of Woman's Suffrage* (Rochester, N.Y., 1881), Volume I, p. 752; *New Northwest*, August 15, 1878; and *Ballot Box*, August 1878. On banking see *Springfield Republican*, October 22, 1872 and *New Century*, May 26, 1876. For attacks on husbands as "wage robbers" see *Ballot Box*, July, August 1877; Epes Sargent, *The Woman Who Dared* (New York, 1869), pp. 218–219; and Lillie Devereux Blake, *Fettered for Life* (New York, 1873), pp. 30–32. See also, for statements on the legalization of the household wage, *Woman's Journal*, July 9, 1870, and April 3, 1875; Lucy Stone, *Woman's Journal*, September 7, 1872; *New Northwest*, August 15, 1878; *Ballot Box*, August 1878; George Walker, "Labor vs. Money," *New Age*, October 21, 1876.

19. Willard, *Sexology* (Chicago, 1867), pp. 336–337; and Willard, "Bill of Rights for Women," *Revolution*, March 16, 1871.

20. "Financial Equality of the Sexes," *Woman's Journal*, April 21, 1875.

21. *Woman's Journal*, February 25, 1871; and "Mutual Justice," *Woman's Journal*, September 9, 1882.

22. "Woman and Peer of Man," *New York World*, January 23, 1871.

23. "The Need for Women Lawyers," *Woman's Journal*, September 12, 1874. Higginson believed that women lawyers should construct this plan and recommended Jane Slocum's thesis, "The Law of Coverture," written for the University of Michigan Law School. See also Higginson, "The Problem of Wages," *Woman's Journal*, March 18, 1870.

24. Blackwell to Stone, June 13, July 2, 1853. See also Blackwell to Stone, May 6, December 22, 1854, and April 2, September 17, 1855, Blackwell Family Papers, Library of Congress.

25. Blake, *Fettered for Life*, p. 379.

26. Sargent, *The Woman Who Dared*, p. 261.

27. Unpublished speech on work, Dickinson Papers, Library of Congress.

28. "Book Notices," *Radical*, January 1868, p. 347; and "Maternity," *Revolution*, February 10, 1870.

29. "The Coming Housewife," *Golden Age*, December 30, 1871. See also "Woman's Ability to Earn Money," *Woodhull and Claflin's Weekly*, September 17, 1870; and "The Coming Housewife," *New Northwest*, July 12, 1872.

30. "Minutes," New York City Sorosis, Sophia Smith Library, Smith College.

31. "Social Aspects of the Readmission of Women Into the Medical Profession," *Papers and Letters Presented at the First Woman's Congress of the AAW* (New York, 1874), pp. 168–178.

32. "The Relation of Woman's Work in the Household to the Work Outside," *Papers and Letters*, pp. 178–184. See also report on the 1873 Woman's Congress, *Woman's Journal*, November 8, 1873 which provided a complete transcript of Blackwell's speech.

33. *History of Woman's Suffrage*, Volume I, p. 728.

34. "Industrial Reconstruction," *Woman's Advocate*, January 1869, pp. 41–44.

35. *Woman's Journal*, May 20, 1874; Blackwell, *The Sexes Throughout Nature* (New York, 1875). See also Blackwell, "Sex and Work," *Woman's Journal*, April 25, June 22, July 4, 1874; and Blackwell, "Work in Relation to the Home," *Woman's Journal*, May 2, May 16, May 23, 1874.

36. *Woman's Journal*, May 2, 1874.

37. Ibid. Blackwell joined the AAW in 1873, served as vice-president in 1875 and 1878 and as the chairman of Committees on Topics and Papers in 1876. She presented papers at four successive congresses in New York, Boston, Syracuse, and Philadelphia. On the formal functions performed by Blackwell for the AAW, see *Association for the Advancement of Women Papers, 1874–1875* (Syracuse, N.Y., 1875) p. 3, Vassar College Library; and Lita Barney Sayles, *History and Results of the Past Ten Congresses of the AAW* (New York, 1882), pp. 12–17.

38. "Report on the Second Congress," *Woman's Journal*, October 31, 1874; *Papers Read at the Fourth Congress of Women* (New York, 1877), pp. 19–25; and *Association for the Advancement of Women Papers, 1874–1875*, pp. 27–35, Vassar College Library.

39. Anna Garlin Spencer to Ednah Cheney, February 10, 1887, New England Hospital Papers, Sophia Smith Library, Smith College. For a biographical sketch of Garlin see Louis Filler, *Notable American Women* (Cambridge, Mass., 1971), edited by Edward T. James, Volume III, pp. 331–332. See also Garlin's speech at the NWSA Boston Convention, *Ballot Box*, June 1881.

40. "The Organization of Household Labor," *Papers Read at the Fourth Congress of Women*, pp. 32–34.

41. "Women and Journalism," *Association for the Advancement of Women Papers*, pp. 41–51, Vassar College Library.

42. "Social Science," report of the second annual meeting of the Illinois Social Science Association, *Inter-Ocean*, October 3, 1879. See also the statements of two other social science women, Elizabeth Boynton Harbert and Julia Ward Howe. "The old idea," Harbert wrote for *Inter-Ocean*, "that literary and professional women were not good housekeepers or efficient mothers must have emanated from the brain of some prejudiced opponent, since a careful and thorough study of the facts proves the statement to be cruelly false" (July 19, 1879). Howe made the following diary entry after hearing Mary Carpenter, a noted prison reformer, speak disparagingly of public work for "mothers" before a social science prison congress: "she shut out mothers and true motherhood, an unfortunate exclusion" (July 13, 1872, Howe Papers, Houghton Library, Harvard University.

43. *Alpha*, January 1, 1878. See also, for other affirmative statements on two careers, *Woman's Journal*, April 3, April 21, 1875; *New Northwest*, April 3, 1874; and *New Century*, July 29, 1876. In an article called "On the Spindle Side," the editor of the *New Century*, Sarah Hallowell, wrote of Indian women who worked with their husbands in a factory built on an Indian settlement in Sault Sainte Marie,

Michigan. "By the protection that the factory system affords them," she wrote, "in solid wages which are paid to themselves, not to their lords, they have grown in self-respect, and are 'the most industrious of womankind.' In this settlement a marked change has taken place in their social relations; no longer the beasts of burden and drudges of their husbands, they live with the respect which their increased social importance demands." These words, of course, say more about Hallowell than they do about th Sault Sainte Marie Indians.

44. Unpublished paper on "organization," Howe Papers, Houghton Library, Harvard University.

45. On Fourierism see Jonathan Beecher and Richard Bienvenu, *The Utopian Vision of Charles Fourier* (Boston, 1971), especially the introductory essay, pp. 1–75. For Fourier's impact on reformist thought see Dolores Hayden, "Two Utopian Feminists and Their Campaign for Kitchenless Houses," *Signs*, Winter 1978, pp. 274–290; Lindsay Swift, *Brook Farm* (Boston, 1899); and Georgiana Bruce Kirby, *Years of Experience* (New York, 1887). The influence of the English socialist, Robert Owen, was also strong.

46. Unpublished paper on female labor, undated, Mary Putnam Jacobi Papers, Schlesinger Library, Radcliffe College.

47. "Reorganization of Society," *Una*, February 1855.

48. "Our Political Allies," *Woman's Journal*, January 20, 1872. See also Phillips's speech at Steinway Hall in Boston, *Woodhull and Claflin's Weekly*, January 6, 1872; Stanton, "Book Notices," *Radical*, January 1868, pp. 350—351; practically any major editorial by Stanton for the *Revolution* from 1868 to 1870; Anna Dickinson, "About Women," *New York World*, October 22, 1871; and Louisa Alcott, *Work* (New York, 1874).

49. *Banner of Light*, July 20, 1867. See also these articles in the *Banner of Light*, which indicate progressive spiritualists' support for cooperative ideas and ventures: "The Cooperative System," June 25, 1868; "Cooperation of Labor," July 13, 1867; "Centripetal and Centrifugal," December 25, 1867; "Cooperative Movement," February 19, 1870; and "The Tendencies of Civilization and Competition," May 1, 1869. Progressive spiritualists also contributed to the *New Age*, a journal designed, in part, to expound a new "cooperative" and "associationist" ideology. See, for example, "The Spiritualist Philosophy of Life," November 6, 1875; "Facts to the Front," November 6, 1875; and "No Progress But By Evolution," June 24, 1876. On the basis of these articles, and on the basis of my analysis of spiritualism in chapter 11, I disagree with R. Lawrence Moore's recent contention that spiritualists remained committed to "democratic individualism" throughout the nineteenth century. See Moore, *In Search of White Crows: Spiritualism, Parapsychology and American Culture* (New York, 1977), p. 100.

50. "A Woman's Cooperative Manufactory," *Woman's Journal*, April 2, 1870.

51. "Humanitarian League," *New York World*, June 6, 1872.

52. See Robert Fogerty's introduction to Marie Howland's novel *Familistère* (New York, 1975), "The Familistère: Radical Reform Through Cooperative Enterprise." Howland was the most important advocate of Godin's ideas in America. See Hayden, "Two Utopian Feminists and Their Campaign For Kitchenless Houses," *Signs*, Winter 1978, pp. 274–290.

53. "Notes and Comments," *Springfield Republican*, November 10, 1874. Stanton gave a series of lectures in 1871 called "Whom to Marry?" and in 1872 lectured before the American Reform League. See *Woodhull and Claflin's Weekly*, October 14, 1871, and May 8, 1872. For a biographical sketch of Bristol see *A Woman of the Century* (New York, 1893), edited by Frances E. Willard and Mary A. Livermore, pp. 123–124.

54. *The Cooperator*, August, September 1881.

55. See Norman Ware, *The Labor Movement in the United States, 1860–1895* (New York, 1929), pp. 320–333.

56. Campbell, "The Working Women of Today," *Report of the International Council of Women* (New York, 1888), assembled by the NWSA, pp. 146–153. Campbell provided a short history of the Sociologic Society.

57. See Howard Quint, *The Forging of American Socialism* (Columbia, S.C., 1953), pp. 79–83. Progressive spiritualists and Theosophists were also members of Nationalist Clubs. On Alice Stone Blackwell's interest in the Communist Party see B. V. to Blackwell, October 3, 1923, Blackwell Family Papers, Library of Congress.

58. Stowe, *Probate Confiscation and the Unjust Laws Which Govern Women* (Boston, 1876), 3rd ed., pp. 54–60; and Caroline Dall, journal entry, December 16, 1879, Dall Papers, Massachusetts Historical Society.

59. On the San Francisco Social Science Association see *Ballot Box*, April 1881. See also Imogene Fales's paper, "Cooperation," delivered before the 1880 Woman's Congress, in Lita Barney Sayles, *History and Results of the Past Ten Congresses of the AAW*, pp. 20–21. See also, for similar statements on the need for cooperation published by feminists throughout the 1870s, Elizabeth B. Harbert, "Cooperation—A Suggestion to Woman's Clubs," *Inter-Ocean*, February 15, 1879; Frederick Hinckley, "Cooperation," *New Age*, April 22, 1876; Hannah Shepard, "Cooperative Houses," *Revolution*, January 6, 1872; "The Coming Housewife," *Golden Age*, December 30, 1871; "Nationalization of Labor on the Basis of Equal Rights," *Woodhull and Claflin's Weekly*, November 18, 1871; "Labor and Capital," *Woodhull and Claflin's Weekly*, August 27, 1870; "Modern Housekeeping," *Woman's Journal*, July 9, 1870; "Cooperation" and "A Horticultural School for Girls," *Herald of Health*, September 1869; "Home and Hotel," *Golden Age*, October 10, 1874; "Glances Toward Social Organization; or, The Science of Society," *Alpha*, December 1, 1878; "Household Problems," *Springfield Republican*, June 3, 1873; "Kitchen vs. Something Else," *Springfield Republican*, November 26, 1874; "Cooperative Housekeeping," *New York World*, December 26, 1875; and "Where Are We Drifting?" *Ballot Box*, August 1877.

60. "Woman," *New Northwest*, November 7, 1873.

61. Speech before the NWSA Convention in Boston, *Ballot Box*, June 1881. See also Abby May's speech before the New England Woman's Club, *Woman's Advocate*, January 1869.

62. Quoted from Schindler's article, "Man and Woman," *Revolution*, December 30, 1871. See also, on Schindler, "Pittsburgh Woman's Suffrage Association," *Woman's Journal*, October 21, 1871; on Fuller, "Manliness and Womanliness," *Laws of Life*, September 1873; and on Sizer, "The Woman Question," *Herald of Health*, October 1869.

63. *College, Market and Court* (Memorial Edition, New York, 1914), pp. 213–214.

64. Blackwell, *The Sexes Throughout Nature* p. 96. See also Hester Pendleton, *Husband and Wife; or The Science of Human Development Through Inherited Tendencies* (Boston, 1863), pp. 113–115, 153–164; Eliza Duffey, *Relations of the Sexes* (New York, 1876), pp. 36–37, 100–101; Elizabeth O. Willard, *Sexology*, pp. 180–181; "Psychology of the Sexes—by Herbert Spencer," *New Northwest*, February 13, 1874; and "Woman and Science," *Radical*, March 1870, pp. 184–185.

65. See Julia Ward Howe, "Paternity," speech to the Fourth Woman's Congress, *New Century*, October 7, 1876; "That Depends," *New Century*, June 24, 1876; "Shall Men Nurse Babies?" *Revolution*, April 23, 1868; Antoinette Brown Blackwell, "Industrial Reconstruction," *Woman's Advocate*, January 1869; and "Warrington's Letters," *Springfield Republican*, April 5, 1875.

66. "Cooperation," *New Age*, April 22, 1876.

67. Bullard, "The Meaning of Marriage," *Revolution*, September 21, 1871; and Eliza Duffey, *Relations of the Sexes*, p. 295. See also "Equalization of the Sexes," *Woodhull and Claflin's Weekly*, August 1, 1874; Jane Croly, *For Better or Worse*, pp. 152, 187; John Neal, *Revolution*, December 2, 1869; and Richard Hume, "The Battle of the Sexes," *Woman's Advocate*, January 1869.

68. "About Women," *New York World*, July 14, 1872. For a sketch of Brackett's life see Norma Green, in James, editor, *Notable American Women*, Volume I, pp. 217–218.

69. "A visit to New York," Caroline Dall observed in the *New Age*, is "never complete to me without a visit to Anna Brackett's school. To hear a recitation con-

ducted by her is refreshing as a play. This time it was a lesson in Comparative Physiology, where the principal point was not to recite, but to think" (*New Age*, January 15, 1876). In a speech before the New England's Woman's Suffrage Association, Mary Livermore singled out Brackett's school as one of woman's "successes" and praised Brackett for "openly speaking of the old prejudices as nonsense" (*Inter-Ocean*, June 9, 1877).

70. See "Analysis of an article on Hegel," *Journal of Speculative Philosophy* 5 (1871), pp. 38–41; and "Margaret Fuller," *Radical* 2 (1872), pp. 25–36.

71. "Organization As Related to Civilization," *Association for the Advancement of Women Papers*, pp. 73–85, Vassar College Library.

72. Ibid., p. 73.

73. Putnam Jacobi, *The Value of Life* (New York, 1879), pp. 199–200, 208–209.

74. *Association for the Advancement of Women Papers*, pp. 75–76, Vassar College Library.

75. Ibid., pp. 78–79.

76. "Woman's Congress," *New York World*, October 18, 1873.

77. "Domestic Legislation," *Ballot Box*, September 1881. See also, for other statements on the theory of equivalence, Antoinette Brown Blackwell, *Woman's Journal*, June 20, 1985; Elizabeth Willard, *Sexology*, pp. 10–15, 207–208; and *Revolution*, February 5, 1868.

78. Stanton, "Sharp Points," *Revolution*, April 9, 1868.

79. "Domestic Legislation," *Ballot Box*, September 1881.

80. For Putnam Jacobi, Peirce, Garlin, and Brackett, see *Papers and Letters*, p. 172; *Papers Read at the Fourth Congress*, pp. 32–33, 35; and Brackett, *The Education of American Girls*, p. 28.

81. *Papers Read*, p. 35.

82. "Enlightened Motherhood," *Papers Read at the Fourth Congress*, pp. 12–13.

83. "Woman's Congress," *New York World*, October 18, 1873. These ideas were later incorporated into the opening pages of the first volume of the *History of Woman's Suffrage* (p. 21).

84. *College, Market and Court*, pp. 315–316.

85. Stanton, "Book Notices," *Radical*, January 1868, p. 351.

86. *Papers Read at the Fourth Congress*, pp. 32–34.

87. "About Women," *New York World*, August 11, 1872.

CHAPTER 9. *The Bee and the Butterfly: Fashion and the Dress Reform Critique of Fashion*

1. Annual Report of the National Dress Reform Association, in *Sibyl*, December 1863.

2. Elizabeth Boynton Harbert, "A Woman's Time," *Inter-Ocean*, April 24, 1880; and Harbert, "Woman's Kingdom," *Inter-Ocean*, August 30, 1879.

3. Herman Melville, *The Confidence Man* (New York, 1968), pp. 187–188.

4. See Frederick Pattee, *The Feminine Fifties* (New York, 1940), p. 266. A Wall Street lawyer published "Nothing to Wear" in *Harper's Weekly* in 1857. A reply to it, called "Something to Bear," appeared in *Home Journal*, May 30, 1857.

5. "Street Yarn," *Life Illustrated*, August 16, 1856; republished in *New York Dissected* (New York, 1936), edited by Emory Holloway and Ralph Adimari, pp. 128–131.

6. Pattee, *Feminine Fifties*, p. 274. See also a long article on "Siberian Gems," *Home Journal*, July 30, 1859.

7. Letter to the editor, unsigned, *Una*, December 1854. The attraction to color merged very easily with the contemporary attraction to the exotic (or to use Werner Sombart's or Mario Paz's equation, with the erotic). See, by way of illustration, William Prescott, *History of the Conquest of Peru* (New York, 1847), Volume I, pp. 25, 95–103, for rich descriptions of Indian costume; Nathaniel Willis, *Health Trip to the Tropics* (New York, 1853), pp. 63, 88–89, for descriptions of Haitian dress; George W. Curtis, *Nile Notes* (New York, 1851), pp. 128–132, for lush descriptions of

Arabian dress; and Bayard Taylor, *The Land of the Saracens* (New York, 1855), pp. 133–148, for voluptuous sketches of Middle Eastern landscapes, colors, and gems. "I revelled in a sensuous elysium," wrote Taylor, "which was perfect, because no sense was left ungratified."

8. Constance Rourke, *Trumpets of Jubilee* (New York, 1927), p. 167; Rossiter W. Raymond, "Mr. Beecher in Private Life," which appears in a book of collected essays on Beecher, *Henry Ward Beecher as His Friends Saw Him* (New York, 1904), p. 68; and Robert Shaplen, *Free Love and Heavenly Sinners* (New York, 1954), p. 24.

9. Beecher to Mrs. Jacob Murray, October 13, 1875, Isabella Beecher Hooker Papers, Stowe-Day Memorial Library. As Rourke, Raymond, and Shaplen point out, Beecher's interest in jewels began in the 1850s.

10. The reference to Willis as a "born shopper" appears in Henry Beers, *Nathaniel Parker Willis* (New York, 1885), p. 127; on Willis's peacocks see the *Home Journal*, May 18, 1859.

11. Elizabeth McClellan, *History of American Costume* (New York, 1927), p. 547; C. W. and P. Cunnington, *The History of Underclothes* (London, 1951), p. 127; and Martha Ash, "The Social and Domestic Scene in Rochester, 1840–1860," *Rochester History* 18 (April 1956), pp. 1–20.

12. Constance Rourke, *American Humor* (1959), p. 27; and James Parton, "The Clothes Mania," *Atlantic Monthly*, May 1869, pp. 531–548.

13. McClellan, *History of American Costume*, p. 547–548; and Cunnington, *The History of Underclothes*, p. 127.

14. Quoted in Arthur H. Cole, *The American Wool Manufacture* (New York, 1926), Volume I, p. 298. See also Parton, "The Clothes Mania," *Atlantic Monthly*, May 1869, p. 545.

15. Beers, *Nathaniel Parker Willis*, pp. 120–124. On Willis's admiration of the dandy see *Prose Writings of Nathaniel Parker Willis* (New York, 1885), edited by Beers, especially "A Dinner At Lady Blessington's," pp. 178–97.

16. *Home Journal*, March 12, September 18, 1859.

17. The quotations are from William G. Brooks, *Diary*, October 23, 1841, Brooks Papers, Massachusetts Historical Society. See also Pattee, *Feminine Fifties*, pp. 293–302; and McClellan, *History of American Costume*, p. 598.

18. McClellan, History of American Costume, p. 598. In a sense, Prescott does not fit here since he was wearing court dress, although his delight in his costume does identify him with an older generation of men. Indeed, Prescott was reputed to have a "large array of shoes for his own use." See *Dress Reform: A Series of Lectures Delivered in Boston* (Boston, 1874), edited by Abba Gould Woolson, p. 250.

19. E. L. Godkin, "The Great Dress Question," *Nation*, April 2, 1868.

20. "Editor's Easy Chair," *Harper's Monthly*, June, October 1853; "Male Bloomers," *Lily*, January 2, 1854, and "The Bee and the Butterfly," *Lily*, December 13, 1855; and Holloway and Adimari, *New York Dissected*, pp. 128–131. See also Esther Shepard, *Walt Whitman's Pose* (New York, 1938), p. 111. Sara Willis, otherwise known as Fanny Fern, used a good deal of her novel, *Ruth Hall*, (New York, 1856), to vilify one of the most "exquisite" gentlemen of the period, her brother Nathaniel, a "be-curled, be-perfumed popinjay . . . who wears fancy neckties, a seal ring on his little finger, and changes his coat and vest a dozen times a day" (pp. 134–135). Elsewhere she wrote, after making one of her jaunts along Broadway: "You should see the young men, with staring velvet vests and mock chains looped over them." If you don't believe me, she said, "take a walk down Broadway and see for yourself" ("Fresh Fern Leaves," *Sibyl*, September 15, 1856).

21. Genio Scott, "Interesting to Ladies," *Home Journal*, February 7, 1857. Scott was a curious and interesting figure, more widely known as the author of *Fishing in American Waters* (New York, 1869), a windy, often humorous study of fishing.

22. *Home Journal*, March 14, 1857; and "Mere Mention," *Home Journal*, February 13, 1858. See also Lucy Hooper, "Fig Leaves and French Dresses," *Galaxy* 18 (1874), p. 506; and McClellan, *History of American Costumes*, p. 463.

23. *Home Journal*, February 13, 1858.

24. "Mammoth Skirt Establishment," *Home Journal*, August 14, 1858; and Nathaniel Willis, "Letter From Idlewild," *Home Journal*, January 30, 1856.

25. "The Red Petticoat Connubially Whip-Up-Alive," *Home Journal*, February 13, 1858.

26. *Home Journal*, February 26, 1859.

27. "The French Silk Trade," *Harper's Bazaar*, April 10, 1869. The United States spent nearly sixty-five million francs on silk in 1860, while Britain spent nearly ninety-three million.

28. "Interesting to the Ladies," *Home Journal*, August 29, 1857.

29. Nathaniel Willis, "Letter From Idlewild," *Home Journal*, January 30, 1856.

30. George Ellington, *The Women of New York or the Underworld of the Great City* (New York, 1869), p. 25.

31. Ellington, *The Women of New York*, p. 37; and Mathew Smith, *Sunshine and Shadow in New York* (New York, 1868), *passim*.

32. "A Word on Fashions," *Home Journal*, February 26, 1859. For material on the spas and watering places see Willis, *Health Trip to the Tropics*, pp. 217–221; Harry Weiss, *The Great American Water-Cure Craze* (New York, 1967), pp. 122–126; and William Mangel, *The Outdoor Amusement Industry* (New York, 1952).

33. May Allison, *Dressmaking as a Trade For Women in Massachusetts* (Washington, 1916), pp. 11–16; James Parton, "History of the Sewing Machine," *Atlantic Monthly*, May 1867, pp. 527–544; Grace Rogers, *The Invention of the Sewing Machine* (New York, 1968), pp. 41–59; "The Sewing Machine Business," *New York World*, April 21, 1872; and Harry Corbin, *The Men's Clothing Industry, Colonial Through Modern Times* (New York, 1970), pp. 42–46. Parton wrote that the "whole number of sewing machines made in the United States up to the close of 1866 was about 750,000. During the quarter ending December 10, 1866, the number of machines made by licensed companies was 52, 219! This was above the rate of 200,000 per annum." The first Howe sewing machine was built in 1845, and in 1850 Isaac Singer built the first really popular machine.

34. Martha Wright wrote Lucretia Mott in 1851 that she thought the sewing machine would be "no great thing for home consumption as it can only stitch" (March 14, 1851, Garrison Family Papers, Sophia Smith Library, Smith College).

35. "Social Matters," *Cincinnati Commercial*, July 2, 1869. On the development of new forms of transportation see George Taylor Rogers, "The Beginnings of Mass Transportation in Urban America: Part I," *Smithsonian Journal of History* 2 (1966), and "Part II," 3 (1967); Alan S. Horlick, *Country Boys and Merchant Princes* (New York, 1975), pp. 33–34; William Fullerton, *The First Elevated Railroads in New York* (New York, 1936); Blake McKelvey, *The Urbanization of America* (New Brunswick, 1963), pp. 75–88; and Charles N. Glaab and A. Theodore Brown, *A History of Urban America* (New York, 1967), pp. 148–150.

36. "Springfield Business," *Springfield Republican*, September 22, 1874; "The Dry Goods Trade," *New York World*, March 21, 1873; "Fashion and Society," *Inter-Ocean*, May 27, 1876; and "Thread Manufacture," *Springfield Republican*, December 29, 1875. There had been, nevertheless, a major retrenchment in buying accompanied by heavy layoffs of shop and sewing girls. See "The Labor Distresses," *New York World*, November 12, 1873.

37. H. Pasdermadjian, *The Department Store: Its Origins, Evolution and Economics* (New York, 1954), pp. 3–4.

38. See Harry E. Ressequie, "Alexander Turner Stewart and the Development of the Department Store, 1823–1876," *Business History Review* 39 (1965), pp. 301–322; and Ressequie, "Alexander Turner Stewart's Marble Palace: The Cradle of the Department Store," *New York Historical Society Quarterly* 48 (April 1964), pp. 131–162. Another important historian of the department store, Ralph M. Hower, argues that American stores (especially Macy's) grew largely out of American conditions and did not faithfully imitate French models. See Hower, *History of Macy's of New York, 1858–1919* (New York, 1943), pp. 15–35 *passim*. Pasdermadjian makes the largest claims for the influence of Bon Marché.

39. Jane Croly, "New York and Paris Fashions for September," *Cincinnati Com-*

mercial, September 1, 1869; and Frances R. Sprague, "Stewart's Hotel for Women," *Cincinnati Commercial*, March 4, 1878.

40. Sprague, "Stewart's Hotel for Women," *Cincinnati Commercial*, March 4, 1878.

41. Ressequie, "A. T. Stewart," *Business History Review* 39 (1965), pp. 319–320; and Parker Pillsbury, "The Largest Store in the World," *Revolution*, September 3, 1868.

42. Ressequie, "A. T. Stewart," *Business History Review* 39 (1965), p. 319.

43. "A. T. Stewart vs. Rapid Transit," *New York World*, March 30, 1871; "Labor and Wealth," *New York World*, December 24, 1871; and "Rapid Transit," *New York World*, February 28, 1873.

44. "Rapid Transit," *New York World*, February 28, 1873.

45. Hower, *History of Macy's*, p. 160.

46. David Montgomery, *Beyond Equality* (New York, 1967), p. 34; and Hower, p. 193. It was not until later in the century that native American young women sought department store work in great numbers. Places like Stewart's were compelled to import many of their saleswomen. See *New York World*, March 25, 1872.

47. "A. T. Stewart vs Rapid Transit," *New York World*, March 30, 1871; and Ressequie, "A. T. Stewart," *Business History Review* 39 (1965), pp 301–322.

48. On vertical integration see Glen Porter and Harold C. Livesay, *Merchants and Manufacturers* (Baltimore, 1971), pp. 132–133. Although this book does not deal with dry-goods and department stores, it is of great use for understanding the evolution of marketing and mercantile trade. The book on nineteenth-century department stores has yet to be written.

49. Croly, "New York and Paris Fashions For September," *Cincinnati Commercial*, September 1, 1869.

50. "Shopping at Stewart's," *Hearth and Home*, January 9, 1869; and "Reflections of a Fashionable Girl," *Hearth and Home*, May 15, 1869.

51. "Reflections of a Fashionable Girl," *Hearth and Home*, May 15, 1869.

52. Karl Polyani, *The Great Transformation* (Boston, 1944), p. 40.

53. "A. T Stewart vs Rapid Transit," *New York World*, March 30, 1871.

54. Pasdermadjian, *The Department Store*, pp. 13–21; John Ferry, *A History of the Department Store* (New York, 1960), pp. 53–58; and Leon Harris, *Merchant Princes* (New York, 1979), p. xvii.

55. See Thomas Bender, *Toward an Urban Vision* (Lexington, Ky., 1975).

56. "Clerks and Salesmen," *New York World*, October 16, 1870; and Bertram Wyatt-Brown, *Lewis Tappan: The Evangelical War Against Slavery* (Cleveland, 1969), pp. 43–80.

57. Croly, "New York and Paris Fashions for September," *Cincinnati Commercial*, September 1, 1869.

58. For attacks on Stewart see Pillsbury, "The Largest Store," *Revolution*, September 3, 1868; "Letters from Mrs. Dall," *New Age*, April 29, 1876; and Chase, "The Tendencies of Civilization and Competition," *Banner of Light*, May 1, 1869. On monopolies and working-class conditions see the platform of the Equal Rights Party (Stanton, Gage, Woodhull, etc.), *Woodhull and Claflin's Weekly*, May 4, 18, 1872; "Army of White Slaves in New York," *Woodhull and Claflin's Weekly*, September 17, 1870; "The Saleswomen's Meeting," *Revolution*, July 21, 1871; "Mendacity of Business Women," *Revolution*, December 20, 1870; and "Social Science," *New Age*, October 14, 1876.

59. Wright to her sisters, October 8, 1868, Garrison Family Papers, Sophia Smith Library, Smith College.

60. Johnson to Dickinson, November 17, 1871, Dickinson Papers, Library of Congress.

61. *New Century*, August 12, 1876. See also "The Lone Woman's Bed and Board," *Golden Age*, April 8, 1871.

62. Sprague, "Stewart's Hotel For Women," *Cincinnati Commercial*, March 4, 1878; and "About Women," *Cincinnati Commercial*, April 13, 1878.

63. Croly, "New York and Paris Fashions for September," *Cincinnati Commercial*, September 1, 1878.

64. Ibid.

65. "Social Science," *New Age*, October 4, 1876; and "A. T. Stewart vs. Rapid Transit," *New York World*, March 30, 1871.

66. "Movement Against the System of Drumming," *Cincinnati Commercial*, January 6, 1878; and Corbin, *Men's Clothing Industry*, pp. 29–30.

67. Robert Twyman, *History of Marshall Field and Company, 1852–1906* (Chicago, 1954), p. 93.

68. Croly, "New York and Paris Fashions For September," *Cincinnati Commercial*, September 1, 1878.

69. Sprague, "Stewart's Hotel for Women," *Cincinnati Commercial*, March 4, 1878.

70. "Shopping at Stewart's," *Hearth and Home*, January 9, 1869.

71. Ibid.

72. *Home Journal*, October 1859.

73. "A New Dry Goods Palace," *New York World*, November 29, 1870; "Spring Opening at Lord and Taylor's," *Hearth and Home*, April 22, 1871; and Ferry, *A History of the Department Store*, pp. 35–51.

74. On Altman's, Stern's, Brooks, and other stores, see "Women's Clothes," *New York World*, October 20, 1872, and September 28, 1873; and "The Suits on the Brooks Brothers Men," *New York Times*, August 15, 1976. Brooks first opened in 1818 and subsequently became, like Stewart's, a major outfitter to the military ("The Suits," p. 29).

75. "Dry Goods Trade," *Woodhull and Claflin's Weekly*, November 12, 1870.

76. Hower, *History of Macy's*, pp. 69, 101–103; and "Women's Clothes," *New York World*, October 20, 1872.

77. "Letters of Mrs. Dall," *New Age*, January 15, 1876.

78. Hower, *History of Macy's*, pp. 166, 193–196. See also, on female buyers, J. L. G., "New York Letter," *New Century*, June 3, 1876. "Some of the largest buyers," the writer declared, "are women who represent firms like Macy's, Daniels and others."

79. "The Case of Mr. Macy," *Revolution*, January 5, 1871.

80. "The Clothing Trade," *New York Independent*, March 9, 1865.

81. See Michael H. Frisch, *Town into City, Springfield, Massachusetts, and the Meaning of Community, 1840–1880* (Cambridge, Mass., 1972), pp. 120–121.

82. "A Magnificent Clothing House," *Springfield Republican*, August 24, 1867.

83. Ibid. See also "Springfield's Business," *Springfield Republican*, September 22, 1874.

84. "A Magnificent Store," *Springfield Republican*, December 14, 1871; and "From Boston," *Springfield Republican*, January 2, 1874. See also Richard H. Edwards, Jr., *Tales of an Observer* (Boston, 1950), an in-house history of Jordan Marsh, reliable only for its dates, p. 5.

85. "From Boston," *Springfield Republican*, January 2, 1874; and "The Great Smuggling Case in New York," *Springfield Republican*, May 15, 1875.

86. Robert Twyman, *History of Marshall Field*, p. 23. See also Lioyd Wendt and Herman Kogan, *Give the Lady What She Wants* (New York, 1952).

87. "They Are Coming Sure," *Inter-Ocean*, August 8, 1876; and Twyman, pp. 51–53.

88. "A Feast of Finery," *Inter-Ocean*, October 2, 1879; and "A Mammoth Establishment," *Inter-Ocean*, March 27, 1880. See also Twyman, *History of Marshall Field*, p. 69. John Wanamaker bought out Stewart's New York store in 1896. See John Appel, *John Wanamaker* (Philadelphia, 1930), pp. 124–126. I have not discussed Wanamaker's in this chapter because it did not sell dry goods or women's clothes until 1877. After that time Wanamaker (like Macy and Stewart) also employed women in many of his departments. See Appel, *John Wanamaker*, pp 103, 343.

89. "A Feast of Finery," *Inter-Ocean*, October 2, 1879; and "A Mammoth Establishment," *Inter-Ocean*, March 27, 1880.

90. "Domestic Manufacture," *Cincinnati Commercial*, August 3, 1869; "The

Exposition of Textile Fabrics," *Cincinnati Commercial*, August 4, 5, 1869; and "Industrial Fair," *Cincinnati Commercial*, November 4, 1869.

91. "The Saleswomen's Meeting," *Revolution*, July 21, 1870; and "Saleswomen," *Revolution*, July 15, 1869.

92. "The French Silk Trade," *Harper's Bazaar*, April 10, 1869.

93. "Our Silk Industry Again," *Inter-Ocean*, March 23, 1877. On the popularity of silk see "The Holiday Season," *New York World*, December 18, 1870; "In Silk Attire," *New York World*, November 30, 1870; "The Dry Goods Trade," *New York World*, March 21, 1873; "Silk Attire," *Inter-Ocean*, October 10, 1879; and "The Great Smuggling Case in New York," *Springfield Republican*, April 6, 1873.

94. "The Communists of New York," *New York World*, December 16, 1872; and "Silk Making and Silk Mills," *New York World*, April 6, 1873.

95. For a wide-ranging discussion of this relationship see Ann Douglas, *The Feminization of American Culture* (New York, 1977).

96. Quoted in *Sibyl*, November 1, 1856.

97. "Spring Fashions," *Springfield Republican*, March 10, 1870.

98. "Clothes and Things," *Springfield Republican*, February 17, 1875.

99. "Les Tableaux Vivants," *Cincinnati Commercial*, May 19, 1872.

100. Appel, *John Wanamaker*, pp. 31–35.

101. "Winter Fashions," *New York World*, December 4, 1870.

102. "Fashion's Dictum," *Inter-Ocean*, September 27, 1879, See also "Fall Openings," *Springfield Republican*, October 5, 1871.

103. *New York World*, March 25, 1872.

104. "The Holiday Season," *New York World*, December 18, 1870.

105. Hower, *History of Macy's*, p. 69.

106. "Women Shopping," *Woman's Journal*, April 26, 1873.

107. Ella Giles Ruddy, *Maiden Rachel* (New York, 1879) p. 21.

108. "Domestic Manufacture," *Cincinnati Commercial*, August 3, 1869; and "The Exposition of Textile Fabrics," *Cincinnati Commercial*, August 4 and 5, 1869.

109. "Shopping at Stewart's," *Hearth and Home*, January 9, 1869.

110. Wendt and Kogan, *Give the Lady What She Wants*, pp. 33–34. The feminization of American stores occurred much earlier in the United States and France than it did in England. Émile Zola called the Paris Bon Marché "*une chapelle élevée au culte des grâces de la femme . . . une vaste enterprise pour la femme, il faut que la femme soit elévée a sa gloire, pour sa jouissance et son triumphe. La toute puissance de la femme, l'odeur de la femme domine le magasin.*" English stores were not really feminized until late in the nineteenth century. Even then, as Thomas Higginson pointed out in 1892, "Paris went to London for its men's clothing and its equipages." See Zola, "Ebauche de 'Au Bonheur des Dames,' " pp. 467–480, appended to *Au Bonheur des Dames*, edited by Maurice le Blond and found in *Les Oeuvres Complètes*, Volume II; Asa Briggs, *Friends of the People: The Centenary of Lewis's* (London, 1956), pp. 19–95; and Higginson, *Concerning Us All* (New York, 1892), p. 33.

111. Hower, *History of Macy's*, p. 193.

112. "A New Dry Goods Palace," *New York World*, November 27, 1870.

113. "Women's Clothes," *New York World*, October 20, 1872.

114. Ressequie, "A. T. Stewart," *Business History Review* 39 (1965), pp. 321–322.

115. Twyman, *History of Marshall Field*, p. 125; and Hower, *History of Macy's*, p. 160.

116. "The Daily Promenade," *Cincinnati Commercial*, November 1, 1869.

117. On the peddlers see Constance Rourke, *American Humor*, pp. 3–33; and Lewis E. Atherton, *The Pioneer Merchant in Mid-America* (New York, 1969) pp. 35–37.

118. "The Case of Mr. Macy," *Revolution*, January 5, 1871; and "Macy's Rookery," *Woodhull and Claflin's Weekly*, March 15, 1871.

119. Pendleton, *Husband and Wife; or The Science of Human Development Through Inherited Tendencies* (Boston, 1863), pp. 135–136.

120. "Female Thieves," *New York World*, December 25, 1870; "About Women,"

New York World, March 31, 1872; and "New York Shoplifting," *New York World*, March 3, 1872. According to the December 25 article, of the 73 women appearing in the rogues' gallery, 27 were shoplifters and 34 pickpockets.

121. See Mary Owen Cameron, *The Booster and the Snitch* (New York, 1964), pp. 159–172.

122. This position was formulated in a conversation with Elizabeth Fox-Genovese. The matter is obviously very complex and demands further study.

123. Frank Mayfield, *The Department Store Story*. (New York, 1949), p. 30; "The Clothing Trade," *Independent*, March 9, 1865; and Cole, *The American Wool Manufacture*, Volume I, pp. 139–146. The ready-made industry, however, would not dominate trade until the 1880s and 1890s.

124. "The Fashionable Colors," *Springfield Republican*, October 16, 1875; and "Corsets vs. Brains," *Woman's Journal*, March 1, 1873.

125. Quoted from *Harper's Bazaar* (New York, 1967), edited by Jane Trahey, p. 2. See also Russell Lynes, "Gaudy to Drab to Gaudy," *Harper's Magazine*, April 1951, pp. 43–48.

126. *Demorest's Monthly*, January 1866; and Trahey, *Harper's Bazaar*, p. 2; and Margaret Walsh, "The Democratization of Fashion: The Emergence of the Women's Dress Pattern Business," *Journal of American History*, September 1979, pp. 219–313.

127. Carrie Hall, *From Hoopskirts to Nudity* (New York, 1938), pp. 3–38; Hall wrote nostalgically about her first *Godey's*, which she received in the 1870s: "The excitement of its arrival was proof positive of the delight I felt in having a fashion magazine of my very own" (p. 7).

128. "New York and Paris Fashions For May," *Cincinnati Commercial*, May 1, 1869; and "Opening of Fall Fashions," *New York World*, September 22, 1871.

129. "Clothes and Things," *Springfield Republican*, February 17, 1875; and "Spring Fashions," *New York World*, March 21, 1871.

130. Mrs. Frank Leslie, quoted in Madeleine Stern, *The Purple Passage, The Life and Times of Mrs. Frank Leslie* (New York, 1953), p. 110.

131. "Fashion's Dictum," *Inter-Ocean*, September 27, 1879; "The Fall Openings," *Springfield Republican*, October 5, 1871; and "Fashion and Society," *Inter-Ocean*, May 27, 1876.

132. *New York World*, April 4, 1873.

133. *Woman's Journal*, February 2, 1870.

134. "Fashion and Society," *Inter-Ocean*, May 27, 1876.

135. For a discussion of the "cylindrical" or "tubular" style, see Bernard Rudofsky, *The Unfashionable Human Body* (New York, 1974), pp. 157–162. See also Quentin Bell, *On Human Finery* (New York, 1976), pp. 107–134.

136. Flugel, *The Psychology of Clothes* (London, 1930), p. 111. See also Carl Kohler, *A History of Costume* (New York, 1963), p. 412.

137. "On Females Growing Old," *Home Journal*, February 13, 1858; "Titcomb's Letters," *Home Journal*, September 18, 1858; and Willis, *Health Trip to the Tropics*, pp. 88–89, 296.

138. Croly, "The Art of Conversation," *Demorest's Monthly*, April 1866; Octavius Frothingham, "Women in Society," *Radical*, June 1867, pp. 598–610; and Elizabeth Boynton Harbert, "Conversation and Conversationalists," *Inter-Ocean*, April 14, 1877.

139. Nathaniel Hawthorne, *Blithedale Romance* (Boston, 1852), p. 34.

140. Willis, *Health Trip to the Tropics*, pp. 88–89. 296.

141. *Godey's*, February 1870.

142. Alfred Malden, "A Man's View of Woman's Dress," *Woman's Journal*, March 16, 1872.

143. H. A. Debille, "American Woman and French Fashions," *Harper's Magazine*, June 1867, pp. 118–120.

144. L. E. Furniss, "The Bondage of the Furbelows," *Appleton's Journal*, January 1873, pp. 75–78.

145. "The Rationale of Fashions," *Nation*, November 21, 1867.

146. "So far," declared the *Home Journal*, "fashion is a great moral benefit, by teaching the fallacy of relying on appearance and yet annihilating classes, so that all

men are equal at the forum and on change" (May 2, 1857). See also *Woman's Journal*, June 22, 1872.

147. "Taste and Dress," *Home Journal*, September 18, 1858.

148. *Godey's*, February 1872.

149. Hale, *Manners; or Happy Homes and Good Society All the Year Round* (New York, 1868), p. 39.

150. "New York Fashions," *Cincinnati Commercial*, January 2, 1878.

151. "Life Sketches," *Sibyl*, October 15, 1856; *Woodhull and Claflin's Weekly*, March 18, 1871; and "A Warning to the Ladies," *Inter-Ocean*, April 17, 1877. The literature on dress reform is not large. See, for example, Robert Riegel, "Woman's Clothes and Woman's Rights," *American Quarterly*, Fall 1963, pp. 87–98; and David Kunzle, "Dress Reform as Antifeminism," *Signs* 2 (1977), pp. 570–579.

152. "Fashion Falsities," *Sibyl*, November 1, 1856.

153. *Sibyl*, July 1863.

154. "Practical Experiences," *Sibyl*, August 28, 1857, and September 15, 1858.

155. "Our Dress," *Lily*, May, June 1851; and Gerrit Smith, letter to the editor, *Sibyl*, June 1, 1857.

156. "The Bloomer and Weber Dresses," *Water-Cure Journal*, August 1851.

157. "List of Dress Reformers," *Sibyl*, June 1858, July 15, 1859; and "Dress Reform Convention," *Sibyl*, July 1863.

158. "List of Dress Reformers," *Sibyl*, June 6, 1858, and July 15, 1859.

159. See Whitney Cross, *The Burned-Over District* (Ithaca, N.Y., 1950), pp. 64–75, 84–88.

160. "Mormonism and Free Love," *Sibyl*, March 15, 1858.

161. *Sibyl*, July 14, 1857; "Mormonism and Free Love," *Sibyl*, March 15, 1858; and "Secretary's Report of the Dress Reform Convention Held at Randolph, Crawford Co., Penn.," *Sibyl*, March 1, 1859.

162. Davis to Stanton, July 20, 1851, Stanton Papers, Library of Congress.

163. Jones, *Woman's Dress; Its Moral and Physical Relations* (New York, 1865), p. 22; and Woolson, *Dress Reform*, pp. vii–xiv.

164. *Sibyl*, December 15, 1855.

165. "Annual Dress Reform Convention," *Sibyl*, May 15, 1861. For other sectarian views see Smith, *Sibyl*, June 1, 1857; Mary Tillotson, address before the Third Annual Dress Reform Convention, *Sibyl*, August 1, 1858; James C. Jackson and Harriet Austin (both presidents of the National Dress Reform Association), *Sibyl*, June 15, 1860; Harriet Austin, "What Is Woman's Rights?" *Laws of Life*, November 1865; and on Mary Walker see Charles McCool Synder, *Dr. Mary Walker, The Little Lady in Pants* (New York, 1962).

166. J. E. B., "Louis Blanc," *Sibyl*, October 1, 1860. See also "Successful Solution," *Sibyl*, January 1, 1861; "Harmonial Healing Institute," *Sibyl*, April 15, 1860; Harriet Martineau, *Sibyl*, December 1, 1859. See also advertisements for the free-love books of M. Edgeworth Lazurus, *Sibyl*, June 15, 1860.

167. "Cayuga Dress Reform Convention," *Sibyl*, September 1, 1857.

168. See circulars distributed by the Friends of Human Progress, Waterloo, N.Y., April 28, 1862, and May 1869, Post Family Papers, University of Rochester Library; and "National Dress Reform Association," *Sibyl*, June 15, 1860.

169. Caroline Severance, "Humanity—a Definition and a Plan," *Una*, April 1, 1, 1853.

170. Stanton to Smith, unpublished letter, c. 1856, Garrison Family Papers, Sophia Smith Library, Smith College; and Smith, letter to the Syracuse Dress Convention, *Sibyl*, June 7, 1857.

171. Jackson, *American Womanhood* (Boston, 1870), p. 42.

172. "Reminiscence of Emily Collins," in *Feminist Papers*, edited by Alice Rossi (1974), p. 423.

173. Address to the 1858 Convention, *Sibyl*, August 1, 1858.

174. Editorial, *Sibyl*, August 15, 1856.

175. Quoted from a phrenological text by Tillotson, letter to the editor, *Sibyl*, October 1, 1858.

176. See Bloomer, "Woman's Dress," *Lily*, March 1851; Ellen Beard, letter to the editor, *Sibyl*, March 1, 1859; J. H. Stillman, "Male/Female Dress," *Sibyl*, March 1863; "Woman's Dress: A Cause of Uterine Displacement," *Water-Cure Journal*, November 1851; and "A Lecture on Woman's Dress," *Water-Cure Journal*, August 1851.

177. "Dress and Its Abuses," *Sibyl*, December 1, 1859.

178. "Errors and Fashions of Society," *Sibyl*, August 1863; and "A Man's View of the Reform Dress," Sibyl, August 15, 1859.

179. Jones, *Woman's Dress*, p. 11.

180. "Our Dress," *Lily*, May 1851. See also "A Word on Dress Reform," *Una*, December 15, 1856; and Amelia Mott Gummere, *The Quaker: A Study in Costume* (Boston, 1901), pp. 4–17.

181. "Errors and Fashions of Society," *Sibyl*, August 1863.

182. Wollstonecraft, *A Vindication of the Rights of Women* (Norton, 1967), p. 84.

183. "Practical Experience," *Sibyl*, September 17, 1858.

184. Stanton to Smith, January 5, 1851, Garrison Family Papers, Sophia Smith Library, Smith College.

185. On Tillotson's social science views see "Dress Reform," *New Age*, August 12, 1876, and on the Dress League see "American Free Dress League," *Woman's Journal*, April 25, 1874.

186. Letter to editor, *Sibyl*, October 1, 1858; and address, *Sibyl*, August 1, 1858.

187. Ward, *Dynamic Sociology* (New York, 1882), Volume I, pp. 607–608.

188. Ibid., pp. 655–656.

189. See Hasbrouck, *Sibyl*, October 1, 1856, and March 1, 1857, for an extreme statement of this position.

190. "Editor's Drawer," *Harper's New Monthly Magazine*, June 1853.

191. See John Doran, *Habits and Men and the Makers of Both* (New York, 1855), p. 10; and Sally Gerusha Stokes, "I Hate a Fool," *Godey's*, October 1851. *Lily*, December 13, 1855.

192. *Lily*, December 13, 1855.

193. "Errors and Fashions of Society," *Sibyl*, August 1863. See also, for another dress reform dialogue, "A Short Catechism," *Lily*, June 1, 1850, in which "fashionable fops" are called "trashy, mushroom excresences of humanity." For a similar statement see "Anecdotes of Fashion," *Una*, September 1854.

194. "What Is A Woman's Right?" *Laws of Life*, November 1868.

195. "Dress Question," *Woman's Journal*, July 20, 1873.

196. *History of Woman's Suffrage* (Rochester, N.Y., 1881), Volume I, p. 48.

197. "Manners and Behavior," *Demorest's Monthly*, June 1878.

198. Speech before the National Association of Woolen Manufacturers, *Cincinnati Commercial*, August 3, 1869; "Domestic Manufactures," *Cincinnati Commercial*, August 5, 1869; and *Woman's Journal*, January 29, 1870.

199. In spite of its admiration for Altman's, *Woodhull and Claflin's Weekly* frequently exposed such conditions, and so did *Revolution*.

200. Bullard, "Women and Raiment," *Revolution*, June 16, 1870; and Moulton, *Women's Journal*, September 20, 1873.

201. Anthony to Stanton, February 14, 1865, Stanton Papers, Library of Congress; and "Letter From Mrs. Dall," *New Age*, January 17, 1876.

202. On Putnam Jacobi see *New Century*, June 3, 1876; on Wells see Kate Gannett Wells, "A Defense of Fashionable Girls," *Woman's Journal*, May 24, 1873; on Howe and Moulton see Moulton, "Cloaks and Costume," *Revolution*, January 13, 1872 and *Woman's Journal*, September 20, 1873; and on Blake see "Woman's Kingdom," *Inter-Ocean*, January 18, 1879.

203. *New Century*, September 2, 1876.

204. Bullard, "Women and Raiment," *Revolution*, June 16, 1870.

205. Unpublished speech on the recruitment of fashionable women, 1870, Stanton Papers, Library of Congress.

206. "Evening Session," report of the NWSA St. Louis Convention, *Ballot Box*, May 1879.

207. Lita Barney Sayles, report of New York City Sorosis Meeting, *Inter-Ocean*, February 12, 1880.

208. Journal entry, April 21, 1862, Dall Papers, Massachusetts Historical Society.

209. On Dickinson's depressions see Chester Girard, *Embattled Maiden, The Life of Anna Dickinson* (New York, 1951), pp. 17, 105–106.

210. Ibid., p. 86, p. 109

211. For an indication of the great warmth these women felt for Dickinson, see their letters to her, Dickinson Papers. For the cited Anthony letter, see Anthony to Dickinson, July 12, 1867, Dickinson Papers, Library of Congress.

212. Girard, *Embattled Maiden*, pp. 110–111.

213. Invitation card from Tiffany, 1871, Dickinson Papers, Library of Congress; and Girard, *Embattled Maiden*, p. 87.

214. *Inter-Ocean*, August 16, 1880.

215. *Springfield Republican*, November 18, 1865. Dickinson was "entitled to the name of the Flamingo of woman's rights, for she appeared on the platform at the Music Hall . . . in a furious red gown."

216. Grays to Post, November 14, 1865, Post Family Papers, University of Rochester Library.

217. Journal entry, June 15, 1867, Dall Papers, Massachusetts Historical Society.

218. "Stage Dressing," *New York World*, September 27, 1872; and "Actresses' Dresses," *Inter-Ocean*, March 8, 1879.

219. Alice Hooker to Dickinson, c. 1872, Dickinson Papers, Library of Congress.

220. Girard, *Embattled Maiden*, pp. 87–88.

221. "Washington Letter," *Cincinnati Commercial*, January 15, 1878.

222. Stanton to Wright, 1873, Garrison Family Papers, Sophia Smith Library, Smith College.

223. "Letter From Washington," *Inter-Ocean*, February 12, 1880.

224. *Woman's Journal*, September 13, 1873.

225. "First Woman's Congress," *Inter-Ocean*, March 31, 1877; "Indiana Social Science Association," *Inter-Ocean*, January 25, 1879; and "Second Woman's Congress," *Woman's Journal*, October 13, 1874.

226. "Minutes," New York City Sorosis, May 19, November 17, 1873, January 19, February 2, 1874, Sophia Smith Library, Smith College; "Report of the New England Woman's Club," *Woman's Journal*, August 1, 1874; and Abba Gould Woolson, "Official Report of the Dress Committee of the New England Woman's Club," *Woman's Journal*, July 25, 1874.

227. "The Dress Question," *Woman's Journal*, July 26, 1873; and "Report of the New England Woman's Club—Dress Reform," *Woman's Journal*, August 1, 1874.

228. *Revolution*, June 10, 1869. See also Booth, *Harper's Bazaar*, October 10, 1869; Eliza Bisbee Duffey, *No Sex in Education* (Philadelphia, 1873), pp. 38–39, 40–41; and Faith Rochester (pseudonym of Frances Russell), "Dot and I—Boy and Girl," *Revolution*, March 17, 1870; Mary Tillotson, "Fashion," *Woodhull and Claflin's* Weekly, May 17, 1873; and Josiah Warren, *Woodhull and Claflin's Weekly*, April 3, 1872. Warren also described Robert Owen's approach to dress at New Harmony in the 1820s: "Mr. Owen believed that a uniformity of dress would have a tendency to allay jealousies and envy . . . and produce a feeling of equality among ourselves."

229. Runkle Calhoun to Dickinson, March 18, 1872, Dickinson Papers, Library of Congress.

230. See Frances B. Johnson, "Proof of the 'Development Theory' or Natural Selection," *Laws of Life*, November 1868; Booth, "Nothing to Do," *Harper's Bazaar*, February 20, 1869 and "The Wholesomeness of Work," *Harper's Bazaar*, February 27, 1869; Croly, "Manners and Behavior," *Demorest's Monthly*, June 1878; and Woolson, *Dress Reform*, p. 244.

231. "Magnificent Beasts and Magnificent Hussies," *Woodhull and Claflin's Weekly*, April 5, 1872. See also "Feminine Men and Masculine Women," *New Northwest*, May 5, 1871. I might add here that attitudes toward the dandy in the "highest" feminist circles were not always so critical. In the early 1880s, for example, Julia Ward Howe invited the "delightful" Oscar Wilde (a dress reformer in his own

right) to her home several times during his visit to the United States. See Deborah Clifford, *Mine Eyes Have Seen the Glory* (Boston, 1979), pp. 227–228.

232. Woolson, editor, *Dress Reform*, p. 34.

233. Ibid., pp. 149–155.

234. Ibid., pp. 157–162.

235. Ibid., pp. 172–173.

236. "The Dress Question—Business Suits for Business Women," *Inter-Ocean*, November 2, 1878.

237. "Washington Fashion and Society," *Cincinnati Commercial*, February 1, 1879.

238. *Woman's Journal*, June 22, August 15, September 25, 1874, October 17, December 7, 1874.

239. Woolson, editor, *Dress Reform*, p. xiv.

240. Ibid., p. 119. See also "Minutes," New York City Sorosis, June 16, 1873, Sophia Smith Library, Smith College; "Woman's Kingdom," *Inter-Ocean*, April 7, 1877; *Woman's Journal*, July 25, 1874; and Elizabeth Stuart Phelps, "Where It Goes," *New Northwest*, September 1, 1871.

241. "Masquerade Ball," *Independent*, June 13, 1867; and G. B. Willcox, "The Roman Pageant," *Independent*, July 11, 1867. "Now a masquerade," wrote Mr. Willcox, "is an arrangement by which a company of young people lay aside, for a time, their identity, and escape the judgement of men on their actions. . . . . It is said that the Ancient Athenians in the theatre, were at times almost thrown into a feenzy of excitement by the sight of the real eyes of the actor flashing through the false face upon them."

Part 3 / The Relation of Feminism to Social Practice

CHAPTER 10. *From Personal Disunion to Social Community: The Feminist Career of Caroline Dall*

1. Dall to Martha Choate, September 20, 1839, Dall Papers, Massachusetts Historical Society. All subsequent citations for this chapter which are not otherwise identified are from this collection. For recent accounts of Dall's life and thought see Stephen Nissenbaum, "Caroline Dall," in *Notable American Women* (Cambridge, Mass., 1971), edited by Edward T. James, Volume I, pp. 428–429; Barbara Welter, "The Merchant's Daughter: A Tale Told from Life," *New England Quarterly*, March 1969, pp. 13–22; and Susan Conrad, *Perish the Thought* (New York, 1976), pp. 162–182.

2. CD to Choate, September 20, 1839.

3. CD to Charles Dall, c. 1844. Caroline quotes her father in this letter.

4. CD to Charles Dall, c. 1844.

5. Journal entry, February 7, 1866.

6. Introduction to Dall's "reconstructed" journal, June 10, 1870. Contrary to Barbara Welter's view in "The Merchant's Daughter," *New England Quarterly*, March 1969, footnote p. 4, Dall's journals are not fully reconstructed or cleaned up, and often reflect badly on her; it was difficult, probably impossible, for her to alter her more than fifty journals, many commonplace books, and occasional notebooks.

7. Introduction to journal, November 1896.

8. On such women as Jane Croly and Mary Putnam Jacobi see chapter 6 of this study; on Emily Howland see Judith Colucci Breault, "The Odyssey of a Humanitarian, Emily Howland, 1829–1929," unpublished dissertation, University of Pennsylvania, 1974; on Matilda J. Gage, *Report of the International Council of Women* (Washington, 1888), p. 347; and on Clemence Lozier, *New Northwest*, June 29, 1872.

9. Journal entry, June 15, 1838.

10. Introduction to journal, November 1896.

11. Introduction to journal, November 1896.

12. CD to Charles Dall, February 15, 1843.
13. Introduction to journal, November 1896.
14. Journal entry, November 30, 1858.
15. Journal entry, September 25, 1865.
16. Journal entry, May 23, 1873.
17. Journal entry, September 23, 1879.
18. CD to Mrs. Gaskell, June 25, 1857. See also CD to George William Curtis, February 4, 1856.
19. Journal entry, November 22, 1862.
20. CD to Charles, February 22, 1843.
21. See chapter 9 of this study, p. 255.
22. CD to her mother, April 9, 1841.
23. CD to Mrs. Gaskell, June 25, 1857.
24. Commonplace notebook, March 1877.
25. Introduction to journal, January 20, 1897.
26. Journal entry, July 22, 1863.
27. Introduction to journal, January 20, 1897.
28. Journal entry, August 1836 and December 5, 1877.
29. Journal entry, September 23, 1867.
30. CD to Choate, July 26, 1839.
31. In this regard Charles Dall resembled such men as John Hooker, Grinfill Blake, Charles Ames, Moncure Conway, David Child, Richard Hildreth, and, to a point, Calvin Stowe, all liberal Protestants, all gentle and given to illness, and all respectful of their wives' abilities. For descriptions of the health, temperaments, religion, and sexual attitudes of these men, see John Hooker, *Some Reminiscences of a Long Life* (Hartford, Conn., 1899), pp. 19–44, 101–114, 280–295; Moncure Conway, *Autobiography* (New York, 1904), pp. 121–129, 287–290; Donald Emerson, *Richard Hildreth* (Baltimore, 1946), 120–148; Grinfill Blake, Katherine Blake, *Champion of Women, The Life of Lillie Devereux Blake* (New York, 1943), pp. 66–70. On David Child see Helene Baer, *The Heart Is Like Heaven* (Philadelphia, 1964), pp. 36–48, 95–124; on Calvin Stowe see Edmund Wilson, *Patriotic Gore* (New York, 1966), pp. 3–70, and Stowe's piece in *Hearth and Home*, "The Woman Question and the Apostle Paul," September 11, 1869; and on Charles Ames see *Herald of Health*, March 1871. A Unitarian minister and close friend of Dall's, Ames encouraged his wife to preach in his place whenever he was sick or away.
32. Charles Dall to parents, January 31, 1849. Barbara Welter mistakenly claims in "The Merchant's Daughter," *New England Quarterly*, March 1969, footnote p. 5, that no letters from Dall's husband exist in her papers. I found nineteen of his letters, most of them long, useful and all chosen by Caroline to shed the best light on her marriage. See Dall's own date list of Charles's letters, written December 5, 1879. Dall destroyed most of her husband's correspondence, but she could not completely edit her own journals, which bear constant witness to the fact that Charles resented her and that her marriage was very unhappy.
33. Charles to parents, October 20, 1850. See also Charles to his mother, c. 1849.
34. CD to Charles, May 1, 1843. It is important to note here that Dall thought Emerson's essay "Compensation" the finest thing he had ever written, while she considered "Self-Reliance" "extravagant" and "unsafe (journal entry, June 3, 1841)." More than Margaret Fuller, Dall feared her own destructive impulses too much to ever feel completely comfortable as a transcendentalist.
35. Journal entry, September 15, 1839.
36. CD to Choate, October 23, 1839.
37. CD to Choate, October 23, 1839.
38. CD to Parker, November 31, 1841.
39. Journal entry, April 25, 1866.
40. James, speech before New England Woman's Club, *Woman's Advocate*, January 1869, p. 37.
41. Journal entries, January 22, 1862, and April 24, 1868.
42. Journal entry, November 22, 1861.
43. Journal entry, June 9, 1866.

44. Journal entry, December 27, 1866.
45. Journal entry, October 17, 1863.
46. CD to A. J. Peabody, June 18, 1846; CD to Charles, June 11, October 1, 1843, and February 4, 1844.
47. Sadie Dall to William Dall, June 1, 1866. See also journal entries, June 1 and 11, 1866; and CD to Thomas Whitridge, October 19, 1861.
48. Journal entries, November 29, 1851, to July 1864.
49. CD to Charles, letter fragment, May 1843.
50. I would speculate here that Dall's struggle is representative of the struggles many feminists experienced in an effort to achieve a feminist position.
51. "Woman's Preacher's Convention," *Woman's Journal*, June 7, 1873.
52. Ann Douglas offers a different and one-sided view of the liberal clergy in her *The Feminization of American Culture* (New York, 1977), pp. 44–80.
53. "The Universalist Convention," *Springfield Republican*, September 24, 1870. See also *Woman's Journal*, July 30, 1870, October 12, 1872, and November 8, 1873.
54. "Revolutions Never Go Backward," *New Northwest*, August 6, 1875.
55. "Women and the American Unitarian Convention," *Woman's Journal*, June 18, 1870. On Unitarianism see also *Woman's Journal*, October 14, December 30, 1871, January 17, 1872, and November 21, 1874; on the other denominations see *Woman's Journal*, January 8, 1870, April 4, 1874, and "Reverend Ann Oliver," *Inter-Ocean*, May 11, 1880.
56. James, editor, *Notable American Women*, Volume I, pp. 407–408.
57. "About Women," *New York World*, June 4, 1871.
58. CD to L.T.S. Mumford, March 17, 1869.
59. Journal entry, September 7, 1868.
60. *Nazareth* (Boston, 1903), originally written in 1866; and *Egypt's Place in History* (Boston, 1868).
61. *Egypt's Place in History*, pp. vii, 20–21.
62. Journal entries, February 5, 1872, April 7, 1873.
63. Journal entry, January 21, 1877.
64. See "Syllabub," *New Age*, April 15, 1876; "Letters from Mrs. Dall," June 1, 1876; journal entry, February 28, 1868; and William Potter to CD, February 8, 1867, and May 7, 1870.
65. *College, Market and Court* (Memorial Edition, New York, 1914), p. xlvi, and CD to Anna Parsons, January 9, 1861.
66. Journal entry, January 19, 1878.
67. *College, Market and Court*, pp. 233–234.
68. On Dall's lecture plans see Samuel May to CD, January 3, 1856; on the Portsmouth antislavery society and fugitive slave relief, see CD to Parker, September 14, 1846. and CD to Ednah Cheney, October 26, 1862; and Dall's article in the *Liberator*, June 2, 1850.
69. Caroline Foster Healey to CD, September 25, 1855.
70. CD to Sumner, June 6, 1856.
71. On the increasingly conservative orientation of Garrisonianism and its connection, in some respects, with a developing industrial system, see Ronald Walters, *The Antislavery Appeal* (Baltimore, 1976), especially chapter 7, "Economy, Toward a Yankee South," pp. 112–128; James McPherson, *The Struggle for Equality* (Princeton, 1964); and David Montgomery, *Beyond Equality* (New York, 1967). The Post Family Papers at the University of Rochester offer further confirmation that abolitionists had considerable stake in a capitalist economy. Garrisonians Amy and Issac Post invested much of their money in Western Union (1867) and West Virginia silver mines (1871).
72. Journal entry, May 12, 1859.
73. "Breakers," *Woman's Advocate*, March, April, and August 1869.
74. Catt to Maud Wood Park, March 3, 1943, NAWSA File, Library of Congress.
75. Stanton, *Radical*, January 1868, p. 345; and C. Clark, "An Average Woman," *Woman's Advocate*, March 1869, p. 298. For a sample of Dall's contributions to *Una*, see her critique of Fanny Fern's *Ruth Hall*, January 1, 1855, her translation of

George Sand's *Spiridion*, July 1855 and on marriage laws, October 15, 1855. See also *A Practical Illustration of Woman's Right to Labor or a Letter from Marie E. Zakrzewska* (Boston, 1860) and *The Life of Anadabar Joshee* (Boston, 1888).

76. Welter, "The Merchant's Daughter," *New England Quarterly*, March 1969, pp. 12–13. Welter's description of the character of Dall's reform career is distorted by her heavy emphasis on the early writings; actually Dall, like most people, changed over the years. Welter tries too hard to place Dall within her own paradigm of "True Womanhood." Dall was really a complicated and conflicted woman, not the shallow and sentimental lady she seems to become in Welter's account. Welter quotes, for example, without citations, from Dall's *Essays and Sketches* (Boston, 1849), in which Dall clearly and rather haughtily disavowed any interest in overt feminism, although even here she insisted upon woman's right to a decent education and on her right to "obtain a livelihood for herself and her children without overtasking the generosity of man" (pp. 83–84). It is true, of course, that the young Caroline Dall had a sentimental side to her personality. She was also a romantic, a fact Welter fails to note. Both these tendencies gave way to a realistic rationalism that dominated her thought.

77. CD to Sedgewick, October 4, 1857.

78. *College, Market and Court*, pp. 6–26.

79. On coeducation see *College, Market and Court*, pp. 380–386, and on Antioch see Mrs. Horace Mann to CD, April 25, May 1, 1862.

80. *College, Market and Court*, pp. 156–157.

81. *College, Market and Court*, pp. 135, 178–179.

82. *College, Market and Court*, p. 157.

83. CD to Sedgewick, October 4, 1857.

84. *College, Market and Court*, pp. 224–225, 245, 363. See also Dall's "Studies toward the life of a 'business woman'—being conversations with Mrs. Rebecca P. Clarke," Dall Papers, Radcliffe College Library. James F. Clarke's mother, Rebecca Clarke, made a fortune in real estate after her husband died. Dall offered her as a model to other women.

85. *College, Market and Court*, p. 364.

86. *College, Market and Court*, p. 266; and CD to Channing, June 15, 1859. Dall also wrote in *College, Market and Court* that "society sets man free from every conceivable family duty, without a word. On the other hand, it binds women down to them with cords of iron, and is pitiless if a single one is snapped. . . . The life of the family does not rest on the mother . . . it belongs to men *and* women" (pp. 366–368).

87. *College, Market and Court*, pp. 208–209.

88. Journal entry, September 15, 1859.

89. Journal entry, May 14, 1859.

90. Journal entry, May 1, 1876.

91. Journal entry appended to an earlier entry, November 5, 1840.

92. Journal entry, November 5, 1840.

93. Elizabeth Peabody to CD, c. 1859.

94. Journal entry, September 6, 1859.

95. Journal entry, August 14, 1869.

96. John Worcester to CD, January 8, January 9, 1868.

97. Worcester to CD, January 9, 1868.

98. Journal entry, April 17, 1873.

99. Journal entry, August 31, 1871.

100. Journal entries, September 1, 2, 1879. In the second entry Dall made comparisons—in the manner of her reformist generation—between this distinguished group of "scientific men" and the "foul tide" of politicians who had just arrived in Saratoga Springs for a convention.

101. See *Historical Pictures Retouched* (Boston, 1860), pp. 10–11; and *College, Market and Court*. See also Susan Conrad, *Perish the Thought*, for a discussion of this kind of reconstruction by other educated women, pp. 108–120.

102. *College, Market and Court*, pp. 125–126, 138.

103. *College, Market and Court*, pp. 149, 154; and "Lodging Houses For Young Women," an address before the ASSA in New York, *The Friend*, December 1867, pp. 376–386.

104. CD to Rev. Wilson, November 29, 1859.

105. CD to George Tichnor, May 22, 1859. See also the *Liberator*, "Woman's Rights Meeting." June 10, 1859: "As one of the most encouraging signs of the times, Mrs. Dall mentioned the formation of the Society for the Cultivation of Social Science in England, a movement which has both men and women of highest social position and culture prominent among its leaders."

106. CD to Tichnor, May 22, 1859. See also "Memorial A, Concerning a Department of Social Science in the City Library."

107. CD to Tichnor, May 22, 1859.

108. Journal entry, October 4, 1873, and notebook II, July 1893.

109. "Social Science in Britain and America," *Springfield Republican*, October 7, 1867.

110. Journal entry, October 14, 1874.

111. "Lodging House for Women," pp. 377–379, 381–382. See also "Report of the Milk Committee," December 1865.

112. CD to Sanborn, September 24, 1865.

CHAPTER 11. *Feminism and Social Science*

1. Progressive feminist spiritualists existed in every part of the country from New York to San Francisco. See, on their meetings and feminist activities, "Annual Convention of Spiritualists," *Cincinnati Commercial*, June 21, 1869; "Secretary's Report of the Spiritualist Grove Meeting Held at Woodburn," *New Northwest*, October 24, 1873; "Spiritualist Camp Meeting," October 17, 1871; *Voice of Truth*, December 14, 1877; and "From Materialization," *New Northwest*, November 9, 1877.

2. All three of these women wrote articles for the *Banner of Light*, the major organ of spiritualism. See, for example, Sayles, "Non-Immortality and Re-Incarnation," May 14, 1870; Davis, "Advantages of the Children's Lyceum System," October 23, 1869; and Chandler, August 20, 1870.

3. For historical material on the Friends of Human Progress, see the following circulars in the Post Family Papers, University of Rochester Library: "To Friends of Spiritual Progress in Western New York," from J. W. Seaver, president of the general association of Spiritualists; circulars from Waterloo, N.Y., April 28, 1862, May 1869, and May 16, 1871. On the National Liberal League see Sidney Warren, *American Free Thought* (New York, 1943), pp. 34–35, 162–168, and *passim.*

4. "A Growl from Our Friends," *New Northwest*, July 9, 1875.

5. *Banner of Light*, October 4, 1867.

6. For a thumbnail sketch of progressive spiritualism, see T. Chapman to Amy Post, November 28, 1874: "I would just say with respect to the spiritual movement here, it progresses just in ratio to the growth, development and harmony of the inhabitants where the ranking discord of sectarian bigotry and religious hallucination have so long had ascendancy over common sense and scientific reason" (Post Family Papers, University of Rochester Library). For historical accounts of spiritualism see R. Lawrence Moore, *In Search of White Crows: Spiritualism, Parapsychology and American Culture* (New York, 1977); Geoffrey Nelson, *Spiritualism and Society* (New York, 1969); and Frank Podmore, *Mediums of the Nineteenth Century* (London, 1963).

7. Colman, *Reminiscences* (Boston, 1891), pp. 25–26. See Colman also for an account of the conflict between upper-class Radical Club types and lower-middle-class reformers in the Liberal League, pp. 79–83.

8. Nelson, *Spiritualism and Society*, pp. 27–28; *Woodhull and Claflin's Weekly*, November 2, 1872; and "Proceedings of the Annual Convention of the American Association of Spiritualists," *Woodhull and Claflin's Weekly*, October 18, 1873.

9. "Spiritual Reformation," *Banner of Light*, September 17, 1857. See also *Banner of Light*, June 29, December 25, 1867.

10. Selden Finney, circular of the Michigan State Spiritual Association, August 9, 1866, Post Family Papers.

11. Colby, "Latter-Day Politics," *Banner of Light*, March 25, 1867.

12. "The Queer Philosophers," *New York World*, October 6, 1870.

13. *Woodhull and Claflin's Weekly*, June 8, 1872.

14. On the emergence of organization see *Banner of Light*, June 29, September 28, 1867, and June 5, 1869; and William Potter, "Orderly and Disorderly Spiritualism," published circular, December 21, 1868: "Probably more Societies and Lyceums have been organized in the last eighteen months than during the whole previous eighteen years. One hundred and twenty societies have been organized in the single State of Michigan" (Post Family Papers, University of Rochester Library).

15. Lois Waisbroker to Amy Post, April 25, 1867, Post Family Papers, University of Rochester Library.

16. *Banner of Light*, November 13, 1869.

17. *Banner of Light*, July 20, 1867.

18. *Banner of Light*, July 13, 1867; and "State Organization," *Banner of Light*, April 20, 1867.

19. *Banner of Light*, May 4, 1867; and Davis, *Children's Lyceum Manual* (New York, 1868), first published in 1865.

20. *Banner of Light*, May 11, June 22, August 3, August 10, 1867, and January 18, April 10, 1869.

21. *Banner of Light*, July 3, 1869.

22. Davis, *Children's Lyceum Manual*, pp. 5–39. On spiritualist enthusiasm for lyceum work, see L. Scott to Amy Post, December 27, 1868, about the Philadelphia lyceum: "I deem it the most necessary work and a work of the greatest importance that the age demands—The work of setting the minds of children free or rather allowing them to remain free and so activating and expanding the mind so that it will not get warped or befogged by any dogma. . . . The *salvation of the Nation depends upon* the training of *youth* and *children* of our land" (Post Family Papers, University of Rochester Library).

23. Davis, *Children's Lyceum Manual*, pp. 32–36.

24. Mary F. Davis, "Advantages of the Children's Lyceum System," *Banner of Light*, October 23, 1869.

25. *History of Woman's Suffrage* (Rochester, N.Y., 1886), Volume III, p. 535.

26. On King see "Woman's Present and Prospective Sphere," *Banner of Light*, April 2, 1870; on Emma Hardinge see "What Our Women Can Do," *Banner of Light*, June 22, 1867; and on Palmer see *Revolution*, May 20, February 11, 1869, and *Springfield Republican*, April 27, 1875. Palmer was president of the NWSA Convention in New York City in 1870.

27. *Banner of Light*, May 13, 1871.

28. Nathaniel Potter to Amy Post, August 18, 1869, Post Family Papers, University of Rochester Library. See also Warren Chase, *The Fugitive Wife: A Criticism of Marriage, Adultery and Divorce* (Boston, 1866): "*Our whole system of training boys,* and mostly girls, is defective, in early life, except in a few families (mostly Spiritualist). In society, in early life, we cultivate exclusively in the boys, the intellect and passions, and crush out the affections as weaknesses, and this almost entirely unfits them for social or married life. In females Nature has planted the affections deeper and stronger, and it is not so easy to root them out. They are better prepared for marriage" (p. 13).

29. Isaac Post, notes in response to a friend's letter, undated, Post Family Papers, University of Rochester Library.

30. See Mary Furner, *Advocacy and Objectivity: A Crisis in the Professionalization of American Social Science, 1865–1905* (Lexington, Ky., 1975), pp. 1–2; and Thomas Haskell, *The Emergence of Professional Social Science: The ASSA and the Crisis of Authority* (Urbana, Ill., 1977), pp. 93–100.

31. Sanborn to William T. Harris, April 10, 1874, Sanborn Papers, Concord Free Public Library. See also Furner, *Advocacy and Objectivity*, pp. 26–27; *Journal of Social Science* 11 (May 1880), pp. 26–37; and *Springfield Republican*, July 20, 1870.

32. Other members were George Snelling, Elizur Wright, Samuel Bowles,

Gamaliel Bradford, Franklin Sanborn, Amassa and Francis Walker, George W. Curtis, Charles L. Brace, Andrew D. White, Enoch Wines, Dorman Eaton, Francis Bird, J. W. Hoyt, Samuel G. Howe, and E. L. Godkin. These names are listed in *Documents Published by the American Association for the Promotion of Social Science* (Boston, 1866), Volume I, pp. 4–70; and Furner, *Advocacy and Objectivity*, pp. 17–19.

33. *Springfield Republican*, December 7, 1865, October 22, 1866, and October 13, 1867.

34. See James Mohr, *The Radical Republican and Reform in New York During Reconstruction* (Ithaca, N.Y., 1973); essays by Felicio Bonadio, Richard Abbott, Philip Swenson, and George M. Blackburn in *Radical Reconstruction in the North* (Baltimore, 1976), edited by James Mohr; William Gillette, *The Right to Vote: Politics and the Passage of the Fifteenth Amendment* (Baltimore, 1965); James McPherson, *The Struggle for Equality* (Princeton, 1964); David Montgomery, *Beyond Equality* (New York, 1967); and Barbara Rosenkrantz, *Public Health and the State, Changing Views in Massachusetts* (Cambridge, Mass., 1972).

35. See Michael Frisch, *Town into City* (Cambridge, Mass., 1972), chapter 10, pp. 202–218 and *passim.*

36. "New England's Women's League," *Springfield Republican*, January 17, 1865. These men were George Snelling, Franklin Sanborn, and Elizur Wright.

37. These people included Isaac Ray, William T. Harris, Mary Eliot Parkman, Henry Barnard, Daniel Coit Gilman, David Lincoln, David Wells, Amassa and Francis Walker, Edward Atkinson, Henry Carey, Octavius Frothingham, Carroll Wright, Theodore Dwight, Theodore Woolsey, and Francis Lieber.

38. See David Brion Davis, *The Problem of Slavery in the Age of Revolution* (Ithaca, N.Y., 1975), pp. 237–240; Alan S. Horlick, *Country Boys and Merchant Princes* (Lewisburg, Pa., 1975), passim; Edward C. Kirkland, *Dream and Thought in the Business Community, 1860–1900* (Chicago, 1964), pp. 15–17; and for the Radical link to industrial capital see Kenneth Stampp, *The Era of Reconstruction* (New York, 1965). pp. 95–100.

39. On Villard see Haskell, pp. 115, 116–121; on Bradford, see *Springfield Republican*, February 26, 1875 and Van Wyck Brooks' preface to the *Journal of Gamaliel Bradford, 1883–1932* (New York, 1933), pp. vii–xiii; on Wright see *Springfield Republican*, November 9, 1875; and on Atkinson see Harold Williamson, *Edward Atkinson* (New York, 1934).

40. Frisch, *Town into City*. Frisch, historian of Springfield, Mass., calls the paper "a reasonably accurate barometer of public opinion" (p. 55).

41. On Bowles see "The Phelps Libel Suit," *Springfield Republican*, May 1, 1875; on Walker, "Death and Amassa Walker," *Springfield Republican*, October 30, 1875; and *Springfield Republican*, October 30, 1875.

42. "The American Public Health Association," *Sanitarian* 1 (July 1873), p. 405.

43. "The Social Sciences, Their Growth and Future," *Concord Pamphlets* (Concord, Mass., 1885), p. 1.

44. Sanborn to Harris, November 18, 1873, Sanborn Papers, Concord Free Public Library.

45. "The Social Science Meeting At New Haven," *Springfield Republican*, October 22, 1866.

46. Scovel, letter to the editor, *Revolution*, August 5, 1869. Scovel was president of the New Jersey State Senate in the early 1870s. See *New York World*, April 5, 1873.

47. H. M. Darlington to Thomas Higginson, September 7, 1854, Dall Papers, Massachusetts Historical Society.

48. Horatio Alger to Lucy Stone, September 22, 1856, Blackwell Family Papers, Library of Congress. On Young Men's Suffrage Leagues, see *Golden Age*, August 16, 1874; *New Northwest*, February 26, 1875; and Martha Wright to Ellen Wright, December 22, 1869, Garrison Family Papers, Sophia Smith Library, Smith College.

49. *Ballot Box*, May 1879. See also Gage to Martha Wright, December 29, 1870, in which she expresses opposition to making men presidents of feminist societies, Garrison Family Papers, Sophia Smith Library, Smith College.

50. *Revolution*, June 3, 1869. See also Stanton and Woodhull to Isabella B. Hooker, December 28, October 19, 1871, Isabella Beecher Hooker Papers, Stowe-Day Memorial Library.

51. *Ballot Box*, November 1879.

52. See, on male executive participation, *Woman's Journal*, January 15, January 29, February 19, March 18, 1870, February 25, October 21, 1871, December 2, 1871, May 16, 1874; *Woman's Advocate* (February 1869), pp. 112–113; *New Northwest*, May 5, 1871, and October 22, 1875; and *History of Woman's Suffrage*, Volume III, p. 594.

53. On the Vermont Woman's Suffrage Association, school committees and the AWSA, see *History of Woman's Suffrage*, Volume III, pp. 351, 382–390, 830.

54. Reference to all these couples can be found scattered throughout the *History of Woman's Suffrage*, with the exception of the Campbells and Jacobis. On these couples see *Springfield Republican*, December 26, 1875; and Linda Gordon, *Woman's Body, Woman's Right* (New York, 1976), pp. 172–173.

55. *New York World*, February 25, 1871.

56. Higginson, *Cheerful Yesterdays* (Boston, 1898), p. 35.

57. Ibid.

58. These men also included Theodore Tilton, Andrew Jackson Davis, Moncure Conway, William Lloyd Garrison, William Gannett Channing, and Henry Ward Beecher.

59. Quoted in Tilden Edelstein, *Strange Enthusiasm* (New Haven, 1968), p. 55.

60. Edelstein, *Strange Enthusiasm*, pp. 52–57.

61. *Banner of Light*, June 25, 1857, and December 15, 1858.

62. See Edelstein, *Strange Enthusiasm*, p. 5.

63. Ibid.

64. Mary Higginson, *Thomas W. Higginson* (Boston, 1941), p. 56.

65. Higginson, *Cheerful Yesterdays*, p. 9.

66. Mary Higginson, *Thomas W. Higginson*, p. 56.

67. Mary Higginson, *Thomas W. Higginson*, p. 7; and Higginson, *Cheerful Yesterdays*, pp. 16–17.

68. For other references to his mother see "The Invisible Mothers," *Woman's Journal*, May 13, 1882; *Contemporaries* (Boston, 1899), pp. 1–2; *Concerning Us All* (New York, 1892), pp. 186–187; and "Home and Its Defenders," *Woman's Journal*, July 21, 1883.

69. Blackwell, *Report of the International Council of Women* (Washington, 1888), pp. 335–336.

70. Higginson, *Outdoor Papers* (Boston, 1863), pp. 34–35. Vain enough about his own manhood, Higginson pasted published descriptions of himself in his diary. See, for example, diary entry for January 17, 1871: "On the other side of the room is Thomas Wentworth Higginson, lusty and healthy in body; not less rugged mentally than physically; a fighter who strikes with Cromwellian directness, directly at the heart of his antagonist's thought" (Diary, Higginson Papers, Harvard University Library). Higginson's fascination with his own image resembled Whitman's, although he would have been the last man to appreciate the resemblance. See also Higginson's *Travellers and Outlaws* (Boston, 1889), especially the chapter on the maroon of Surinam.

71. See Higginson, *Army Life in a Black Regiment* (Boston, 1962), first published in 1869, pp. 4, 16–17, 44, 83, and *passim;* and *Cheerful Yesterdays*, pp. 152–162.

72. Higginson, *Army Life in a Black Regiment*, chapter 6, "Night in the Water," pp. 152–156.

73. Higginson, *Oldport Days* (Boston, 1873), pp. 246–249. See also Edelstein, *Strange Enthusiasm*, pp. 95, 162; Mary Higginson, *Thomas W. Higginson*, p. 124; and *Outdoor Papers*: "There is, or ought to be, in all of us a touch of untamed gypsy nature, which should be trained, not crushed. We need, in the very midst of civilization, something which gives a little of the zest of savage life. . . . Physical exercises give to energy and daring a legitimate channel, supply the place of war, gambling, licentiousness, highway robbery, and office seeking" (pp. 137–138).

74. Higginson, *Cheerful Yesterdays*. pp. 18–19, and *Outdoor Papers*, p. 30. See also his description of the Victoria Regia Water Lily: "Can this be the virgin Vic-

toria—this thing of crimson passion, this pile of pink and yellow, relaxed, expanded, voluptuous;" and of the humming bird: "a winged drop of gorgeous sheen and dross, a living gem, poised on its wings" (*Outdoor Papers*, pp. 288, 296–297). Higginson's diary records responses to all forms of natural beauty, to sunsets and sunrises, beautiful days, and long walks by the sea. See, for example, diary entry for February 7, 1869: "The ocean was a royal purple and so was the Narrangansett-golden sky of sunset chowder;" and for December 18, 1868: "All over the emerald sea there were occasional jets and sparkles of white;" or, for February 19, 1869: "Clearing suddenly into a wild golden sunset." Higginson also collected insects, especially butterflies, wrote and dreamed about them. "I grow tired of pictures—never of a butterfly." For references to his dreams see *Army Life in a Black Regiment*, p. 146; *Oldport Days*, pp. 246–249, 61, 91–92, 199.

75. Higginson, *Part of a Man's Life* (Boston, 1906), p. 191.

76. Edelstein, *Strange Enthusiasm*, pp. 98–99.

77. Higginson, *Outdoor Papers*, p. 139.

78. Higginson, "Finer Forces," *Women and Men* (New York, 1888), pp. 134–135; and *Concerning Us All* pp. 186–187. See also "Woman and Man," *Woman's Journal*, January 22, 1870; "Celery and Cherubs," *Woman's Journal*, October 1, 1870; "Shy Graces," *Women and Men*, pp. 308–310; and *Oldport Days*, pp. 226–227.

79. Higginson, "Unmanly Manhood," *Women's Journal*, February 4, 1882; and *Woman and the Alphabet* (Boston, 1858), p. 196. See also *Book of the Heart* (New York, 1899): "All strong novels involving illicit love are necessarily tragedies, not vaudeville; and nowhere is this more true than in French literature. . . . The violation of the domestic tie is often punished with cruel severity, even by the most tolerant novelists" (p. 506).

80. "Fact of Sex," *Woman's Journal*, March 23, 1870.

81. "Unmanly Manhood," *Woman's Journal*, February 4, 1882.

82. Edelstein, *Strange Enthusiasm*, p. 64, 313. How else can we understand the way Higginson exploited for literary purposes his romantic attraction for the feminine and darkly handsome southerner William Hurlbut, who appears like some fascinating modern Alcibiades in Higginson's novel *Malbone* and in his short story "The Haunted Window?" Or how else can we accept the ingenuousness with which Higginson's second wife, Mary, included a long account of Higginson's feelings for Hurlbut, mostly in Higginson's words, in her biography of her husband. "I have never loved but one male friend with passion," he wrote to an inquiring admirer, "and for him my love had no bounds—all that my natural fastidiousness and cautious reserve kept from others I poured on him; to say that I would have died for him was nothing. I lived for him; it was easy to do it, for there was never but one such person." Far from confirming Higginson's homosexuality, this innocently candid revelation denies its existence. The fact is, Higginson was not a homosexual, because he never believed he was.

83. Higginson to Dall, March 27, 1854, Dall Papers, Massachusetts Historical Society.

84. Higginson to Dall, October 5, 1859, Dall Papers, Massachusetts Historical Society.

85. Anthony to Stone, July 18, 1857, NAWSA File, Library of Congress. Later, Higginson would disappoint the NWSA leadership with his obsessive emphasis on suffrage.

86. Twenty of Higginson's thirty-five books dealt directly or indirectly with sexual relations. Beyond the books and essays already mentioned, Higginson helped Lucy Stone edit a *Woman's Almanac* in 1857; in 1862 he edited a series of women's rights pamphlets; and throughout his career he helped other feminists write and publish books, giving freely of his knowledge and money. On *Woman's Almanac* and pamphlets see Higginson to Stone, May 20, 1857, NAWSA File, Library of Congress; and Higginson to Dall, March 16, 1862, Dall Papers, Massachusetts Historical Society. Virginia Penny constructed her famous *Cyclopedia of Women's Employment* (New York, 1863) partly from material gathered by Higginson. See Higginson to Dall, June 20, 1867: "I furnished her much matter which she worked over thoroughly," Dall Papers.

87. Higginson to Hooker, February 19, 1859, Isabella B. Hooker Papers, Stowe-Day Memorial Library.

88. See, in this order, Helen Gray Cone to Higginson, February 1891; Viola Roseboro to Higginson, March 15, 1890; Emma Lazarus to Higginson, November 4, 1876; Kate Louise Brown to Higginson, July 27, 1898, Higginson Papers, Boston Public Library; and Emily Dickinson to Higginson, written in 1862 and quoted by Anna May Wells, *Dear Preceptor, The Life and Times of Thomas W. Higginson* (Boston, 1956), p. 151. See also Rose Terry Clark to Higginson, November 17, 1880: "It is very gracious of you to imply that I am still a whole author, at this point when I feel that I am not more than one-sixteenth of one;" and Alicia Brown to Higginson: "I have never forgotten the encouragement you sent me . . . when I first began to write for the *Atlantic* so many years ago" (Higginson Papers).

89. For Higginson's attack on the positivists see "Womanly Brains," *Woman's Journal*, March 12, 1870; and "Comte's View of Woman," *Woman's Journal*, October 6, 1883. For his other viewpoint, see "Fact of Sex," *Woman's Journal*, March 23, 1870.

90. *Woman's Journal*, February 15, 1873; and "The New Celibacy," *Woman's Journal*, August 23, 1884.

91. See Higginson to Dall, November 18, 1860, Dall Papers, Massachusetts Historical Society, in which he refers to Stanton's letter of November 16, 1860, in the *Liberator*: "In her letter there seems an unusual logical confusion between the consequences of the present unequal marriage and the consequences of marriage as such." Actually, no confusion whatever existed in Stanton's letter. No one believed more than she did in the social importance of marriage. She did assert, however, that women should "repudiate marriage utterly and absolutely, until our tyrants shall reverse their canons . . . and by the talisman of justice transform the 'femme covert' into an equal partner." Higginson also believed in marital egalitarianism but, unlike Stanton, refused to discuss it at conventions—a refusal that in fact provoked his letter to Dall.

92. Edelstein, *Strange Enthusiasm*, pp. 314–315; and diary entry, February 20, 1869, Higginson Papers, Boston Public Library.

93. Mary Higginson, *Thomas W. Higginson*, p. 129.

94. Quoted in *Sketches and Reminiscences of the Radical Club* (Boston, 1880), edited by Mrs. John T. Sargent. The Five Points area, in the slums of New York's Lower East Side, was considered the most crime-ridden in America.

95. See Christopher Lasch, *The New Radicalism in America* (New York, 1965).

96. Higginson, "Drawing the Line," *Woman's Journal*, November 12, 1870; and Paulina W. Davis to Isabella Beecher Hooker: "This year Higginson had the control here and the convention did not make a ripple on the surface—everything was shut out but suffrage" (November 29, 1870, I. B. Hooker Papers, Stowe-Day Memorial Library).

97. *Journal of Social Science* 5 (1873), pp. 34–47.

98. *Journal of Social Science* 6 (July 1874), pp. 1–13.

99. *Woman's Journal*, January 4, 1873; "Woman and the Bible," *Springfield Republican*, March 12, 1872; and "Literary Anniversaries," *Springfield Republican*, June 27, 1872. Bascom's wife, Emma, was also one of the vice-presidents of the American Association for the Advancement of Women in the late 1870s. See Lita Barney Sayles, *History and Results of the Past Ten Congresses of the Association for the Advancement of Women* (New York, 1882), p. 16.

100. For the 1860s see Vincent Bowditch, *Life and Correspondence of Henry Ingersoll Bowditch* (Boston, 1902), pp. 212–213. For the 1870s and 1880s see "Proceedings of the Suffolk District Medical Society," *Boston Medical and Surgical Journal* 105 (July–December 1881), pp. 109–110; and "The Medical Education of Women," *Boston Medical and Surgical Journal* 105 (July–December 1881), pp. 289–293.

101. "Woman's Club," *Springfield Republican*, April 22, 1872.

102. On Jarvis see *Springfield Republican*, October 10, 1866; and on Bowles and Robinson see "Woman's Suffrage," *Springfield Republican*, March 21, 1872. See also, on general ASSA support for the introduction of women as equals with men

into the state boards of health, "Women's Work in Public Institutions," *Occasional Papers of the ASSA* 1 (May 1868), pp. 4–5. See also John Stuart Mill to Edward Jarvis: "I am happy to find so complete sympathy on your part in my opinions in favor of the equality of women. . . . America is . . . at the head of the movement" (August 22, 1869, Jarvis Papers, Free Concord Public Library).

103. "Woman Suffrage Party," *Springfield Republican*, September 19, 1870; and "Woman's Suffrage," *Springfield Republican*, January 27, 1871. The wives of both men—Harriet Robinson (formerly Hanson and once a Lowell Mill girl) and Mrs. Samuel Bowles—belonged respectively to the NWSA and AWSA. See "The Suffrage Agitation," *Springfield Republican*, May 25, 1870, and Harriet Robinson, *Massachusetts in the Woman's Suffrage Movement* (Boston, 1881), a book biased heavily toward the NWSA. Bowles's daughter, Ada, also belonged to the AAW. See *Papers and Letters Presented at the First Woman's Congress* (New York, 1874), p. 188.

104. Walker to Stone, February 26, 1867, NAWSA papers, Library of Congress. Walker also spoke in defense of woman's suffrage at a New England Suffrage Association meeting in 1871. See *New York World*, May 31, 1871.

105. "The Wages of Women," *Woman's Journal*, July 18, 1873.

106. Sanborn to Samuel Eliot, November 5, 1870, Sanborn Papers, Concord Free Public Library.

107. "General Intelligence," *Journal of Social Science* 2 (1870), p. 201; "The Work of Social Science," *J. of S.S.* 6 (July, 1874), pp. 41–43; and list of AWSA officers, *Revolution*, December 2, 1869.

108. On Peabody, see "The Voting of Women in School Elections," *J. of S.S.* 11 (December 2, 1869), pp. 42–54.

109. Sanborn, "Illiteracy of Women in the United States," *J. of S.S.* 3 (1871), pp. 208–209. "The average is from 140 to 150 females to every 100 males, among the illiterate. . . . These are significant figures and they point even more directly than the needle to the pole to the necessity of taking a fresh start in the so-called development of women."

110. *Documents Published by the American Association for the Promotion of Social Science* (Boston, 1867), pp. 77–78.

111. Journal entry, October 9, 1867, Dall Papers, Massachusetts Historical Society.

112. *History of Woman's Suffrage*, Volume III, p. 306.

113. *J. of S.S.* 2 (1870), p. 25.

114. *Documents Published by the ASSA* 1 (July 1866), pp. 6–7.

115. *J. of S.S.* 6 (July 1874), pp. 3–5.

116. Talbot to Dall, June 14, 1877, Dall Papers, Massachusetts Historical Society.

117. "Women and Prison Reform," *Women's Journal*, June 6, 1874.

118. "The National Prison Congress," *Springfield Republican*, October 18, 1870.

119. *J. of S.S.* 2 (December 1879), pp. 1–3; and on Thompson see *Notable American Women*, (Cambridge, Mass., 1971), edited by Edwin T. James, Volume II, p. 453.

120. *Ballot Box*, May 1881. On Zakrzewska and Sewall see Dall to Sanborn, September 14, 1865, Dall Papers, Massachusetts Historical Society.

121. "The Work of Social Science," *J. of S.S.* 6 (July 1874), p. 37.

122. Quoted in Thomas Haskell, *Emergence of Professional Social Science* (Urbana, Ill., 1977), p. 137.

123. M.C.P., "A Letter From Minnesota," *New Century*, November 4, 1876.

124. Ella Giles Ruddy, *Maiden Rachel* (New York, 1879), pp. 305–306.

125. "Glances Toward Social Organization or the Science of Society," *Alpha*, December 1, 1878.

126. Willard, *Woman and Temperance* (Chicago, 1883), pp. 345–356; *Annual Report of the AAW* (New York, 1880), pp. 1–2. See also Mary Earhart, *Frances Willard* (Chicago, 1944), pp. 186–187; Willard's speech before the International Council of Women, in Blackwell, *Report of the International Council of Women* (Washington, 1888), p. 222; and on Willard's early work in social science see Julia Ward Howe, diary entry, March 5, 1875, Howe Papers, Houghton Library, Harvard University.

127. *Revolution*, December 29, 1870. *New York World*, January 23, 1871.

128. "Notes," *Inter-Ocean*, January 25, 1879; and Sayles, *History and Results of the Past Ten Congresses*, pp. 19–20, Vassar College Library.

129. "Humanitarian League," *New York World*, June 6, 1872.

130. Sayles, *History of Results of the Past Ten Congresses*, pp. 19–20.

131. "Report of the General Secretary," *J. of S.S.* 11 (May 1880), pp. v–vii; Caroline Dall, *College, Market and Court* (Memorial Edition, New York, 1914), pp. 379–380; *Ballot Box*, July 1878, April 1880; *The Cooperator*, June, July 1881; "Notes," *Woman's Advocate*, September 20, 1869 and November 1869; *Radical* 1 (1866), pp. 227–230; Caroline Dall to the Social Science Association of Hopedale, Mass., December 15, 1865, Dall Papers, Massachusetts Historical Society; *History of Woman's Suffrage*, Volume III, p. 710; and, for a mine of invaluable information on social science associations, see Elizabeth Boynton Harbert's scrapbook, Mary Earhart Dillon Papers, Schlesinger Library, Radcliffe College. An NWSA organ edited by Matilda Joslyn Gage, the *Ballot Box* declared in 1877 that the work of the Woman's Congresses demonstrated "the vast dormant legislative, administrative, legal, literary and scientific ability possessed by women" (November 1877).

132. "Report of the General Secretary," *J. of S.S.* 11 (May 1880), pp. vi–vii.

133. "Domestic Legislation," *Ballot Box*, September 1881.

134. "Minutes" of the Social Science Club, Cornell University Library; and *Ballot Box*, March 1881.

135. See *Philadelphia Social Science Association* (Philadelphia, 1886) for a list of officers and members, pp. iv–x.

136. *The Cooperator*, August, September 1881.

137. Report on the organization of the Denver, Colorado, and Southern Illinois Associations, *Inter-Ocean*, March 8, 1879.

138. *Ballot Box*, April 1881; and see *History of Woman's Suffrage*, Volume III, for a short biographical sketch of Harbert, pp. 592–593.

139. *Inter-Ocean*, July 13, 1878. For reports on the Illinois Association, see July 13, 1878 to October 6, 1883. I do not know when this organization ceased to exist.

140. Ruddy, *Maiden Rachel*, p. 239.

141. "Sanitary Science," *Inter-Ocean*, November 23, 1878. See also Mrs. S. Van Benshoten, "Practical Social Science," *Inter-Ocean*, February 15, 1879.

142. *Documents Published by the Association for the Promotion of Social Science* 1 (Boston, 1867), pp. 67–69.

143. Ruddy, *Maiden Rachel*, pp. 131, 241–242, 266–267; and "The Genesis of Crime," *Papers Read at the Fourth Congress of Women* (New York, 1877), pp. 117–120. For a short biographical sketch of Giles, see Ruddy, *Maiden Rachel*, pp. 313–314.

144. *J. of S.S.* 1 (June 1869), p. 2; and *Springfield Republican*, April 8, 1872.

145. Putnam Jacobi, "Social Aspects of the Readmission of Women into the Medical Profession," in *Papers and Letters, Presented at the Third Woman's Congress of the American Association for the Advancement of Women* (New York, 1877), p. 113; and Putnam Jacobi, *The Value of Life* (New York, 1879), pp. 172–173.

146. Ruddy, *Maiden Rachel*, p. 239.

147. See Mrs. Helen Shed, "Our Prison System and the Reformation Considered," *Inter-Ocean*, August 30, 1878; "General Notes," *Inter-Ocean*, December 14, 1878, report of the Indiana SSA; "Social Science," *Inter-Ocean*, August 30, 1879, program of the 1879 Illinois SSA Conference; and "Education of Vagrant Girls," *Inter-Ocean*, October 3, 1879, account of the second Indiana SSA.

148. M.C.P., "Letter From Minnesota," *New Century*, November 4, 1876.

149. *Inter-Ocean*, October 26, 1878.

150. "Sanitary Science," *Inter-Ocean*, November 23, 1878.

151. "Child-Birth Made Easy," *Inter-Ocean*, November 23, 1878; *Woman's Journal*, September 23, 1873; and "Mistress and Maid," *Woman's Journal*, September 9, 1872.

152. Bristol, "Enlightened Motherhood," *Papers and Letters*, pp. 10–11.

153. Brown, "Social Aspects of the Readmission of Women into the Medical Profession," *Papers and Letters*, p. 173; and Blackwell, "Comparative Mental Power of the Sexes Physiologically Considered," *Papers Read at the Fourth Woman's Congress* (New York, 1877), p. 20.

154. "Woman's Congress," *New Century*, October 7, 1876.

155. "Woman's Congress," *New Century*, October 14, 1876.

156. "Social Science," *Inter-Ocean*, August 30, 1879, program of 1879 conference; *Inter-Ocean*, October 3, 1879, October 8, 1879; *Woman's Journal*, October 23, 1875, report of the Third Woman's Congress; and *Ballot Box*, November 1878, report of the Illinois SSA conference.

157. *Inter-Ocean*, September 6 and 13, 1879.

CHAPTER 12. *A Unified World View: The American Social Science Association*

1. For the most recent accounts of this association see Thomas Haskell, *The Emergence of Professional Social Science, The ASSA, and the Crisis of Authority* (Urbana, Ill., 1977); and Mary Furner, *Advocacy and Objectivity: A Crisis in the Professionalization of American Social Science, 1865–1905* (Lexington, Ky., 1975). For an earlier discussion see Luther and Jessie Bernard, *Origins of Sociology* (New York, 1943), pp. 527–567.

2. *Journal of Social Science* 2 (1870), pp. 33–34.

3. Wines to Dall, September 22, 1868, Dall Papers, Massachusetts Historical Society.

4. See, for example, the statement of the first president of both the association and MIT, William Rogers, before this body in 1865, *Documents Published by the American Association for the Promotion of Social Science* (Boston, July 1866), p. 16. "It has too long been shown," said Rogers, "that the man of science who confines himself to a specialty, who does not, at the very least, conquer the underlying principles of the other branches of scientific inquiry, is necessarily misled and cannot avoid frequent mistakes. To have any perception of the perspective of his subject, he must see it in relation to ther subjects."

5. *Springfield Republican*, August 3, 1870, and September 20, October 5, 1871.

6. Kate McKean, *Manual of Social Science* (New York, 1872), pp. 523–525.

7. "International Industrial Competition" (Philadelphia, 1870), p. 8.

8. "Interdependence," *Springfield Republican*, December 28, 1871. See also Haskell, for a discussion of the concept of interdependence, especially chapter 2, pp. 24–47.

9. David Wells, "Influence of the Production and Distribution of Wealth on Social Development," *J. of S.S.* 8 (May 1876), p. 8.

10. Samuel Eliot, address before the ASSA, *Documents Published by the American Association for the Promotion of Social Science*, pp. 67–68.

11. David Wasson, "The International," *J. of S.S.*, 1873, pp. 109–121; Robert T. Davis, "Pauperism," *J. of S.S.* 7 (July 1874), pp. 74–78; and Samuel Eliot, opening address, *Documents Published*, pp. 75–76.

12. "Organization," *Radical*, December 1866, p. 220. "Hundreds of men and women in the future may be engaged in the administration of economies of large scale," Stephen Pearl Andrews wrote in *Science and Society* in the 1850s. Carefully indicating that such change would come "naturally" or socially and not through political imposition, he added that "some of these administrators would be at the head of the various departments, but all . . . will be rigidly subordinate to the grand design of the projector who will be despot of his dominion exercising, nevertheless, a beneficent despotism." In 1871 Andrews called for a "New Catholic Church—the unified civil government of mankind under the direction of science." Positivist Albert Brisbane wrote similarly: "The Social System is . . . to be considered as a Whole, composed of subordinate parts or branches like other Wholes—like the human body, for example, which is composed of subordinate organs, such as the brain and nervous system. . . . To the Social Whole, called the Social System or Order, the term Organism may . . . be justly applied." See Andrews, *Science of Society* (New York, 1853), pp. 181–182, and *Golden Age*, April 29, 1871; and Brisbane, *Modern Thinker* 1 (1870), pp. 225–248.

13. For a discussion of interdependence as a reality men and women had to accept and adjust to, and not as a concept of the good society they also consciously projected, see Haskell's long footnote, pp. 30–31, and *passim*, in *The Emergence of Professional Social Science*. I also disagree with Haskell on two other counts: one, that

"transcendental idealism" dominated ASSA thinking (p. 190) and, two, that the ASSA collapsed because it "failed" both to alter its backward-looking "gentry class" orientation and to offer a new generation of social scientists an adequate theory of "casual attribution" (pp. 241–242). As Haskell himself admits elsewhere, by the post–Civil War period radical individualism had ceased to inform reform thought in any compelling way, although bourgeois Americans continued to employ the rhetoric of individualism through to the end of the Progressive period (as they still do, to some extent, today). Yet Haskell still insists upon the influence of Emersonian idealism on reform thought when that, too, had taken on a Hegelian character. On the second count, the ASSA did not survive as an organization because its reform base had been cut away by the existence of newer reform groups, made up of members of the same class, which the ASSA itself had helped, in fact, to generate. It collapsed, in other words, as a reform and not as a scientific association, which it had never really been in the first place. Moreover, because Haskell sees the ASSA statically as a collection of idealistic "gentry" rather than as representative of a progressive bourgeoisie, he does not understand how it may have dynamically created changes within itself, which did not lead so much to its demise as an organization as to its transformation into more effective and more organized expressions of class dominance.

14. Charles F. Adams, "The Protection of the Ballot in National Elections," *J. of S.S.* 1 (1869), pp. 105–106. "Excitable natures," Adams wrote "rarely strengthen free institutions."

15. Z. R. Brockway, "Reformation of Prisoners," *J. of S.S.* 4 (1875), pp. 144–159.

16. The term "advanced industrial society" appears in Francis Walker's public lecture, "Hard Times," *Springfield Republican*, February 25, 1875.

17. *Springfield Republican*, May 14, 1875.

18. Brockway, "Reformation of Prisoners," *J. of S.S.* 4 (1874), pp. 144–159. For similar statements see "The Summer's Social Science," *Springfield Republican*, June 6, 1870; Edward Jarvis, "Common Sense as Applied to Insanity," *Springfield Republican*, March 15, 1872; and Theodore Tilton, "The Philosophy of Punishment," *Independent*, January 21, 1869.

19. *J. of S.S.* 3 (1874), pp. 220–221.

20. Alfred Carroll, *J. of S.S.* 7 (September 1874), pp. 236–269.

21. A. B. Palmer, *Documents Published by the American Association for the Promotion of Social Science* 1 (July 1865), p. 40.

22. "Doctors and Teachers," *Springfield Republican*, July 10, 1875.

23. "Great Subjects—Disease," *Congregationalist*, March 29, 1882.

24. "Are Yankees Dying Out?" *Springfield Republican*, October 6, 1866.

25. William C. Robinson, "The Diagnostics of Divorce," *J. of S.S.* 14 (September 1881), pp. 136–150. For a similar statement see Francis Walker, "Some Results of the Census" and "The Causes of Retardation in the National Increase Between 1860–1870," *J. of S.S.* 5 (1873), pp. 73–97.

26. "Human Life and Its Destroyers," *Springfield Republican*, October 15, 1867.

27. For the best discussion of this sanitary assault, see Barbara Rosenkrantz, *Public Health and the State, Changing Views in Massachusetts* (Cambridge, Mass., 1972). See also Franklin Sanborn, "The Work of Social Science in the United States," *J. of S.S.* 6 (July 1874), pp. 36–45.

28. George Bayle, "Disposal of the Dead," paper read before the New York Public Health Association, *Sanitarian*, June 1874, pp. 97–98. For the same analysis see Mary Safford Blake, *New Age*, February 10, 1876.

29. "Important Sanitary Movement," *New York World*, September 17, 1872. For a report on the Nashville convention see "American Public Health Association," *Nashville Republican Banner*, November 12, 1878.

30. Elizabeth Peabody, "A Letter," *New Century*, May 13, 1876.

31. "The New Century for Women," *New Century*, May 13, 1876.

32. Ibid., May 13, 1876.

33. "Kindergarten," *New Century*, May 13, 1876. See also William T. Harris, Emily Talbot, and Henry Barnard, "Kindergarten System," *J. of S.S.* 12 (September 1880), pp. 8–10; Anna Hallowell, "The Care and Saving of Neglected Children,"

J. of S.S. 1 (September 1880), pp. 117–120; an article by an ASSA member, Elizabeth Boynton Harbert, "Kindergarten," *Inter-Ocean*, November 23, 1878; and Emily Talbot, *Papers on Infant Development* (published by ASSA Education Department, 1882). See also, for an excellent recent account of the rise of the Kindergarten system, Marvin Lazerson, "Urban Reform and the Schools: Kindergartens in Massachusetts, 1870–1915," *History of Education Quarterly* 11 (Summer 1971), pp. 115–142.

34. "The Social Element of the Kindergarten," *New Century*, May 27, 1876. The reform approach to the kindergarten remained fixed throughout the century. See, for example, Sarah Cooper's speech before the International Council of Women in 1888: "Better, far better, that we plant kindergartens and organize industrial schools and educate the young for work, than to let them grow up in such a manner as to form Jacobin clubs and revolutionary brigades, which will be the beginning of the end of our greatness and prosperity. . . . Take the very little child into the kindergarten and there begins the work of physical, mental and moral training; develop his faculties; unfold his moral nature; cultivate mechanical skill in the use of the hands, give him a sense of symmetry and harmony; a quick judgement of numbers, measures and size; stimulate his inventive faculties; make him familiar with the customs and usages of well-ordered lives . . . and thus equipped mentally, morally and physically, send him forth to the wider range of study, which should include within its scope some sort of industrial training" (*Report of the International Council of Women* [Washington, 1888], p. 72)."

35. *J. of S.S.* 15 (1882), p. 88.

36. "Report on a Developing School and School Shop" (Boston, 1877), pp. 1–17. The following quotations come from this report and will not be cited separately.

37. See Dorman Eaton's paper delivered before the ASSA in 1873 on "Municipal Government," which presents the same argument in a different context. "Everything has become so complex," Eaton said, "and nearly every branch of administration must be treated as the sole business of city officers . . . Not one person in five hundred understands the duties of the higher offices, and not one in five thousand knows who is at fault for the neglect and abuses manifest on every hand" (*J. of S.S.* 5, 1873, pp. 1–2). See also Caroline Dall, speech before the ASSA, "Social Science," *Springfield Republican*, October 10, 1866; Dr. Abraham Jacobi, "Proceedings of the New York Medico-Legal Society," *Sanitarian*, August 1874, pp. 205–206; and George Beard, "Hysteria and Allied Affections," a lecture at the Long Island Hospital, June 1872, George Beard Papers, Yale University Library.

38. Quoted from Harry Braverman, *Labor and Monopoly Capital* (New York, 1974), p. 78.

39. "American Public Health Association," *Sanitarian*, July 1873, pp. 413–428.

40. *J. of S.S.* 7 (September 1874), p. 236.

41. "Sanitary Organization of Nations," *Boston Medical and Surgical Journal* 101 (July–December 1879), pp. 50–66; and 102 (January 1880), pp. 25–30. See also "Preventive Medicine and the State and the Doctor of the Future," *Springfield Republican*, March 24, 1874; and *Springfield Republican*, May 22, 1874. Dorman Eaton made the same proposal in 1873. See Eaton, "An Address on Sanitary Legislation," *New York World*, November 14, 1873. See also Ezra Hunt, "The Need of Sanitary Organization and Rural Districts," a paper delivered before the APHA, *New York World*, November 14, 1873. Hunt lamented the absence of sanitary "care" and "authority" in these districts. We need, he wrote, "some discreet and widespread sanitary power that should extend its protection to the whole country under the rule that no town might cherish such a condition of things as should endanger its own health or that of the country."

42. Wharton, "International Industrial Competition," p. 4.

43. *Boston Medical and Surgical Journal* 102 (January 1880), p. 25. See also Enoch Wines, *J. of S.S.* 2 (1870), p. 241; "It has been doubted whether the public mind is sufficiently aware of the dangerous elements around us; whether the connection between filth and disease is as yet proved to the public satisfaction. . . . It is thought . . . that local and private interests have often been so strong as to paralyze the action of the Health authorities." The ASSA claimed credit for "nationalizing" this question. See *J. of S.S.* 11 (May 1880). For biographical material on

Bowditch see Vincent Bowditch, *Life and Correspondence of Henry Ingersoll Bowditch* (Boston, 1902), Volume II, pp. 84–103; and Rosenkrantz, *Public Health and the State*, pp. 51–65.

44. *J. of S.S.* 11 (May 1880), pp. vi–vii.

45. Letter to an unidentified New York Journal (probably Wendell Phillips's *Anti-Slavery Standard*), January 28, 1865. Other quotations are from *Documents Published by the American Association for the Promotion of Social Science* 2 (Boston, 1867), p. 67; and Henry Villard, *J. of S.S.* 1 (1869), p. 2.

46. Franklin Sanborn, "Training School for Nurses," *J. of S.S.* 7 (September 1874), pp. 294–298; Sanborn's report to the ASSA, *J. of S.S.* 11 (January 1878), pp. 5–6; George F. Marhoe, "Legislation in Relation to Pharmacy," *J. of S.S.* 5 (1873), pp. 122–135; Alfred Carroll, *J. of S.S.* 7 (September 1874), pp. 236–269; Andrew D. White, "The Relation of National and State Governments to Advanced Education," *J. of S.S.* 7 (September 1874), pp. 299–321; and L. P. Alden, response to a paper delivered by Charles L. Brace, "What is the Best Method For the Care of Poor and Vicious Children?," *J. of S.S.* 11 (May 1880), pp. 98–99. Principal of the Michigan State Public School for Poor Children, Alden praised Brace's "placing out" system as a good method for dealing with such children; at the same time, however, he disagreed that most parents had the "expertise" to discipline children. "The great majority of even respectable well-to-do families," he said, "do not understand child-life, how to train and discipline it" (p. 100). Sanborn responded to Alden's words with shock. If "few parents are fit to bring up their own children," he declared, "who then is fit? . . . Even a bad family . . . is not always the worst place for children; a great many good children have grown up in wretched families (p. 102). For positions similar to Alden's, see Caroline Dall, "Social Science," *Springfield Republican*, October 10, 1866: "We should have laws that would sanction the taking away of children of vicious parents, that they may be reclaimed to respectability and virtue;" and "Unconsidered Murders," *Sanitarian*, June 1873, in which the writer claims that no man's house is his "castle" when it comes to public health: "Fathers and mothers are absolute; and provided there is not malice and no direct action, they are able to kill their children indirectly and by ignorance, without prevention or rebuke" (p. 111). For recent accounts of the systematic rationalization of American professional life, see Burton Bledstein, *The Culture of Professionalism* (New York, 1976); Michael Katz, *Class, Bureaucracy and Schools: The Illusion of Educational Change* (New York, 1975); Roy Lubove, *The Professional Altruist* (Cambridge, Mass., 1965); Perry Miller, *The Life of the American Mind* (New York, 1965); Robert Wiebe, *The Search for Order* (New York, 1967); Thomas Haskell, *The Emergence of Professional Social Science;* Mary Furner, *Advocacy and Objectivity;* Lawrence Vesey, *The Emergence of the American University* (Chicago, 1965); William Rothstein, *American Physicians in the Nineteenth Century* (Baltimore, 1972); and R. Jackson Wilson, *In Quest of Community* (New York, 1968). For a brilliant critique of the cult of expertise see Christopher Lasch, *Haven in a Heartless World* (New York, 1977).

47. *Springfield Republican*, December 25, 1875. See also Sanborn, "The Work of Social Science," *J. of S.S.* 11 (September 1879), pp. 23–29; Dorman Eaton, "Municipal Government," *J. of S.S.* 5 (1874), pp. 1–35; "The Experiment of Civil Service Reform in the United States," *J. of S.S.* 8 (May 1876), pp. 77–80; and Charles F. Adams, "The Protection of the Ballot in National Elections," *J. of S.S.* 1 (1869), pp. 108–109.

48. "The Relation of State and Municipal Governments, and the Reform of the Latter," *J. of S.S.* 11 (January 1878), pp. 140–149. Social scientists pleaded for an end to the "sect-spirit," the "scattering" or "squandering of resources," and the inefficiency of "local control," all of which, they believe, subverted individual as well as public welfare. In their place they put "harmonious systems" of subordination and superordination under "singe-headed supervision," founded on "systematic principles of order and economy," and run by "systematically instructed" men and women. See on charities and boards of health, Sanborn, "The Supervision of Public Charities," *J. of S.S.* 6 (July 1874), pp. 85–86; 7 (September 1874), p. 210; 2 (1870), p. 241; and 11 (May 1880), pp. vi–vii. On education see Andrew D. White, "The Relation of National and State Laws to Advanced Education," *J. of S.S.* 7 (September

1874), pp. 299–322; John Philbrick, "Inspection of County Schools," *J. of S.S.* 1 (1869), pp. 11–23; and Daniel C. Gilman, "American Education," *J. of S.S.* 11 (December 1879), pp. 4–16. On libraries, museums, and prisons see Sanborn, *J. of S.S.* 11 (May 1880). See also "The Social Science Meeting at New Haven," *Springfield Republican*, October 22, 1866. For the best account of the attack on local control, see Michael Katz, *The Irony of Early School Reform* (Boston, 1968).

49. Haskell, *The Emergence of Professional Social Science*, pp. 57–62. See also George M. Fredrickson, *The Inner Civil War* (New York, 1965).

50. "Social Sciences, Their Growth and Future," *Concord Pamphlets* (Concord, Mass., 1885), p. 4.

51. Ibid., p. 5.

52. Sanborn to Harris, April 26, 1866, December 29, 1873, February 2, 1874, January 24, 1876, Sanborn Papers, Concord Public Library.

53. Sanborn and Harris, *A. Bronson Alcott, His Life and Philosophy* (Boston, 1893), Volume I, p. 545. On Harris see Henry Pochman, *New England Transcendentalism and St. Louis Hegelianism* (Madison, Wis., 1948), p. 17; and Loyd Easton, *Hegel's First American Followers* (Athens, Ohio, 1966), pp. 16–19.

54. Louise Olson, "Contemporary Criticisms of the Concord School of Philosophy and Literature," M. A. Thesis, Columbia University, June 1930, p. 17.

55. Haskell, *The Emergence of Professional Social Science*, pp. 141–143.

56. Curti, *The Social Ideas of American Educators* (New York, 1935), pp. 313–314.

57. "Social Science and Social Conditioning," *Concord Pamphlets* (Concord, Mass., 1885), p. 22.

58. E. L. Godkin, "Legislation and Social Science," *J. of S.S.* 3 (1871), pp. 122–123.

59. Wines to Dall, September 22, 1868, Dall Papers, Massachusetts Historical Society.

60. Dall to the Social Science Association of Hopedale, December 15, 1865, Dall Papers, Massachusetts Historical Society.

61. On emergence of new forms of social science thought at the end of the century, see Reba N. Sofer, *Ethnics and Society in England: The Revolution in the Social Sciences, 1870–1914* (London, 1978), pp. 1–68; and Haskell, *The Emergence of Professional Social Science*, pp. 190–256.

62. Frederic Harrison, "The Positivist Problem," *Modern Thinker* 1 (1870), p. 248.

63. "General Intelligence . . . Social Reform Movements in the United States," *J. of S.S.* 2 (1870), p. 201.

64. "Influence of the Production and Distribution of Wealth in Social Development," *J. of S.S.* 8 (May 1876), pp. 8–21; and *Woman's Journal*, July 18, 1874. For a good account of the economic thought of Wells, Francis Walker, Edward Atkinson, and Carroll Wright, see David Horowitz, "Genteel Observers: New England Economic Writers and Industrialization," *New England Quarterly*, March 1975, pp. 65–83.

65. Sanborn, "General Intelligence, *J. of S.S.* 2 (1870), p. 201; and Elizur Wright, "Life Insurance for the Poor," *J. of S.S.* 8 (May 1876) pp. 147–164.

66. See Horowitz, "Genteel Observers," *New England Quarterly*, pp. 72–74.

67. Samuel Eliot, opening address before the ASSA, *Documents Published by the American Association for the Promotion of Social Science* 2 (Boston, 1867), pp. 67–69; David Wasson, "The International," *J. of S.S.* 5 (1873), pp. 109–121; R. T. Davis, "Pauperism in the City of New York," *J. of S.S.* 6 (July 1874); and William Robinson, "The Diagnostics of Divorce," *J. of S.S.* 14 (1881), pp. 136–150.

68. *History of Woman's Suffrage*, Volume III, p. 411.

69. For a listing of the names of these women see Sayles, *History and Results of the Past Ten Congresses of the Association for the Advancement of Women* (New York, 1882), *passim*; "Woman's Congress," *New York World*, October 13, 1873; and *New Century*, October 14, 1876.

70. On this shift see Warren Susman, " 'Personality' and the Making of Twentieth-Century Culture," in *New Directions in American Intellectual History*. (Baltimore, 1979), edited by John Higham and Paul Conkin, pp. 212–226.

71. On Progressive communitarian theory see Jean B. Quandt, *From the Small Town to the Great Community* (New Brunswick, N.J., 1970); and on the New Nationalism see Charles Forcey, *The Crossroads of Liberalism* (New York, 1961).

72. On the sociology of marriage and parenthood see Christopher Lasch, *Haven in a Heartless World*, pp. 10–11, 107–110.

73. Himmelfarb, *On Liberty and Liberalism* (New York, 1974), p. xx.

EPILOGUE: *Feminism and Reform*

1. For Mead's early history see her autobiography, *Blackberry Winter* (New York, 1975).

2. Mrs. Frank Crocker to Isabella Beecher Hooker, May 31, 1871, Isabella Beecher Hooker Papers, Stowe-Day Memorial Library; Elizabeth Stuart Phelps, *Woman's Journal*, August 12, 1871; and Howe, unpublished speech on finance to the Association for the Advancement of Woman, 1873, Howe Papers, Harvard University Library.

3. Gilman, *Women and Economics* (New York, 1966), p. 265.

4. On sexual patterns in the 1920s see Paula Fass, *The Damned and the Beautiful* (New York, 1976), pp. 25–40; and on planned parenthood see Linda Gordon, *Woman's Body, Woman's Right* (New York, 1976), pp. 341–390.

5. Rice to Miss Moore, January 9, 1909, Rice Papers, Wesleyan University Library.

6. Mary Ruth Walsh, *"Doctors Wanted: No Women Need Apply": Sexual Barriers in the Medical Profession, 1835–1975* (New Haven, 1977), pp. 181–239.

7. Sheila M. Rothman, *Woman's Proper Place* (New York, 1978), p. 136.

8. Ibid., pp. 142–143.

Index